Handbook of Applied Thermodynamics

Author

David A. Palmer, D.Sc.
Research Associate
Amoco Chemicals Company
Naperville, Illinois

CRC

CRC Press, Inc.
Boca Raton, Florida

Library of Congress Cataloging-in-Publication Data

Palmer, David A.
Handbook of applied thermodynamics.

Bibliography: p.
Includes index.
1. Thermodynamics. I. Title.
TJ265.P27 1987 660.2′969 87-781
ISBN 0-8493-3271-0

This book represents information obtained from authentic and highly regarded sources. Reprinted material is quoted with permission, and sources are indicated. A wide variety of references are listed. Every reasonable effort has been made to give reliable data and information, but the author and the publisher cannot assume responsibility for the validity of all materials or for the consequences of their use.

Direct all inquiries to CRC Press, Inc., 2000 Corporate Blvd., N.W., Boca Raton, Florida, 33431.

International Standard Book Number 0-8493-3271-0
Library of Congress Card Number 87-781
Printed in the United States

FOREWORD

This book is for the practicing engineer or scientist involved in process development and design. The emphasis is on applied thermodynamics and for this reason, the text is organized with respect to the stage of development of a process rather than according to logical development of thermodynamic principles. Therefore, it is assumed that the reader has some familiarity with concepts of ideality, activity coefficients, fugacity, chemical potential, etc.

The use of thermodynamics in process development and design has become a very useful tool since the advent of computers, because computers allow the use of detailed information which would otherwise not be feasable. Therefore, the use of thermodynamic principles is becoming an integral part of process development and design activities; and this book will be useful to all engineers and scientists involved in these activities. The applied principles presented here can be used in most areas of industry including oil and gas production and processing, chemical processing, power generation, polymer production, food processing, synthetic fuels production, specialty chemicals and pharmaceuticals production, bioengineered processes, etc.

Grant M. Wilson

PREFACE

This book was written for engineers and chemists who have previously studied thermodynamics for two or more semesters in college. The writer therefore assumes that the reader is already familiar with activity coefficients, fugacity coefficients, and equations of state. However, some uncertainty on the part of the reader as to their significance and use is acceptable. The primary audience is the practicing engineer or chemist. This handbook may also fill a useful niche as an auxiliary textbook for college courses. This book is not organized like a typical thermodynamics textbook. Rather, it is organized according to the types of problems encountered in industry. They range from *probing research*, through *process assessment*, and finally to *process development*. Even projects to improve operating processes fall into those three basic categories. Because of the approach taken, concepts are not always developed in a logical sequence and the book contains no derivations.

Topics included in this book include: an overview of the importance of physical properties and thermodynamics, the use of thermodynamics to predict the extent of reaction in proposed new chemical combinations, and the use of special types of data and prediction methods to develop flowsheets for probing projects. To cover process assessment there are two chapters. One give sources of critically evaluated data and divides the published works into three categories, depending on quality. Methods of doing one's own critical evaluation of literature data are included. Process development usually requires experimental data. Therefore, there is a listing of known North American contract experimentalists with the types of data measured by each. Another chapter gives methods for measuring phase equilibrium data. Two difficult examples are solved to illustrate the implementation of concepts previously presented. Some thermodynamic concepts to carry out process optimization are given in the final chapter.

This handbook is intended to be a working tool and a helpful source of information. Sometimes the best resource is not an equation but a person or organization which has experience in a particular area of technology. Therefore, their telephone numbers and addresses are included. It is thought far more helpful to include these specific numbers, than to not include them at all. Most should remain correct for a number of years.

The author is indebted to the many physical property experts with whom he has associated over the years. These include professors in college, his dissertation advisor B. D. Smith, and fellow students. At Amoco Chemicals, he is indebted to Dr. P. G. Thornley for reviewing and encouraging the writing of this manuscript. Special thanks also to J. E. Myers for reviewing the entire manuscript and making valuable suggestions. The author is likewise indebted to M. A. Albright, Dr. T. B. Selover, Dr. C. Tsonopoulos, Dr. R. P. Danner, Dr. G. M. Wilson, and Dr. W. H. Seaton for reviewing various portions of the manuscript. The author, nevertheless, bears full responsibility for the material and opinions expressed herein. Many other industry experts, participants in the Design Institute for Physical Property Data, have had a direct impact on the material chosen for presentation here. The friendship and assistance of all these gracious people is gratefully acknowledged.

NOMENCLATURE

B	Second virial coefficient
f_i^v	Vapor phase fugacity
f_i^L	Liquid phase fugacity
g_{ii}	NRTL activity coefficient equation parameter
ΔG_f^0	Gibbs free energy
G^E	Excess Gibbs free energy of mixing
G_{ij}	NRTL parameter
ΔH^v	Molar enthalpy of vaporization
H^E	Excess enthalpy of mixing
H_{ij}	Henry's constant
K_i	Vapor-liquid equilibrium constant
K_p	Equilibrium reaction constant in terms of partial pressures
MW	Molecular weight
$perm_i$	Permeability of component i
P	Pressure
P^*	Vapor pressure
P_c	Critical pressure
P_r	Reduced pressure (P/P_c)
q_i	Permeation rate of component i
q_i	Shape factor for UNIQUAC equation
R	Universal gas constant
S^E	Excess entropy of mixing
SG	Specific gravity
T	Absolute temperature
V_c	Critical volume
V_i^L	Liquid molar volume of component i
x_i	Liquid mole fraction of component i
y_i	Vapor mole fraction of component i
Z	Compressibility factor
Z	Coordination number for UNIQUAC equation
Z_c	Critical point compressibility factor
α	Relative volatility
α	NRTL parameter
γ_i	Liquid phase activity coefficient of component i
γ_i^∞	Infinite dilution activity coefficient
η	Viscosity
η_{II}	Second Law efficiency
λ	Thermal conductivity
λ_{ij}	Interaction energy parameter
Λ_{ij}	Wilson equation parameter
ν_i	Area fraction of component i in mixture
ρ	Density
σ	Surface tension
τ_{ij}	NRTL parameter
ϕ	Volume fraction of component i in mixture
$\hat{\phi}_{i,P}$	Fugacity coefficient of mixture at system pressure
ϕ_{i,P_i^*}	Pure component fugacity coefficient at its vapor pressure

THE AUTHOR

David A. Palmer has played a key role in cooperative research on applied thermodynamics. He was one of a small group of people behind the formation of a research organization devoted to thermodynamics and physical properties in general. He was the one who saw the wisdom of having that effort under the aegis of the American Institute of Chemical Engineers (AIChE), and promoted that association. Today, the Design Institute for Physical Property Data, better known as DIPPR, is well known and recognized. It is an integral and important part of the AIChE. Dr. Palmer has been in charge of the DIPPR Technical Committee from almost the inception of the organization. In that position, he has initiated a number of programs which have already been successfully completed. The work of the DIPPR has involved regular contact with practically all of the American and some of the important foreign industry experts in thermodynamics. It has also involved considerable interaction with potential and actual contractors in universities. Some of the products of that effort will be mentioned in this handbook.

Palmer received his high school education in Lethbridge, Alberta, Canada. His bachelor's degree in chemical engineering was received from Brigham Young University, Provo, Utah. His master's and doctor's degrees were received from Washington University in St. Louis, Missouri. His specialization and doctoral dissertation were in fluid phase equilibrium thermodynamics, both experimental and theoretical.

Dr. Palmer is employed by the Amoco Chemicals Company, where he has had a very broad range of experience in process design work. For example, he develops ideas for exploratory work, tests the progress of exploratory probing work, and has followed ideas through the process development stage. He has also been involved in development and application of ideas for improvement of operating processes. At the time of publication of this book he had received 13 patents and several foreign patents. Palmer is also chairman of the Amoco Chemicals Physical Properties Committee, which shares information and conducts courses for other staff members. He has also operated a physical properties laboratory at the Amoco Chemicals Company.

Palmer's hobbies are the Spanish language and Mesoamerican archaeology. Therefore, he has maintained considerable contact with both Mexico and Guatemala. In 1981 he published a book on the archaeology of that area as it related to ancient historical documents.

TABLE OF CONTENTS

Chapter 1

THERMODYNAMICS AND PHYSICAL PROPERTIES IN PROCESS INNOVATION

TABLE OF CONTENTS

I. INTRODUCTION

New process innovation does not end when a new product or process chemistry is discovered. Design of the reactor and subsequent downstream processing must also receive careful attention. It is often necessary to do a significant amount of inventing in order to extract economically the desired product from the reactor effluent, recycle unreacted raw materials, and minimize pollution problems. Otherwise good chemistry sometimes has been abandoned for lack of creative application of separations technology. Sometimes, application of phase equilibrium technology is required just to make the chemistry work.

In this book the issue of phase equilibrium will be carefully considered because of its great impact on the feasibility and cost of chemical processes. Its evaluation requires use of thermodynamics; likewise, thermodynamics can be used to evaluate the reactions which form the basis of most chemical processes. Those who specialize in these areas must of necessity also be knowledgeable in physical properties. Physical properties of the pure compounds and mixtures are at the root of the phase equilibrium characteristics of systems; for example, the vapor pressure is extremely important in distillation. Other physical properties used for process design are thus the natural province of those who specialize in thermodynamics.

II. IMPORTANCE OF PHYSICAL PROPERTIES

It sometimes seems that national security must be threatened to obtain concerted effort by a democratic nation. It was primarily the progress of science that finally ended the war with Japan. An almost forgotten element of the development of the atomic bomb relates to the race between the Allies and Germany to bring it to fruition. There were several ways to develop such a bomb. The short way required knowledge of a physical property of graphite. The Germans guessed at the answer, and guessed incorrectly. As a result they were led on the longer path of development of heavy water in Norway. The Americans knew of the German guess, but measured the physical property. They found that it was substantially different from the guess of the Germans, and were able to devise a much faster route to production of a nuclear weapon.[1,9] If this scenario had been reversed, the U.S. might have become a province of the Third Reich!

Accurate physical properties are at the cornerstone of physics, chemistry, and engineering. Those disciplines have created the comfortable standard of living which many people in the world now enjoy. Petroleum and petroleum products fueled great changes during the first half of this century. Most people overlook the fact that the tractor freed the manpower resources of this nation to concentrate on manufacturing and service industries. Development of nitrogen fixation, which required an understanding of thermodynamics of ammonia, made possible dramatic increases in land productivity. Further great increases in agricultural productivity were due to the now maligned but absolutely essential pesticide industry. Applied chemistry has spurred other important industries: plastics of every variety, synthetic fibers, pharmaceutical chemicals, adhesives, microelectronics, advanced composites, and others. American leadership in these and other technologies has made the U.S. the strongest nation on earth.

It is a common perception that foodstuffs are our primary export, yet chemical exports in 1981 more than doubled in value the food exports![2] Statistics from the same source on national income arising from manufacturing show the following for chemical and related industries.

	($ billions)
Chemicals and allied products	41
Petroleum and coal products	34
Plastic and rubber products	17
Textiles and apparel	30

Together, these industries represented 22% of total U.S. industrial production in the reporting year of 1982. A National Academy of Sciences report noted that the chemical industry provided a $12 billion positive international trade balance — second highest of all commodity groups.[10]

The direction of growth among U.S. industries is also of significance. The data indicate that the fastest-growing major industries are chemicals and related products. An industrial production index using a base year of 1967 was established; then the first half of 1983 was used to rate the various industries. Does it come as any surprise that the primary metals industry had declined by 21%, or that the fabricated metal products industry had increased only 13% in the 15 years of this study? All durable goods had increased only 27% in that time span. Nondurables, by contrast, increased by 61%; most chemical products fall into that category and grew by 105%! Leading all industry groups was the plastics/rubber segment, which expanded by 173%.

The chemical and petroleum industries are leaders in capital investment. In 1982 they were responsible for 45% of the $99 billion in manufacturing investment. Much of that investment is related to energy conservation and replacement of inefficient processes with more efficient ones. This can be illustrated best by a table of energy savings per constant dollar of value added. The time frame studied was 1971 to 1977, and it is quite certain that the trend has continued.

	(%)
Paper	2
Cement	2
Petroleum refining	9
Aluminum	9
Steel	10
Inorganic chemicals	17
Organic chemicals	30

It is clear from these data that the chemical industry has been a leader in energy conservation. This has been made possible through process improvements, increased capital investment, and — at the root of all these changes — physical property data. Yet there is much potential improvement remaining.

The chemical industry generally recognizes that "the road to successful innovation and efficiency is paved with pieces of accurate physical property data cemented together with good correlations". This involves more than fine tuning and optimization; it goes to the heart of process development. It involves the choice of unit operations as well as the size of equipment. It can be the key to development of a breakthrough process or failure to do so. Several examples follow for the gas processing and chemical industries.

A. Gas Processing Industry Examples

M. A. Albright (see References 3 and 4) of Phillips Petroleum Co. has given several good examples of costly failures due to use of inadequate vapor-liquid equilibrium data.

Twenty years ago, gas processing recovered about half the propane plus the heavier hydrocarbons. But the demand for LPG was growing. The lean oil absorption process then pervasive wouldn't do better. So, plants were built using refrigerated oils to recover more propane. As the level of refrigeration went colder, however, the plants recovered far less propane than predicted by design. Why? Because they were based upon projections from warm data to cold which were wrong. So the pressure was increased. This helped sometimes but not always and helped less than predicted. Again — wrong extrapolations. Why? Because the closure of the dew point curve and the bubble point curve to a critical [point] was either not recognized or was done so poorly. Once the data were measured and the problems recognized, it was shown that absorption could *never* accomplish the design goals.

The next step was the development of cryogenic plants where propane plus [heavier hydrocarbons] were recovered by condensation. All went well for a while. Then the market for ethylene for polymers exploded. It became desirable to recover ethane. So, colder plants were designed. Suddenly these did not meet design goals. Why? Because the effects on the phase behavior below the methane critical point were not recognized. Fortunately, data measurements were in progress and this problem was soon resolved . . .

Interestingly, the first big LNG plants, built in Algeria, were failures and required much redesign and rebuilding. Why? Because improper phase relation predictions caused an attempt to violate the second law of thermodynamics in the heat exchange train.

A recognition of the strides that could be made by having improved physical property data led the gas processing industry to form a cooperative research program within the Gas Processors Association (GPA). It has used the leverage of cooperative research dollars to pay for measurement of a significant amount of data. It has also developed to industry standards several equations of state. Industry experts believe that this work has resulted overall in an increase of 60% in extraction plant capacity.

Some gas and oil production companies are achieving enhanced recovery of petroleum using carbon dioxide flooding. The prediction of the effects of such miscible gas flooding on reservoir properties has required the use of sophisticated equations of state. Data for determination of the interaction parameters have been measured by several organizations, including but not limited to the National Bureau of Standards (NBS), Texas A & M University, and the American Institute of Chemical Engineers' (AIChE) Design Institute for Physical Property Data (DIPPR).

A study by Air Products and Chemicals Inc. for a nitrogen rejection plant indicates that equipment size is very sensitive to the accuracy of the thermodynamic values.[5,8] An uncertainty of just 5% in the K value of methane affects the heat exchanger size by 33%. Similarly, a 10% uncertainty in the nitrogen heat of vaporization resulted in a $2 million/year product loss in a standard-size plant. Inaccuracy of available data on light gases resulted in considerable oversizing of a hydrogen recovery plant built in 1961. Subsequent efforts to measure data accurately and develop better correlations have allowed reductions of 40% in capital cost for the same size of plant.

B. Chemical Industry Examples

The chemical industry faces a different set of challenges due to the great variety of molecules that it must process. When one considers the enormity of the problem facing engineers with respect to mixtures, the successes enjoyed thus far are remarkable. Consider, for example, the successful design and operation of units to produce high-octane aviation gasoline and synthetic rubber during World War II. Much of the credit would have gone to data measured by Standard Oil of New Jersey and M. W. Kellogg Company, as well as to "new and improved" correlations[6] of that time.

The successful design of plants for the extractive distillation of butadiene also has required considerable expertise and judgment. Consider the complexity of the feed to one of these plants. A total of 153 important binary mixtures can be identified, and composition effects are important. Then there are the thousands of combinations of multicomponent interactions. Thermodynamics experts have made phase equilibrium measurements on selected systems and used good correlation tools to solve an otherwise intractable process simulation problem. The process development relied more on physical properties work than on unit operation testing. Such men as Dr. Cline Black and Dr. Carl Deal played leading roles in development of the important sulfolane aromatics extraction process and the acetonitrile/water mixed-solvent extractive distillation process licensed by Shell Development Co. Development of models for such complex systems is illustrated in Chapter 8.

Dr. D. Zudkevitch[6] of Allied Corp. reports that in the early 1960s a major engineering company captured a large share of the ethylene plant business. One of the tools used was the development of a physical property data base better than others in the industry. The company was able to redesign, revamp, and reorient the way that gases are handled coming out of the cracking section reactors. As a result, the cost of the recovery section was reduced, making an enormous difference in the overall plant cost. Mr. S. B. Adler was a key contributor, not only in developing the basis for those designs, but also in leading the way to development of methods for free energy minimization.

A process which won the Kirkpatrick award of *Chemical Engineering* was the carbonylation of methanol to acetic acid. It now represents the majority of installed acetic acid plant capacity. Monsanto Co. can be justly proud of the new technology. Its engineers approached the development from a fundamental point of view. They skipped the semiworks phase, and had a 100,000:1 scale-up. The plant was started in 1 day at full design capacity! There was more than good chemistry involved in this development. A great deal of data was found in the literature or measured, and then critically evaluated. The engineers developed complete simulations of the plant material and energy balances. One key problem that they successfully took into account was the dimerization of acetic acid in the gas phase. As part of Dr. J. R. Fair's engineering technology group, Dr. H. R. Null (author of *Phase Equilibrium in Process Design*) implemented an approach which handled this problem successfully. It took into account the chemical equilibria between the monomer and dimer in the gas phase. A failure to account properly for this phenomenon would have resulted in erroneous design not only of the separations section, but also of the reactors.

In contrast to that success, a major U.S. chemical company paid a very large licensing fee to a Japanese company for rights to a process on which the U.S. company held the basic patents. What put the U.S. company in that strange position? In development of the process, a reversible chemical equilibrium with a separating agent was involved. The U.S. technologists were unable to design a method for breaking the complex without isomerizing the extracted chemical. Subsequently, Japanese chemical engineers found a simple solution based on phase equilibrium principles. As is the case in many situations, a knowledge of the chemistry is insufficient by itself. A workable design requires knowledge of phase equilibrium, and the correct data to back up the ideas which flow from the basic principles.

C. Physical Properties in Data Bases

A large data base is a necessary and important part of process simulation programs sold commercially. The combination of the data base and the thermodynamic algorithms is usually a key feature of these programs and is touted in the advertising. However, each of those data banks and algorithms has limitations. Some of the data may be in significant error, and all correlations have regions where they perform poorly. There is a slang phrase used in the computer industry, "Garbage in — garbage out." It is very true in this case. Small errors in some of the data can result in very costly design mistakes.

D. Importance of Experimental Physical Property Data

One of the purposes of this book is to help identify which new data should be measured during development of proprietary products or processes. The limitations of correlations are also evaluated, with recommendations for the best methods to be used. Care in this endeavor can reduce both the capital and operating cost of new plants. By comparison with probable savings, the cost of physical property data is insignificant. Collection and measurement of accurate physical property data needed to do the design correctly can prevent costly reconstruction or even abandonment of new chemical plants.

E. Summary

Physical properties have considerable commercial significance and, on occasion, strategic importance. The chemical process and related industries represent about 22% of total industrial production in the U.S. Even a small improvement in energy or raw material utilization can have a large economic impact; the organic chemical industry has been the leader in energy reduction. Further, capital costs are very sensitive to some of the physical properties used for process design. The history of problems in the design of gas processing plants has been traceable largely to inadequate phase equilibrium data. Large reductions in capital have been possible due to improved data. Complicated chemical processes have been successfully designed and operated without first building demonstration plants; this would have been impossible without a solid basis of physical property data and thermodynamic correlations.

III. ORGANIZATION OF THE BOOK

This book is intended for engineers, chemists, and others who already have had a course in basic thermodynamics. They will already understand vapor pressures, Gibbs free energies, activity coefficients, and fugacity coefficients. However, the strategy of how to use information from college thermodynamics books is seldom taught. The following are questions of great industrial importance:

- *When* is it appropriate to use a given technique?
- *When* are experimental data required?
- *Where* does one go to get information in the literature?
- *Which* literature sources can be trusted?
- *Who* can be called on for contract data measurements?
- *How* are the more common types of data measured?

These are all practical questions, and the object of this book is to answer these and many more. The reader will look in vain for derivations herein. In fact, an attempt has been made to err on the side of too few equations, with compensation by appropriate reference to the basic papers and textbooks in the field.

There should be considerable information herein of interest to chemists. However, the writer is a chemical engineer and the primary thrust of this book is toward those concerned with "process engineering". The term is used in the broadest sense, and includes the entire span of activities from conception of an idea through plant start-up and optimization. It is important to recognize that just as the activities of the chemist and engineer proceed along a spectrum, so too do the data needs.

The very organization of this book is intended to help determine the nature of effort that should be expended for a given set of problems. The time spent on the physical property aspects of a research program should be neither out of proportion nor inadequate to the needs at a given point in time; otherwise, resources are not used optimally. For convenience, the different stages in a project are divided into *probing research, process assessment,* and *process development*. Most companies make similar distinctions but may call these stages by different names. In order to familiarize the reader with this nomenclature, the following definitions are offered.

A. Section I: Probing Research

This is the stage of tentative experimentation and process evaluation which helps to assess the feasibility of new ideas. Early thermodynamic computations can be particularly useful to chemists, screening out some proposed reactions and finding optimal conditions for others. At this stage a preliminary flow sheet should be developed by the engineer associated with the project. The recovery/purification train proposed at this time can make the difference in whether the project is pursued or dropped. This stage requires very quick access to a few fundamental pieces of information. Because of the very preliminary status of the project, new physical property data almost never are measured unless a key data point makes the difference or unless the process innovation is itself a separation process.

Probing research is covered in two chapters: in Chapter 2 the thermodynamics of reactions, with the objective of calculating equilibrium compositions as a function of temperature, and in Chapter 3 the broader range of physical properties concerns, including examination of different types of phase behavior such as azeotropes and phase splitting. Predictive methods are presented. The concepts discussed are appropriate to the needs of a probing project.

B. Section II: Process Assessment

Success in the probing phase leads to a need for more in-depth information. Many questions remain to be answered. At this point there is a need for an engineering assessment of the technical and economic feasibility of the project, to justify the increased manpower that will be invested. During this phase there should be continual communication between the experimentalist(s) and the process engineer who is working on the flow sheets, material balances, and energy balances. All of the major obstacles to development should be identified during this effort, and many of them solved. This interchange results in a steadily improving design and therefore a more realistic assessment of the economic potential of the proposed new technology.

The engineer probably will have access to a sophisticated process simulator. At the very least, he or she will have computer programs for the individual unit operations. The accuracy of these programs, assuming they converge, is not usually limited by the unit operation algorithms. It is limited by the data input accuracy and thermodynamic correlations. Care in assembling the data is justified at this point. Not all data are created equal in accuracy. A professional data evaluator from Britain made this pithy comment: ''I am used to plotting out data and seeing a scatter of perhaps +/− 25 percent, where all the data themselves are claimed to have individually accuracies of +/− 2 or 3 percent.'' Further, a former director of the NBS made this telling comment: ''Half or more of the numerical data published by scientists in their journal articles are unusable because there is no evidence that the researcher accurately measured what he thought he was measuring or no evidence that possible sources of error were eliminated or accounted for.''[7]

For the reasons cited, this book gives a detailed bibliography, and distinguishes between the different types of sources. That is, the user of this book is led first to the sources that represent critically evaluated data and especially those sources that indicate the probable error in the numbers given; this is done in Chapter 4 for both pure component and mixture property data. The concept of testing (''qualifying'') data is also presented, and there is discussion of utilization of limited amounts of mixture data to predict broader ranges of composition, using the power of thermodynamic relationships to best advantage.

Quite often, the needed data cannot be found anywhere in the literature, or contradictory data are found. At this stage it is necessary to apply prediction and correlation techniques for both pure component and mixture properties; this is the subject of Chapter 5. There is nothing fundamentally wrong with making such predictions, provided that the margin of expected error does not materially affect the overall attractiveness of the process or product being developed. In Chapter 5, methods for correlating and predicting phase behavior data of many types are explained. These include use of equations of state and activity coefficient equations. Both low- and moderate- to high-pressure phase equilibria are covered, as are gas solubility, use of electrolytes, polymer-solvent equilibria, solid-liquid equilibria, and properties of petroleum and synfuel fractions.

C. Section III: Process Development

If the technical, economic, and business/marketing factors are all favorable at the end of the evaluation phase, the next step is process development. It is the final phase of research before construction of a plant. A team of people is assigned to evaluate every aspect of the new process and to provide all the data necessary for a final process design. When the product is new, the team usually must have a second objective of providing product for customer samples. It is necessary at this stage of development to determine, by analysis of economic sensitivity to data accuracy, which elements of the physical property package are most important. Replacement of predictions with physical property measurements then may be advised. As with other aspects of the development program, this is the most costly stage in the physical properties work-up. It is not by accident that the measurement program is covered in the third phase. Too many eager researchers want to start expensive data measurement programs before the basic reaction is even demonstrated.

Data measurements can be either contracted out or measured in-house. A discussion of contract data measurements is given in Chapter 6, and a unique listing of the people and organizations willing to take such data is included. This information is based on a survey by the author of those actively involved in data measurement activities in the U.S. An indication is given of the various types of data measured by each investigator, and of his or her level of experience.

Sometimes there is a justifiable desire to have in-house capability to measure physical properties. Unfortunately, all too often this is left to the discretion of each new project team. The result is almost without exception a set of useless data. The only really practical way for such data to be taken is to have one or more people responsible for the data, regardless of the project for which they are being developed. In Chapter 7, information is given on various types of phase equilibrium data measurements which can be reasonably carried out by experienced technical people.

D. Section IV: Applications

Two major examples in Chapter 8 utilized the various principles taught throughout the preceding chapters. In one example, a problem is solved which involves simultaneous chemical and phase equilibria. In the other, the physical property model necessary to simulate the butadiene extraction process is developed. That system is challenging because of large departures from ideality and many discrete chemical components.

A subject which is very useful for the process optimization area is called "availability", "exergy", or "second law" analysis. It is a fundamentally sound approach to the analysis and improvement of chemical processes. Unfortunately, it tends to be made overcomplicated. It is possible to use a few simple rules and apply practical considerations to arrive at the same conclusions with much less agony and effort. These simple approaches are brought together in Chapter 9. That chapter concludes with a thermodynamic overview of choice of Rankine cycle fluids, which also can play a role in process optimization.

In spite of all the progress made and all the data taken, there is much left to be accomplished. Events of the last decade have shown that cooperative efforts and determination can dramatically improve on otherwise random developments in thermodynamics and phase equilibria. The future can be determined by those willing and motivated to do something about it.

REFERENCES

1. National Research Council, National Needs for Critically Evaluated Physical and Chemical Data, Washington, D.C., 1978.
2. U.S. Bureau of the Census, Statistical Abstract of the United States: 1984, 104th ed., Washington, D.C., 1983.
3. **Williams, C. C., III and Albright, M. A.,** Energy Saving Through Better Thermodynamic Data, Preprint 06-76, presented at 41st Midyear Meet., American Petroleum Institute Refining Department, Los Angeles, May 11, 1976.
4. **Albright, M. A.,** Thermodynamic and physical property data — current needs, in *Thermodynamic Needs for the Decade Ahead: Theory and Experiment,* Proc. NSF Workshop, Sandler, S. I., Ed., National Science Foundation, Washington, D.C., 1983, 146.
5. **Klotz, H. C., Copeman, T. W., Vines, H. L., and Miller, E. J.,** What's the impact of bad data on NRU Designs?, *Hydrocarbon Process.,* 62, 84, 1983.
6. **Zudkevitch, D.,** Importance of thermodynamic and phase data to the economics of the chemical industry, in *Thermodynamic Needs for the Decade Ahead: Theory and Experiment,* Proc. NSF Workshop, Sandler, S. I., Ed., National Science Foundation, Washington, D.C., 1983, 186.
7. **Shaw, J. A.,** Data for engineering design, in *Awareness of Information Sources,* Selover, T. B. and Klein, M., Eds., AIChE Symp. Ser. No. 237, Vol. 80, American Institute of Chemical Engineers, New York, 1980, 43.
8. **Miller, E. J.,** Needs for energy intensive technologies, in *Thermodynamic Needs for the Decade Ahead: Theory and Experiment,* Proc. NSF Workshop, Sandler, S. I., Ed., National Science Foundation, Washington, D.C., 1983, 67.
9. **Sengers, J. V. and Klein, M.,** Ed., Technological Importance of Accurate Thermophysical Property Information, Spec. Publ. 590, National Bureau of Standards, Washington, D.C.
10. **Agres, T.,** U.S. chemistry deals from strength, *Res. Dev.,* December 31, 1985.

Section I: Probing Research

The first phase in evolution of new product or process technology is sometimes called probing research. The approaches appropriate for this phase of research are covered in two chapters. Chapter 2 reviews the computations appropriate for single and multiple reactions. These computations can be most valuable in eliminating infeasible reactions before trying them at the bench. It can also help to set appropriate reaction conditions for thermodynamically feasible but reversible reactions. Application of elevated temperature may be necessary, as in the steam cracking of hydrocarbons to ethylene and propylene. Other examples include the dehydrogenation of ethylbenzene to styrene and the cracking of methane to acetylene. High pressures are often necessary to overcome equilibrium constraints, as in the synthesis of ammonia or methanol. Sometimes continuous withdrawal of a product is necessary, as in catalytic distillation processes.

Chapter 3 explores the scoping design techniques that are used to evaluate the technical and economic potential of newly discovered reactions. These techniques also can be used to scope new separation processes before any data are obtained. It is presupposed in this section that no new physical property data will be obtained. Rather, literature data and the rapidly improving predictive methods are called upon. This stage is characterized by rules of thumb, analogies, and creation of a variety of new ideas, some of which will form the basis for subsequent development work.

Chapter 2

NEW CHEMICAL REACTIONS

TABLE OF CONTENTS

I. INTRODUCTION

Development of new industrial chemistry to create new chemicals or to make existing chemicals more efficiently consumes a high proportion of exploratory research spending in the chemical industry. It is important to use every possible tool to increase the probability of success in the chemical reaction phase of research. The use of thermodynamics is just one tool, but it is certainly an important one. Through computation of the chemical equilibrium constant for desired reactions, it is possible to determine the temperature range in which a reaction may be favored thermodynamically. This does not mean that the product in fact will be made, but it does imply that the reaction is possible. If the equilibrium constant is extremely small under any practical operating conditions, then it is a waste of time even to try the experiment. Thermodynamics not only can screen out unfavorable reactions, but also may help to determine which of several competing products might be made in the greatest yield.

The key, then, is to test every reaction for thermodynamic viability before laboratory effort is begun. This has five benefits: (1) chemical equilibrium computations help to prevent effort expended on reactions that are unfavorable at reasonable conditions; (2) for reactions that run reversibly, they help to define reasonable temperature conditions; (3) they help development work by predicting the thermodynamic limit on conversion; (4) they stimulate thought as to other possible reactants; and (5) they give the heat of reaction which is necessary for reactor design.

II. COMPUTATION OF CHEMICAL EQUILIBRIUM

Computation of the expected chemical equilibrium requires the Gibbs free energy of reaction. It is computed from the sum of Gibbs free energies of formation of the products minus the sum for the reactants. If the value is negative the reaction is promising. If slightly positive, it will be equilibrium-limited but may be possible, depending on reaction conditions and stoichiometry. However, most reactions with positive free energies of reaction exceeding 10 kcal/g mol should be considered impractical and not worth attempting at the temperature used in the computations, although it may be possible to find a temperature and pressure regime in which a reaction would be feasible which appears infeasible at room temperature.

The corresponding heat of reaction is computed in a manner analogous to the free energy change of reaction. That is, the sum of heats of formation of the products minus the sum of the heats of formation of the reactants gives the overall heat of reaction. That quantity is important not only for design, but also because it relates to the change in chemical equilibrium with temperature.

A simple way to calculate thermodynamic properties for common substances is to use the tables created at the Thermal Laboratory of the Dow Chemical Company.[1] Heat and Gibbs free energies of formation are tabulated for a large number of substances, for many of which the temperature variation of those properties is also given. As an example, consider a simple reaction such as the gas phase esterification of acetic acid.

A. Example

From pages 229, 423, 449, and 451 of Reference 1 the following data are obtained.

	Ethanol	+ acetic acid ⟷	water	+ ethyl acetate
Enthalpies at 500 K (kcal/g mol)	−58.31	−105.42	−58.28	−108.58
Total heat of reaction = Summation of products minus summation of reactants = −3.31 kcal/g mol (exothermic)				
Gibbs free energies at 500 K	−28.80	−80.16	−52.36	−58.74
Total = −2.41 kcal/g mol				
Equilibrium constant $K_p = \exp(-\Delta G/RT) = 8.62$				
Gibbs free energies at 700 K	−16.73	−69.88	−49.92	−38.49
Total = −1.80 kcal/g mol				
Equilibrium constant $K_p = 3.64$				

$$= \frac{P_{water} * P_{ethyl\ acetate}}{P_{ethanol} * P_{acetic\ acid}}$$

B. Basis of the Tables

The heats and free energies of formation are all developed on a standard basis, with the enthalpy and free energy of formation of the elements defined as zero at all temperatures. Therefore, the heats and free energies of formation of both elements and molecules can be added and subtracted to obtain the values for overall reactions. The American Petroleum Institute (API) enthalpy data base, by contrast, does not have any such relationship between the molecules and should not be used for computation of either heat of reaction or chemical equilibrium. Note that the National Bureau of Standards (NBS)/Dow tables were developed with a basis of 1.0 atm. Conversion of these tables to a basis of 1.0 bar has been undertaken by the NBS.

The heats of formation are usually obtained by bomb calorimetry at 298.15 K. The products of combustion are analyzed to assure that complete combustion took place. Knowing the heats of formation of the products, such as carbon dioxide and water, it is possible to work backward to the heat of formation of the substance being tested. The heat of combustion is an integral part of the computation. Heats of formation can also be obtained indirectly by measurement of enthalpy changes in other reactions where the heats of formation of all other chemicals involved are previously known. Finally, a less accurate method uses the change with temperature in equilibrium constant of a reversible reaction. The heats of fusion and of vaporization are used to convert heats of combustion measured for one phase to those of another.

Free energies of formation are somewhat more difficult to obtain. In addition to the enthalpy of formation, the entropy must also be known. This usually involves measurement of enthalpy changes in the substance in cooling from 298.15 to 4.0 K. It is particularly difficult to measure accurately the heat involved in phase transitions. Spectroscopic predictions are also used, but are of questionable value for the complex molecules now of interest to the chemical industry. A final method of obtaining the entropy is to measure the chemical equilibrium at two temperatures, assuming that all the other free energies of formation are previously known.

C. Effect of Pressure

Pressure has almost no effect on the equilibrium in the reaction of the preceding example. That reaction takes place entirely in the gas phase and the moles on each side of the reaction are the same. When the number of moles decreases with reaction, increasing pressure is favorable to equilibrium. An example would be the high-pressure hydrogenation of liquids. Similarly, when the number of moles increases with reaction, increasing pressure is unfavorable to equilibrium. An equilibrium constant in terms of moles is related to the equilibrium constant in terms of partial pressure and the increase in moles $\Delta\nu$:

$$K_n = K_P\{P/n\}^{-\Delta\nu} \tag{1}$$

The solution for the number of moles of each component at equilibrium is obtained by algebraic formulas related to the stoichiometry.

Partial pressures are adequate when the ideal gas law is obeyed. However, that approximation is inadequate at higher pressures (above about 5 atm) and when there is gas phase association. Returning to the previous example, note that acetic acid does associate. In such cases the partial pressures must be converted to fugacities. This is accomplished by multiplying each partial pressure by a fugacity coefficient. Either of these quantities can be obtained through use of equations of state described in Chapter 5.

D. Effect of Temperature

Where the effect of temperature on the thermodynamic properties of all the reactants and products is known, those values are used directly. In Reference 1 there are tables of the enthalpy and Gibbs free energy of formation as a function of temperature for about 600 substances. The computations should be done with all the values for reactants and products at the same temperature. However, for many substances the Gibbs energy of formation and the heat of reaction are only available at 25°C. Fortunately, the van't Hoff equation gives a ready estimate of the temperature variation.

$$\left[\frac{\partial \ln K_P}{\partial T}\right]_P = \frac{\Delta H_r}{RT^2} \tag{2}$$

E. Effect of Liquid Reactants or Products

The partial molar free energies of all components are identical in all phases at equilibrium. Thus, the free energy of a pure component does not change when there is a shift in phases. However, it is necessary to adjust for the fact that the liquid vapor pressure is not at 1 atm. The equation used is

$$\Delta G_f^0\,(l) = \Delta G_f^0(g) + RT \ln P^*\,(\text{atm}) \tag{3}$$

To illustrate the effect this has, consider the Gibbs energy of formation of liquid acetyl chloride. The free energy of the gas is -50.59 kcal/g mol. The vapor pressure is 0.394 atm at 298 K. The correction is small, resulting in a shift to -51.14 kcal/mol for the liquid.

In the equilibrium expression, the activity of the liquid should be entered. For a pure liquid it will be unity. For liquid mixtures it will be the mole fraction in the liquid times its activity coefficient. That will be true for each liquid constituent. An example of the computation of a complex liquid phase chemical and phase equilibrium problem is given in

Chapter 8. Further discussion of the thermodynamics of condensed phases may be found in Reference 18.

III. PREDICTIONS OF CHEMICAL EQUILIBRIA IN THE ABSENCE OF DATA

What can be done when the chemicals of interest are not part of the thermodynamic tables? The best solution in that case is to use group contribution methods. Rather than conduct a thorough search of the literature for the appropriate thermodynamics, it is more efficient to use a good-quality group contribution method to give the required value. It usually will be sufficiently accurate for screening purposes. The literature search approach is likely to consume more time and quite often gives no conclusive result.

A. Thermochemical Prediction Methods Based on Group Contributions

There are a number of methods available for making group contribution predictions of thermochemical data.[2] The *Manual for Predicting Chemical Process Design Data*[19] of the Design Institute for Physical Property Data (DIPPR) also reviews such methods. The Handrick method[16] should be considered for hand calculations. However, the method most accepted by chemical industry experts for filling in missing thermochemical data is the Benson method.[3] The method is second order in that it takes into account not only the groups themselves but also the effect of nearest neighbors.

Fortunately, the American Society for Testing and Materials (ASTM) has developed a computerized version of the method called CHETAH.[4] The program calculates heats and free energies of formation, overall values for balanced chemical equations, heats of combustion, vapor heat capacities, and hazard potential parameters. The user can specify any temperature. The program, which includes all the associated data files, can be ordered at a very modest charge from

ASTM, Attn. P. Lively
1916 Race Street, Philadelphia, PA 19103
(215) 299-5536

At this writing the version distributed was #4.3. The program comes with a copy of the manual and some data input forms. This version contains 584 groups and 117 correction factors. Nevertheless, a little practice in working with the Benson method makes the program easy to use. Committee E-27 of the ASTM is working on an expanded and improved version.

B. Use of the CHETAH Program

Consider a seemingly complex molecule such as maleic anhydride. It is constructed from two doubly bonded carbons, two carboxyl groups, and one oxygen atom. However, it is necessary to specify the nearest neighbors (singly attached ligands, at least one of which must be polyvalent) in order to arrive at the correct answer. The nearest neighbors are given in parentheses. Thus, the molecule is specified as follows.

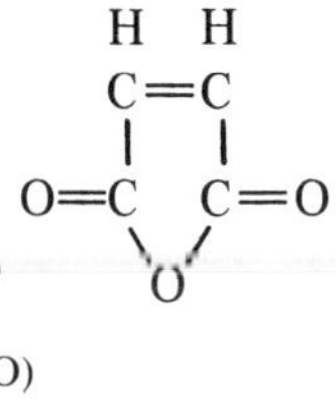

Group	*n*
C_D–(H) (CO)	2
CO–(C_D) (O)	2
O–$(CO)_2$	1
Ring structure	1

C_D–(H) (CO) is the group specified for two of the carbons, indicating that it is coupled to one carbon through a double bond, to a carbonyl group through a single bond, and to one hydrogen atom.

CO–(C_D) (O) is the designation for a carbonyl group coupled to an oxygen atom and a doubly bonded carbon. There are two of this group.

O–$(CO)_2$ represents the oxygen which is coupled to two carbonyl groups. In addition, the program allows for inclusion of a correction factor for the effect of ring strain.

A significant feature of the CHETAH program is inclusion of molecules as well as groups. Molecules not included in the list can be created by adding and subtracting groups from similar molecules which are on the list.

As an example consider terephthalic acid, a molecule not included in the original data bank. It can be constructed by taking two molecules of benzoic acid and subtracting one molecule of benzene, leaving one aromatic ring with two carboxyl groups. The enthalpy of formation is thus computed as −158.5 kcal/g mol. This compares well with the experimental value of −161.4 kcal/g mol. A similar approach can be used to modify other molecules, adding or subtracting groups as well as molecules.

IV. FREE ENERGY MINIMIZATION PROGRAMS

One disadvantage of the CHETAH program is that it is applicable only to single balanced equations or to coupled reactions. Computation of complex chemical equilibria, such as one might encounter with ethane steam cracking, requires more sophisticated software. In essence, free energy minimization techniques are required. Several computer programs have been used by industry for this purpose.

A. The Rand Method

The M. W. Kellogg Company used a program developed by Adler (see Dluzniewski and Adler[5]). Some of the computational methods[6-8] were developed by the Rand Corporation. An actual industrial example of the application of this program was the hydration of propylene to isopropanol in phosphoric acid. Two liquid phases can form between water, isopropyl ether, and isopropanol; there is a complex vapor-liquid and chemical equilibrium established. Reference 17 presents a lengthy FORTRAN computer program using the Rand method. It is applicable to systems with liquid and vapor phases in equilibrium, or either phase separately. The same reference gives another FORTRAN program for the algorithm developed by Villars, Cruise, and Smith. The authors suggest that this second algorithm is well suited to multiphase programs.

The actual program used by Rand can be obtained from

The Rand Corporation, Computer Services Department
1700 Main Street, Santa Monica, CA 90406
(213) 393-0411

It would be appropriate to ask for "Chemist — The Rand Chemical Equilibrium Program". Useful accompanying documents are #RM-5404-PR and #RM-5426-PR. This program requires the user to input Gibbs free energies for each component at the specified temperature of interest. Only isothermal computations are permitted.

B. NASA-Lewis Program

A program developed by the NASA Lewis Research Center is distributed freely to organizations working on U.S. Government projects which can take advantage of its capabilities. For 62 compounds of interest to the propellant and aerospace industries, thermodynamic and propulsion characterizations can be made for multicomponent reacting mixtures. There is no modeling of gas phase nonidealities; ideal gas phases are required. Phases other than the gas phase are restricted to pure compounds. Both thermodynamic and transport properties are available at high temperatures.[9-11] Both isothermal and adiabatic reactions can be specified. The computer programs, designated LEW-14166 and LEW-11740, are available commercially from

COSMIC Computer Center
112 Barrow Hall, University of Georgia, Athens, GA 30602
(404) 542-3265

C. Simulation Sciences Process Simulator

Process simulator programs provide a convenient environment for doing reaction equilibria computations by free energy minimization. The Simulation Sciences Inc. flow sheet simulator can handle this problem, although it is limited to a single phase, either vapor or liquid; solid components are not allowed. The program can draw upon a data bank which is on the order of 1000 components. A practical example is given in the Simulation Sciences manual, for combined methanation and shift reactions. In the methanation reaction, synthesis gas combines to form methane and water. In the shift reaction, the water combines with carbon monoxide to form carbon dioxide and hydrogen. The flows in moles per hour are given below. The reaction was specified to have occurred at 650°F and at a pressure of 90 psia.

	Feed (mol/hr)	**Product (estimate)**
Water	50	4,000
Carbon monoxide	4,150	50
Hydrogen	12,600	200
Carbon dioxide	80	100
Methane	1,420	6,000

Simulation Sciences Inc.
1051 W. Bastanchury Road, Fullerton, CA 92633
(714) 879-9180

D. Chemshare Method

The Chemshare Corporation has introduced a generalized equilibrium reactor computation module. It is restricted to gas phase reactions; either isothermal or adiabatic. As data input, heat and entropy of formation at 25°C are required. Standard equations of state can be used to model the gas phase nonidealities.

Chemshare Corporation
P.O. Box 1885, Houston, TX 77251
(713) 627-8945

E. University of Pennsylvania Method

In the search for better free energy minimization methods there are opportunities to reduce the failure rate of computational algorithms. It is in this area that the University of Pennsylvania has made its contribution. Improvements have been made in computation of complex reaction equilibria, liquid-liquid-liquid equilibrium (LLLE), vapor-liquid equilibrium (VLE), and electrolyte equilibria.[12-14] These techniques are incorporated into the ASPEN program and have been found by industry to work very well.

The ASPEN process simulator was developed at the Massachusetts Institute of Technology with funding from the U.S. Department of Energy. One of the advantages of the simulator is the ability to handle solids. There is a public domain version available at reasonable cost from

National Energy Software Center
9700 South Cass Avenue, Argonne, IL 60439
(312) 972-7250

There are also various commercial operations selling enhancements to the public domain version. Two of the more important companies are

J. S. Dweck, Consultant, Inc.
6000 East Evans Avenue, Building I, Suite G-20, Denver, CO 80222
(303) 758-6862

ASPENTECH
251 Vassar Street, Cambridge, MA 02139
(617) 497-9010

A microcomputer program for free energy minimization has also been developed. CHEMEQU is written in FORTRAN and runs under UNIX. Unfortunately, it is somewhat machine-specific, having been developed on a FORTUNE computer requiring 512 K of RAM and a 10-megabyte hard disk. It is advertised[15] at very nominal cost.

Prof. Ralph W. Pike, Department of Chemical Engineering
Louisiana State University, Baton Rouge, LA 70803
(504) 388-1426

V. SUMMARY

The application of thermodynamic principles gives exploratory researchers many powerful tools. The most important of these is the ability to screen out reactions which have unfavorable chemical equilibria. A large amount of basic thermodynamic data, highly accurate and reliable, has been published. Where data are missing, there are predictive methods available. The Benson method is recommended, and it is available in an inexpensive computer program. For equilibria in multiple reactions there are a number of commercial computer programs available. These involve sophisticated free energy minimization algorithms. Some are able to take into account the changes which occur on condensation into one or more liquid and even solid phases. Those same programs can be applied, in the absence of reaction, to the prediction of phase splitting.

REFERENCES

1. **Stull, D. R., Westrum, E. F., and Sinke, G. C.,** *The Chemical Thermodynamics of Organic Compounds*, John Wiley & Sons, New York, 1969.
2. **Reid, R. C., Prausnitz, J. M., and Sherwood, T. K.,** *The Properties of Gases and Liquids*, McGraw-Hill, New York, 1973.
3. **Benson, S. W. et al.,** Additivity rules for the estimation of thermochemical properties, *J. Am. Chem. Soc.*, 69, 279, 1969.

4. **Seaton, W. H., Freedman, E. H., and Treweek, D. N.,** CHETAH, the ASTM Chemical Thermodynamic and Energy Release Evaluation Program, ASTM DS5, Manual and Computer Program, American Society for Testing and Materials, Philadelphia, 1984.
5. **Dluzniewski, J. H. and Adler, S. B.,** Calculation of complex reaction and/or phase equilibria problems, Int. Chem. Eng. Symp. Ser. No. 5, *Int. Chem. Eng.*, 4, 21, 1972.
6. **Boynton, F. P.,** Chemical equilibrium in multicomponent polyphase systems, *J. Chem. Phys.*, 32, 1880, 1960.
7. **Oliver, R. C., Stephanou, S. E., and Baier, R. W.,** Calculating free energy minimization, *Chem. Eng.*, 69(4), 121, 1962.
8. **Balzhiser, R. E., Samuels, M. R., and Eliassen, J.,** *Chemical Engineering Thermodynamics*, Prentice-Hall, Englewood Cliffs, N.J., 1972.
9. **Zeleznik, F. J. and Gordon, S.,** Calculations of complex chemical equilibria, *Ind. Eng. Chem.*, 60(6), 27, 1968.
10. **Gordon, S. and McBride, B. J.,** Computer Program for Calculation of Complex Chemical Equilibrium Compositions, Rocket Performance, Incident and Reflected Shocks, and Chapman-Jouguet Detonations, NASA SP-273, Washington, D.C., 1976.
11. **Gordon, S., McBride, B. J., and Zeleznik, F. J.,** Computer Program for Calculation of Complex Chemical Equilibrium Compositions and Applications, Supplement I — Transport Properties, NASA 86885, Cleveland, 1984.
12. **Gautam, R. and Seider, W. D.,** Computation of phase and chemical equilibrium, *AIChE J.*, 25, 991, 1979; II. Phase splitting, *AIChE J.*, 25, 999, 1979; III. Electrolytic solutions, *AIChE J.*, 25, 1006, 1979.
13. **White, C. W., III and Seider, W. D.,** Computation of phase and chemical equilibrium. IV. Approach to chemical equilibrium, *AIChE J.*, 27, 466, 1981.
14. **Chakravarty, T., White, C. W., III, and Seider, W. D.,** Computation of phase equilibrium: optimization with thermodynamic inconsistency, *AIChE J.*, 31, 316, 1985.
15. Applications Software Survey for Personal Computers, 1984, American Institute of Chemical Engineers, New York, 1984.
16. **Handrick, G. R.,** Heat of combustion of organic compounds, *Ind. Eng. Chem.*, 48, 1366, 1956.
17. **Smith, W. R. and Missen, R. W.,** *Chemical Reaction Equilibrium Analysis: Theory and Algorithms*, John Wiley & Sons, New York, 1982.
18. **Walas, S. M.,** *Phase Equilibria in Chemical Engineering*, Butterworths, Reading, Mass., 1985.
19. **Danner, R. P. and Daubert, T. E.,** *Manual for Predicting Chemical Process Design Data*, American Institute of Chemical Engineers, New York, 1983.

Chapter 3

SCOPING PROCESS DEVELOPMENT AND DESIGN

TABLE OF CONTENTS

I. INTRODUCTION

At a very early stage in the development of a new chemical process, some thought must be given to phase equilibrium issues. The chemist, if running a liquid phase or gas-liquid reaction, must usually give at least some consideration to the pressure and temperature relationships. The author has examined pressurized reactions, claimed to be entirely in the gas phase, where the product was actually a liquid under reactor conditions; therefore, the residence time was very much greater than had been thought. The phase equilibrium of the reaction itself can be important not only for scale-up and final design, but also sometimes to make the reaction proceed properly. Therefore, an engineering assessment of a reaction can influence the course of the early experiments.

A specific example of the importance of phase equilibria to chemical reactions is in the liquid phase oxidation of *o*-xylene to phthalic anhydride. An acid/anhydride equilibrium exists in the reactor and of course it is strongly influenced by the water concentration.

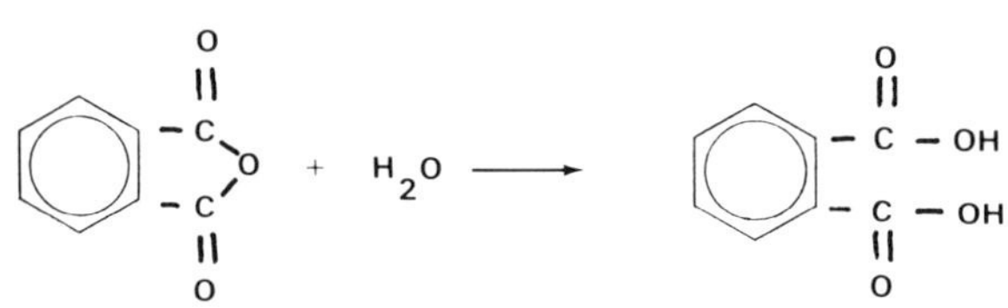

(1)

As a further complication, too much water deactivates the catalyst. On the other hand, too little water results in excessive anhydride, which gives catalyst precipitation.[2] Until the phase equilibrium of the system was understood and the pressure properly set to adjust water concentration, it was not possible to make the reaction proceed properly. The approach used to solve this problem is described in Chapter 8.

Shortly after success with a probing experiment, a scoping process design and economic analysis should be performed. It determines the viability of future research based on the newly demonstrated chemistry. At this stage the overall process is conceptualized, including the product recovery and purification sections. Notwithstanding the importance of this evaluation, the needed data are usually sparse. Laboratory data beyond a reactor yield are rare. If the laboratory product has been recovered, it is usually by methods that are expensive commercially. By use of physical properties, more economical recovery methods often can be devised. However, with probing project status, physical property measurements can seldom be justified. This chapter addresses some of the simple things that can be done to overcome these obstacles in creation of scoping designs.

Phase behavior does not just govern the size of the equipment. The ease or difficulty of potential separations can be used *a priori* to eliminate some proposed new chemical processes. In fact, the separations process in some cases may determine the best sequence of chemical reactions. Thus, thermodynamic/phase equilibrium data are important because they are an integral and indispensable part of the inventive process. The following example will illustrate.

A. Example: Combining Reaction with Phase Equilibria

A knowledge of thermodynamic and phase equilibrium relationships was absolutely essential to the development by Chemical Research and Licensing Co. of their process for methyl-*tert* butyl ether (MTBE). The following thermodynamic computations show that the reaction is exothermic and that it does not go to completion under "normal" reaction conditions.[1]

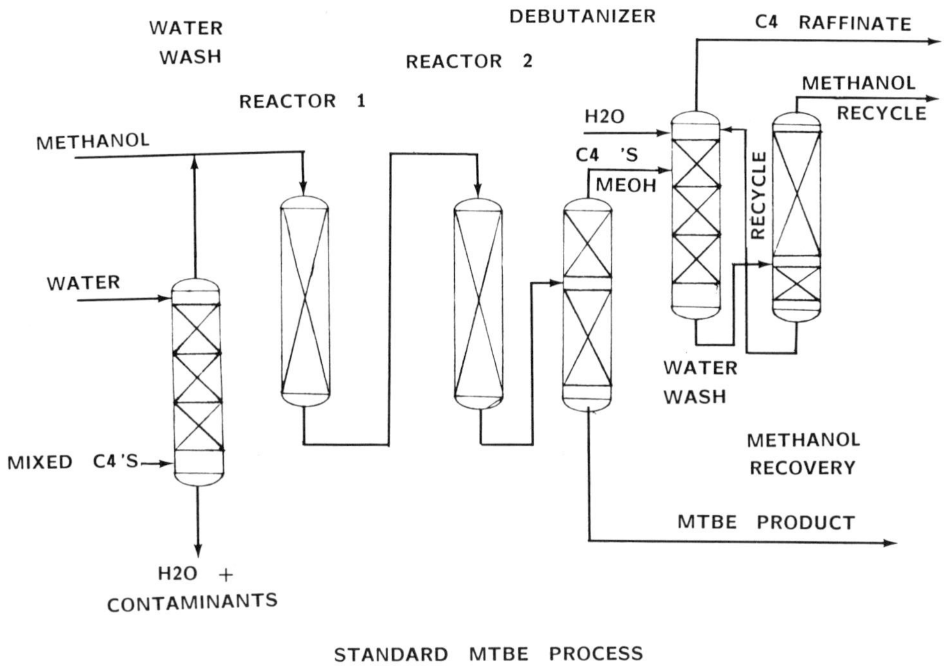

FIGURE 1. Flow sheet for a conventional MTBE process. Note the use of a water wash on the C_4 feed, two reactors which do not extract more than the equilibrium amount of isobutylene or methanol, a debutanizer, and a methanol recovery section that can be quite large. (Courtesy of and schematic representation by Chemical Research and Licensing Co. of Houston, Tex. Redrawn with permission.)

	Methanol +	isobutylene ↔	MTBE
Heats of formation (400 K)	−48.95	−5.43	−72.18

Heat of reaction = −17.8 kcal/g mol (exothermic)

Free energies of formation	−35.54	20.23	−15.97

Free energy change of reaction = −0.66 kcal/g mol (2)

Given a reaction temperature of 400 K, the equilibrium constant for vapor phase reactions is 2.3.

Although the reaction is reversible, it can be driven to completion by carrying out an external distillation and recycling the unreacted feeds. Such a process is illustrated in Figure 1.

Advances in the art were made by combining the MTBE reaction with the distillation. This simultaneously solves the problems of reversibility and heat removal. The distillation tower is constructed with packing instead of trays. The packing consists of catalyst pellets contained in baskets. The heat of reaction reduces the reboiler steam requirement. Because the operation takes place in a distillation column, product can be removed from the bottoms free of methanol or isobutylene. In effect, the equilibrium limitation is completely by-passed. Figure 2 illustrates this new concept known as the CR&L process.[1] In Figure 3 the internal configuration of the reactor/distillation tower is illustrated. For further information, contact

Mr. Larry Smith, President
Chemical Research and Licensing Co.
P.O. Box 34687, Houston, TX 77034
(713) 943-0246 and (713) 941-4646

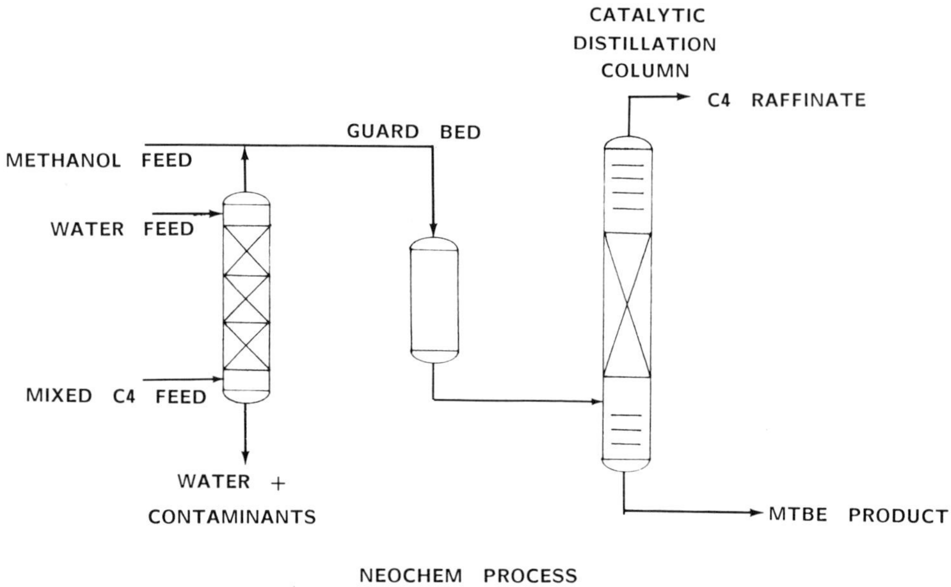

FIGURE 2. Flow sheet for the CR&L MTBE process. A guard bed is used to protect the catalyst from water. The reactive distillation column removes the MTBE product continuously from the bottom. Methanol is continuously removed from the top of the reactor in small concentration (3 to 6%) as an azeotrope with the butylenes and is recovered in a smaller distillation section. (Courtesy of and schematic representation of technology owned and licensed by Chemical Research and Licensing Co. of Houston, Tex. Redrawn with permission.)

The same company has applied reactive distillation to recovery of isoamylenes.[58] These olefins are normally recovered by extraction with sulfuric acid. According to company claims, the reactive distillation process reduces energy consumption by half, and purity is improved from 95 to 99%. The process involves a selective reaction with an alcohol to form an ether. Only the isoamylenes react. Since the reaction is carried out with simultaneous distillation, the equilibrium problems are overcome. The heavy ether is removed continuously from the bottom of the column, while the isoamylene-lean fraction (raffinate) is withdrawn from the top. The ether then goes to a second catalytic distillation where the process is reversed. The alcohol is removed from the bottom of the tower and recycled to the first tower, while the product isoamylene is removed as a distillate. A sulfonic acid catalyst contained in a glass fiber and stainless steel bundle serves as both tower packing and catalyst in both towers.

There are undoubtedly other reactions where this concept could be profitably employed; for example, successive reactions where a certain molecular weight product is needed could potentially be run in such a process. However, simulation of such systems is difficult. It requires simultaneous solution of the equations for vapor-liquid equilibrium (VLE) and chemical equilibrium. One computer program capable of simulating such processes is FRACHEM, developed by OLI Systems Inc. It is designed to handle distillation of electrolyte systems where either chemical equilibria or chemical kinetics are important throughout the distillation column. Nonelectrolyte chemical equilibria are handled in a straightforward manner. The computer tape can be purchased; 90-day trials of the computer program can also be arranged for a reasonable cost. Industry response to this computer program has been favorable.

OLI Systems Inc.
52 South Street, Morristown, NJ 07960
(201) 540-0291 and (201) 540-0353

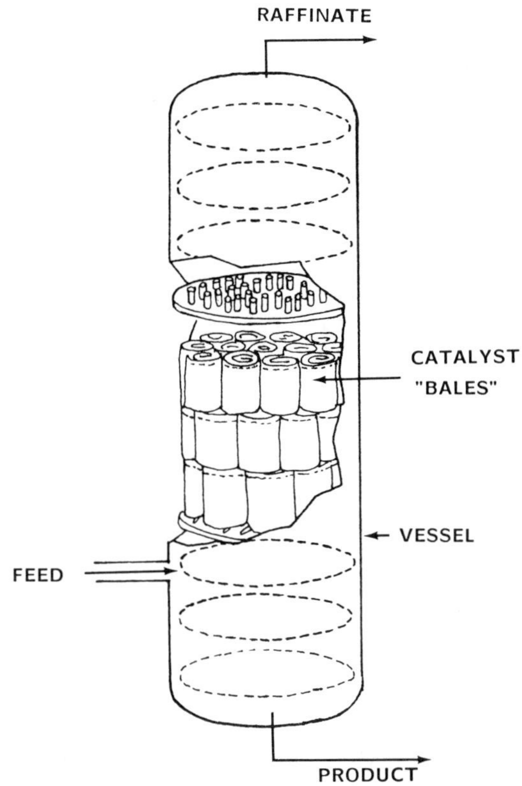

FIGURE 3. CR&L reactor internals showing catalyst bundle. The catalyst "bales" are configured like rolls of metal mesh. They have an open structure to permit passage of gas and liquid, but the mesh has openings small enough to retain the catalyst. The section containing catalyst operates as both a reactor and a packed bed in a distillation column. There are trays above and below the packed bed to remove methanol and isobutylene, respectively. (Courtesy of and schematic representation of technology owned and licensed by Chemical Research and Licensing Co. of Houston, Tex. Redrawn with permission.)

In 1985 a theoretical framework for assessment of reactive distillation was published.[72] It includes as an example the separation of meta-xylene from para-xylene using organosodium compounds with crown ether chelating agents. A modest attempt to demonstrate the predictions experimentally also was reported.[73]

Major improvements in chemical processes for the commodity chemicals are more difficult to achieve than in previous decades because of the maturity of the industry. This generally accepted fact increases the importance of accurate physical property information in the choice of the best separation process and subsequent minimization of overdesign and energy consumption. The author hopes that this chapter will assist in the creative process by giving a better understanding of nature. A qualitative feel for the interactions between molecules can expedite choice of potential entrainers or liquid extraction agents for difficult phase separations. It is also helpful to have the insight to spot optimal combinations of unit operations, such as was accomplished with the CR&L process. Such insights have the potential to enhance the productivity of research, including the first scoping economic appraisals of new processes.

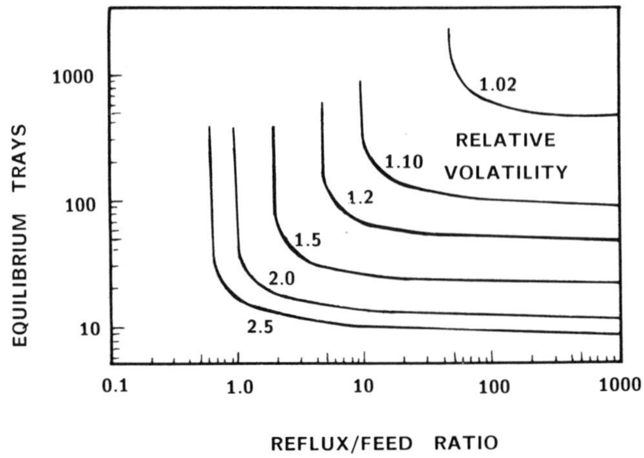

FIGURE 4. Effect of relative volatility, for ideal binary mixtures, on reflux ratio and equilibrium stage requirements in a distillation column. A 50/50 molar feed ratio is assumed. The light product purity specification is 99.5%. The recovery specification for the light component is 98%. Note the enormous increase in distillation requirements as the relative volatility declines. (Adapted from Evans, H. D. and Sarno, D. H., *Seventh World Petroleum Congress Proceedings*, Vol. 5, Elsevier, New York, 1967, 259.)

II. TYPES OF VAPOR-LIQUID PHASE BEHAVIOR

A. Ideal Systems

An ideal system in the context of VLE is one that obeys Raoult's Law. In those rare instances when it applies, the partial pressure of a pure component in a vapor mixture is equal to its mole fraction in the liquid times its vapor pressure at the system temperature.

$$y_iP = x_iP_i^* \tag{3}$$

For such mixtures the relative volatility (α) does not change with composition. The shape of the VLE curve can be predicted knowing only the vapor pressures. For a binary mixture,

$$y_i = \frac{\alpha\ x_i}{(1 + x_i(\alpha - 1)} \tag{4}$$

Figure 4 illustrates the effect of volatility on separation requirements for systems which are close to ideal.[3] The cost of separating close-boiling mixtures is extremely sensitive to the relative volatility. Based on a binary mixture with a 50/50 molar feed ratio, the computations assume 98% recovery at 99.5% purity of the light component. At a relative volatility of 1.10, the separation requires 200 theoretical plates and a reflux ratio of 11/1. If the relative volatility is as low as 1.02 the separation is all but impossible, with a requirement of 700 theoretical plates and a reflux ratio of 90/1! On the other hand, an increase in relative volatility to 2.0 reduces the theoretical plate requirement to 20 and the reflux ratio to 1.5. Addition of another constituent which increases the relative volatility of the key components thus can improve drastically the technical and economic feasibility of the separation. Judicious choice of this separating agent, deliberately introducing nonideal behavior, can often lead to development of a patentable process.

Neither azeotropes nor multiple liquid phases can exist in ideal mixtures. Molecules in ideal mixtures distribute themselves randomly. This is achieved only with molecules that are the same size and shape and that have identical intermolecular forces, criteria which are virtually never met. Obviously, then, we are not dealing with a law at all but rather with an ideal from which the real world deviates.

There are some circumstances when ideality can be assumed. At low pressure, if high accuracy is not essential, the xylene isomers can be considered to form ideal mixtures. A heptane + octane mixture might also be considered ideal as a first approximation. In either of these cases, the error involved could be as much as a couple of percent in vapor mole fraction. As a rule of thumb, one can consider that intermolecular force differences are usually most important, size differences are next, and shape differences are least important.

Even if the molecules behave ideally in the liquid phase, increasing pressure creates nonideality in the gas phase. This is taken into account using fugacity coefficients (ϕ_i). They are calculated from equations of state such as the Redlich-Kwong equation.[4] The virial equation, the BWRS, or other equations of state also can be used. The procedures are given in Chapter 5. For highly nonideal systems of the type usually encountered in scoping designs, the liquid phase nonidealities will grossly outweigh the effect of vapor phase nonidealities so the vapor phase nonidealities are often assumed to be unity at this stage.

B. Nonideal Systems

In the real world, the different functional groups on molecules create very strong nonrandomness in the liquid mixture. Energy effects created by size and shape differences also contribute to nonideality. These effects can be taken into account by using the "activity coefficient" (γ_i) as a multiplier on the vapor pressure in Equation 3. The activity coefficient is not a "fudge factor", but a true thermodynamic quantity. It is related to the excess free energy G^E by Equation 5.

$$G^E = RT \sum_i x_i \ln \gamma_i \tag{5}$$

A rigorous expression for the K value using activity and fugacity coefficients is given in Equation 6.

$$K_i = \frac{\gamma_i P_i^* \phi_{i,P_i^*}}{P \hat{\phi}_{i,P}} \exp\left[\frac{(P - P_i^*) V_i^L}{RT}\right] \tag{6}$$

One common standard state assumed here for calculating activity coefficients is the pure liquid compound at the system temperature and pressure. Accordingly, the activity coefficient for a pure liquid is unity. The activity coefficient of a component in a mixture approaches unity tangentially as the concentration of that component approaches 100%. Activity coefficients are almost always at a maximum at infinite dilution. Figure 5 illustrates this concept for the acetonitrile + benzene system at 45°C.[11] The characteristic sought in activity coefficient equations is the ability to predict correctly the shape of the activity coefficient curve. Such equations also should predict the effect of temperature and the multicomponent behavior from information on binary activity coefficients.

For a thorough discussion of the derivation of the thermodynamic equations, the reader should refer to a number of standard textbooks.[5-10] As promised, the author has avoided cluttering this discussion with derivations. Only the working equations are given, with an explanation of when and how to use them.

C. Effects of Nonideality

Some effects of nonideal liquid mixtures are illustrated in Figure 6, a diagram of four

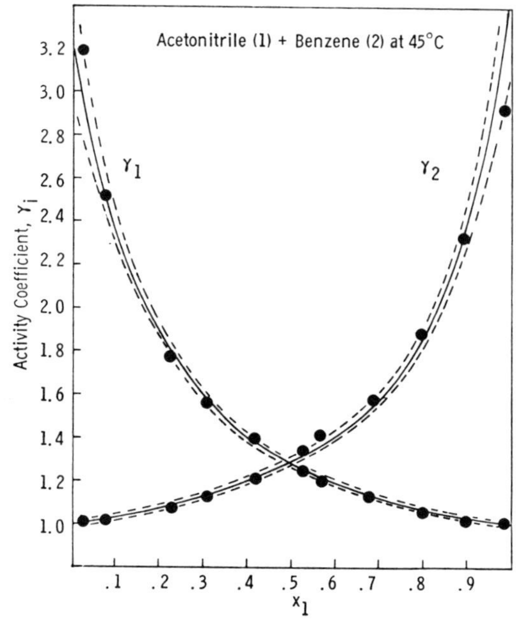

FIGURE 5. Activity coefficients for acetonitrile (1) + benzene (2) at 45°C. The dashed lines indicate the probable range of experimental uncertainty based on composition measurements, temperature, and pressure uncertainty. (Reproduced from Palmer, D. A., Measurement and Correlation of Thermodynamic Excess Properties for the Acetonitrile + Benzyne + *n*-Heptane System, Ph.D. thesis, Washington University, St. Louis, 1971.)

binary VLE. Traditionally, the more volatile component is on the right-hand side, so that the right-hand side of the diagram represents the rectification or top part of a distillation column. The stripping or bottom part of the column is represented on the left-hand side. Curve a illustrates Raoult's law, with unity activity coefficients. The shape of the curve is essentially determined by the relative vapor pressures of the pure components. Curve b shows that as the activity coefficients increase, stripping becomes easier, but rectification becomes difficult or impossible. Increasing the activity coefficients even more creates an azeotrope in Curve c. Finally, in Curve d, the activity coefficients are so large that a horizontal line develops which spans a region of two liquid phases.

From this illustration it is possible to make some important points. Azeotropes are a simple consequence of activity coefficients — usually due to activity coefficients greater than unity. Thus, azeotropes can be predicted by correlation of activity coefficients. Partial miscibility likewise occurs as a result of activity coefficients; in this case, they must exceed unity.

Activity coefficients exceed unity when the molecules are dissimilar and tend to aggregate more with each other than with the dissimilar molecules. Thus, pockets of high concentration of one component or the other are created on a molecular level. The greater the aggregation, the higher the activity coefficient and the greater the positive value of the excess Gibbs free energy. The logical extension of excessive aggregation is mechanical splitting into two or more liquid phases. The *n*-octanol + water mixture is an example of such a system. The higher the activity coefficient for a component, the greater the partial pressure it will exhibit. However, the partial pressure cannot exceed the vapor pressure under any circumstances.

Occasionally, the forces attracting unlike pairs of molecules are stronger than forces

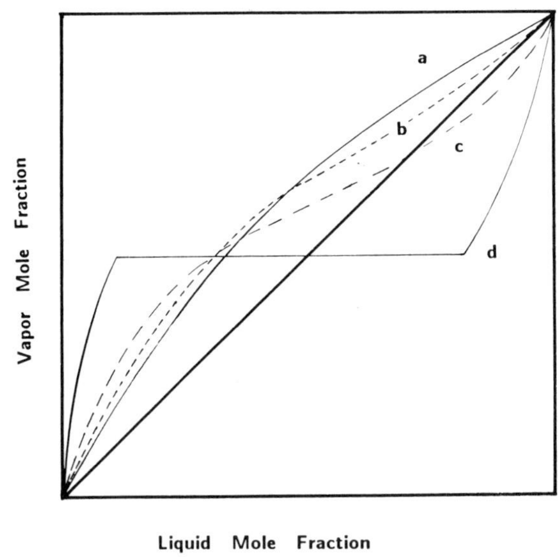

FIGURE 6. Effect of positive deviations from ideality on a binary VLE diagram. Curve a indicates Raoult's law for a mixture with a constant relative volatility of 2.0. Curve b indicates a similar mixture but with activity coefficients exceeding unity. Curve c indicates fairly large activity coefficients, resulting in formation of an azeotrope. Curve d has activity coefficients so large that two liquid phases have formed.

between like molecules. In these cases the excess Gibbs free energy may be negative, and activity coefficients will be less than unity. This negative deviation from ideality is normally caused by formation of chemical complexes; the ammonia + water system is one such example. In some regions the formic acid + water mixture also exhibits negative deviations from ideality.

What are the molecular properties which contribute to the activity coefficients? The first approach to this problem is to see an example for a real system. Data measured[12] on the acetonitrile extraction system for butadiene will illustrate. Activity coefficients were measured for butadiene and for the range of other constituents in the feed to a butadiene extraction unit. First, the effect of *molecular shape* will be examined. The activity coefficients relative to that of 1-butene (at infinite dilution) were as follows.

1-Butene	1.0
cis-2-Butene	0.96
trans-2-Butene	1.0
Isobutene	0.93

It is obvious that the effects of molecular shape are measurable and important for final design. However, they probably could be safely neglected in a scoping design. *Molecular size* is somewhat more significant. In the following numbers, the activity coefficient of a molecule in acetonitrile is compared to that of a smaller molecule having the same amount of unsaturation.

Isobutane	1.0
Isopentane	1.26
1-Butene	1.0
3-Methyl-1-butene	1.36

Propadiene	1.0
1,2 Butadiene	1.11
Propyne	1.0
1-Butyne	1.24

It is evident that the addition of just a single carbon makes a marked difference. That difference is equal to the entire relative volatility in some large commercial separations. When there is an extremely large difference in size of alkanes, two liquid phases can form.

A different order of effect can be seen with changes in *molecular forces* due to increasing unsaturation. There is an attraction between the very strong dipole of acetonitrile and the excess electrons in the unsaturated molecules. The following data were reported on volatilities at infinite dilution in acetonitrile, relative to 1,3 butadiene.

n-Butane	3.7
trans-2-Butene	2.0
1,3 Butadiene	1.0
2-Butyne	0.76
Vinyl acetylene	0.52

These data show why extractive distillation works for butadiene separation. The major diluents are olefins and paraffins whose volatility is enhanced from 2 to 3.7 times relative to butadiene by the presence of the solvent. The acetylenes must be removed in a separate step, since their volatility is decreased relative to butadiene and they will come out the bottom of the tower with butadiene.

Another very polar molecule is dimethyl formamide. It is not surprising to find that it dissolves acetylene, butadiene, and isoprene. It is also a good solvent for the acid gases sulfur dioxide and hydrogen chloride. Having a basic character, it is a poor solvent for ammonia.

There are of course many functional groups to consider. Many technologists have an intuitive feel for the fact that a molecule with a high oxygen content will be immiscible with hydrocarbons. But what of the interaction of a nitrile with an alcohol? As a general rule of thumb, mixtures of highly polar with nonpolar materials will exhibit large positive deviations from ideality. In some cases there will be two liquid phases or an azeotrope. Sometimes the dipole moment can be used as a numerical measure of the degree of polarity.[13]

Mixtures of polar materials will exhibit widely varying degrees of nonideality. Asymmetric activity coefficient and excess free energy curves are common in such systems. Ultimately, what is needed to correlate data on such systems is an analytical equation which accounts for size, shape, and functional group interaction energies. Such an equation (UNIFAC) is examined later in this chapter. To obtain a qualitative feel for the mixing characteristics in a highly nonideal mixture there are other possibilities.

Critical solution temperatures gives a qualitative idea of intermolecular compatibility.[14] The higher the critical solution temperature, the more nonideal the mixture. At the critical solution temperature the infinite dilution activity coefficients will be on the order of 7.4 (when the activity coefficient curves are symmetrical). Examination of tables of azeotropes is also most helpful.[15] A pair of activity coefficients can be derived from such data because the liquid and vapor mole fractions are the same at the azeotrope. The compilations of VLE and liquid-liquid equilibria (LLE) by DECHEMA referenced in Chapter 4 are indispensable to preliminary evaluations. Information on the distribution of solutes between water and organic solvents such as octanol can be helpful. This information is available on-line from Technical Database Services (TDS), as discussed in Chapter 4. If a solute distributes more into the aqueous phase it is generally more polar and/or has oxygen in the molecule.

Conversely, the less polar and/or hydrocarbon-type molecules will distribute into the organic phase.

D. Azeotroping Miscible Systems

Azeotropes are a natural consequence of positive deviations from ideality (increasing activity coefficients). Although various empirical approaches have been developed, they should be ignored in favor of thermodynamic treatment. In fact, it is possible to derive activity coefficients from such data. Simply convert the weight fractions usually listed to mole fractions. Since y = x at the azeotrope, and since the temperature and pressure are given, by computation of the vapor pressures the activity coefficients are calculated approximately by equation 7.

$$\gamma_i = P/P_i^* \tag{7}$$

Having the activity coefficients for both components at one composition, the parameters in the UNIQUAC, Wilson, or Van Laar equations can be computed by regression analysis. The azeotrope data are not always sufficiently accurate to justify the computation of fugacity coefficient corrections.

If the vapor pressures of two components are very different, a high degree of nonideality is required to cause an azeotrope. On the other hand, two chemicals with almost identical vapor pressures will azeotrope with just a small degree of nonideality. Virtually all azeotropes are minimum-boiling. That is, the bp of the mixture is lower than either pure component. This is a consequence of almost all systems exhibiting positive deviations from ideality. Such minimum-boiling azeotropes are usually encountered at the top of distillation columns.

The effect of temperature on azeotropes can be correlated (not predicted) with considerable accuracy. The correlating equations[53] are

$$\ln P = A_1 + B_1/T + C_1T \tag{8}$$

where the mean fit is within 0.5% (average) of the experimental pressure data. Also, the azeotropic composition is represented within 1.5% (average) by

$$x_i = A_2 + B_2T \tag{9}$$

Parameters for 108 binary systems are given by Tamir.[53]

Although azeotropes are usually considered a problem, they also can represent a solution. The best known azeotrope problem is with the ethanol + water system. It is now a factor of considerable commercial significance with development of a large industry producing anhydrous ethanol for addition to gasoline. The challenge is the removal of the 5% of water in the azeotrope. This is done by adding another agent such as pentane.[16] It forms an azeotrope with water at a temperature of 34.6°C. Being lower-boiling than the ethanol + water azeotrope at 78.2°C, it effectively removes water from the mixture. Because the pentane is immiscible with water, it can be decanted and recycled to the column. Thus, one azeotrope is used to break another. Less well known is the purification of acetic acid contaminated with by-product formic acid and water. A ternary azeotrope at 107°C is broken by using a benzene entrainer. It carries the formic acid and water overhead at about 71°C and 1 atm.

Thus, if faced with a group of chemicals difficult to separate, the situation is usually not hopeless. Often water creates the problem, azeotroping with a number of chemicals one would like to separate. As a strategy, therefore, consider whether any of the products or by-products could be used as an entrainer if added in larger amount. If not, is there an inert hydrocarbon that could be added to separate out the water? Or, if there is no water present,

Table 1
PREDICTION OF HETEROGENEOUS AZEOTROPES, BASED ON PURE COMPONENT VAPOR PRESSURES AT THE AZEOTROPE TEMPERATURE

Hydrocarbon	P (calcd)	y (calcd)	y (exptl)
Cumene	765	42.1	43.8
Octane	764	25.0	25.5
Ethylbenzene	763	32.9	33.0
Styrene	764	40.5	40.9
Heptane	763	12.9	12.9
Hexane	763	5.3	5.6
Benzene	762	8.9	8.9

perhaps its addition may actually help in performing a separation. Wherever possible, it is best not to add components that are not already present in the mixture. Overall, the strategy is to modify the liquid phase environment to give the desired separation in the fewest possible steps.

E. Azeotroping Immiscible Systems

Hydrocarbon + water systems have such low solubility in each other that they can be considered immiscible. They are of considerable importance in both the refining and chemical industries. Volatile hydrocarbons azeotrope with water. The only nonazeotropes in hydrocarbon + water systems occur when there is an extreme difference in vapor pressures. An example of such a case would be use of steam for stripping in a heavy-oil vacuum tower. The water is indeed immiscible, but there is no azeotrope formed.

The solubilities, however low, are in fact known for many hydrocarbons. The American Petroleum Institute (API) *Technical Databook — Petroleum Refining* is one source, and there are many others. Because of the very low solubility, the inverse of the solubility, as a mole fraction, gives the activity coefficient of the solute directly. This procedure is inadequate, however, for other substances which exhibit partial miscibility.

Most of the heterogeneous hydrocarbon/water azeotrope data can be predicted *a priori* using vapor pressure data. This simple and otherwise overlooked fact can be of considerable importance when the literature is full of contradictory observations. As long as there are two liquid layers, both water and the hydrocarbon exhibit an activity (mole fraction times activity coefficient) very close to unity. As a result, each exerts its vapor pressure independently. The sum of vapor pressures at a given temperature gives the equilibrium pressure at that temperature. The ratio of vapor pressure of one component to the total pressure gives the mole fraction of the azeotropic composition. This can be easily demonstrated, taking data at random from Horsley's compilation.[15] For each of the hydrocarbon + water mixtures the observed azeotropic pressure was 760 mmHg. In most cases the calculated pressure is a few millimeters high, but usually within 0.5% of the true pressure. The computed mole fractions of water are remarkably close to the stated values. Table 1 presents predicted heterogeneous azeotropes in terms of weight fractions, to be consistent with Horsley's compilation. Vapor pressures were determined from the Design Institute for Physical Property Data (DIPPR) Data Compilation at the experimental bp temperatures.

Table 1 demonstrates the power of the simple use of vapor pressures for computing heterogeneous azeotropes. Having this tool in hand, contradictory data can be critically examined. For cyclohexane, two measurements were listed. One gave 8.4 wt% water and the other 9.0%. This computational procedure showed that the correct value is really 8.4

Table 2
CORRELATION OF HYDROCARBON SOLUBILITY IN WATER

Component	A	B	C	Solubility at three-phase critical pressure T(K)	x_{hc}
Benzene	−170.0402	6,922.912	24.3987	541.7	0.02528
Cyclohexane	−209.1169	8,325.49	29.8231	529.4	0.001748
n-Hexane	−367.9850	16,128.64	52.8208	496.7	0.000498
Ethylbenzene	−185.1695	7,348.55	26.3452	568.1	0.00583
n-Octane	−343.1497	13,862.49	49.2461	539.1	0.000461
Ethylcyclohexane	−334.2468	14,105.21	47.9310	561.4	0.00335

wt%. A predicted pressure for the first experiment of 763 mmHg but only 748 for the second confirmed that value. The real utility in the procedure will be in applying it to situations where there are no data, including the above systems at other pressures.

These conclusions about additivity of the vapor pressures have been confirmed in a study on high-temperature mutual solubilities of hydrocarbons and water.[46,51] The authors concluded that up to 423 K, the three-phase equilibrium pressure is equal to the sum of the pure component vapor pressures. Going up to the critical end point they found that the critical pressure exceeded the sum of the vapor pressures by only 5 to 10%.

In the same study, correlation parameters were presented for the solubility of C_6 and C_8 hydrocarbons in water. Because of the industrial importance of these mixtures, it is worthwhile reprinting the values (Table 2). The correlating equation is

$$\ln x_{hc} = A + B/T + C \ln T(K) \tag{10}$$

In References 46 and 51 the Henry's constants for these hydrocarbons in water are also given. On the other hand, the solubility of water in hydrocarbons is much greater, so that Henry's constants do not apply up to the three-phase critical point. However, the Redlich-Kwong Zudkevitch-Joffe equation[48] does an excellent job of representing that behavior.

Adler[59] has pointed out that the activity coefficients themselves are often required for design of plants with water-hydrocarbon mixtures. Results obtained from unreliable data can be misleading. After review of the data previously measured,[60-67] Adler concluded that a very useful correlation with respect to composition and temperature can be made using just one equation. This correlation uses the Watson Characterization Factor (WCF), which finds widespread use in the petroleum industry. Values of that factor for pure component hydrocarbons can be found in the API *Technical Databook — Petroleum Refining*. The equation is simply

$$\begin{aligned}\log_{10} \gamma_{H_2O} &= 2.3303 - 0.015837\,T + 0.000018963\,T^2 \\ &\quad + 0.062999\,(WCF) + 0.0015614\,(WCF)^2\end{aligned} \tag{11}$$

F. High-Pressure Phase Behavior

Wichterle[56,57] discusses at length the complicated phase diagrams that can be encountered at high pressure. Significantly above the critical pressure, gas-gas equilibria can exist when there is more than one component. The relationships between such equilibria and ordinary VLE are complex, so the reader should consult the original references for more information.

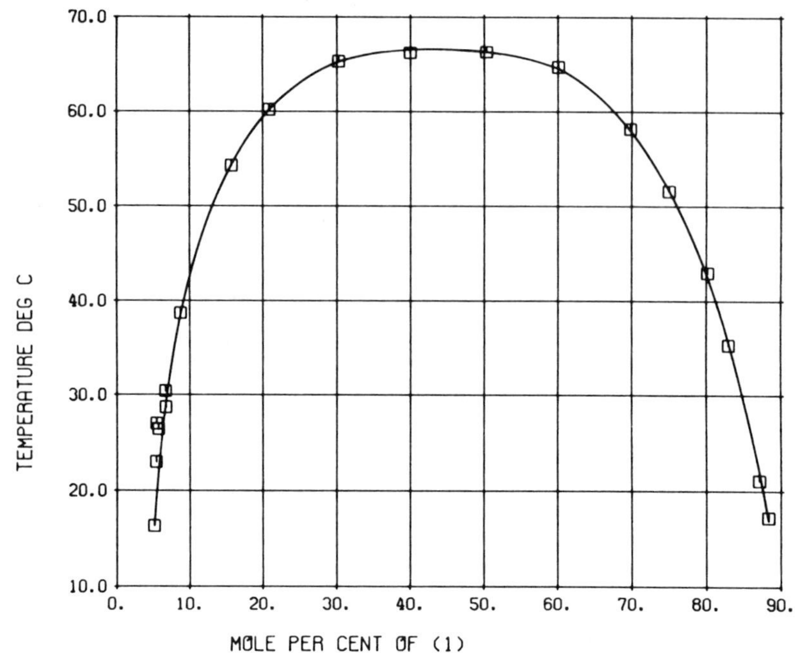

FIGURE 7. Effect of temperature on the miscibility of the furfural (1) + cyclohexane (2) binary mixture. This illustrates the fairly common upper critical solution temperature which can be observed at about 67°C. There is a large miscibility gap at room temperature. (Reproduced from Sorenson, J. M. and Arlt, W., *Liquid-Liquid Equilibrium Data Collection*, Vol. 1 to 3, DECHEMA, Frankfurt, 1979, 1980. With permission.)

III. TYPES OF LIQUID-LIQUID PHASE BEHAVIOR

LLE are of considerable importance in process design. Sometimes, multiple liquid phases are induced deliberately to effect a separation. An example is liquid extraction of aromatics from paraffins and naphthenes using polar molecules such as sulfolane. A useful overall survey of the field of LLE is given in Reference 17.

A. Types of Phase Diagrams

Binary systems exhibiting partial miscibility are relatively easy to understand. Figure 7 shows the effect of temperature on the miscibility of the furfural + methyl cyclohexane system. It exhibits an upper critical solution temperature at 67°C. As is typical, the mutual miscibility declines as the temperature drops. The critical solution temperature is analogous to the vapor-liquid critical temperature, and the same kinds of thermodynamic criteria apply. A less common system is shown in Figure 8. For cyclohexyl-2-pyrrolidone + water there is a lower critical solution temperature; increasing temperature decreases the solubility. Some systems, although rare, exhibit both lower and upper critical solution temperatures. This is illustrated by the nicotine + water system in the range from 60 to 235°C, as shown in Figure 9. There is no explicit equation which correctly predicts the shape of such curves. However, activity coefficient equations can reproduce such curves if the parameters are allowed to vary with temperature. Perfect equations would exhibit no parameter variation with temperature.

Ternary systems fall into common classifications, noted as I, II, or III depending on how many of the binary systems are partially miscible. As is typical for binary systems, raising temperature usually increases the regions of miscibility. When the mixture composition falls

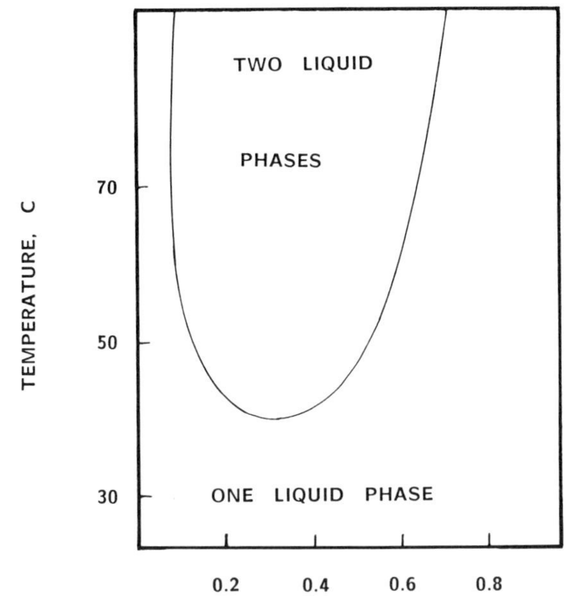

FIGURE 8. Effect of temperature on the miscibility of the cyclohexyl-2-pyrrolidone + water mixture. It illustrates the rather rare occurrence of a lower critical solution temperature. (Adapted from the GAF Corporation. With permission.)

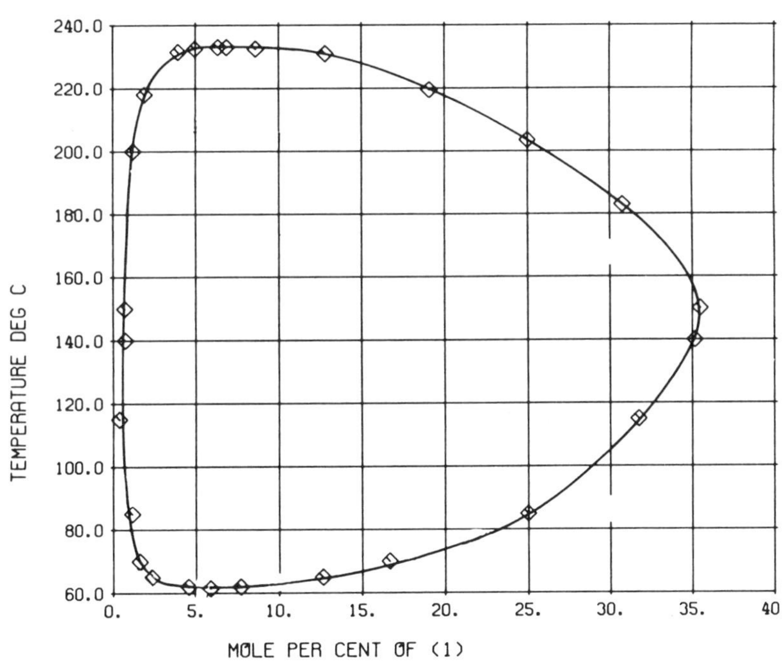

FIGURE 9. Illustration of a system containing both lower and upper critical solution temperatures. The system is nicotine (1) + water. (Reproduced from Sorenson, J. M. and Arlt, W., *Liquid-Liquid Equilibrium Data Collection*, DECHEMA, Frankfurt, Vol. 1 to 3, 1979, 1980. With permission.)

inside the region of partial miscibility, the composition of the two phases in equilibrium falls on the solubility envelope at ends of a "tie line" which passes through the feed composition. In a type I system, the tie lines become progressively shorter with addition of the miscible component until a single point is reached. It is called the "plait point" and represents another type of critical phenomenon. One can usually expect a ternary azeotrope at that point. Some ambitious student may even be able to prove that it is required. Note that these phase diagrams are essentially independent of pressure, under normal pressure conditions.

Sometimes, normally miscible liquids can be "salted out" and caused to phase-separate. Most technologists are quite aware that inorganic salts can do this in aqueous mixtures. It may not be quite so obvious that supercritical gases such as ethylene can accomplish the same thing.[18]

B. Semiquantitative Rules for Phase Splitting

It has been assumed that the reader has a working knowledge of the meaning of activity coefficients, which represent thermodynamically the liquid phase nonidealities. The formation of more than one liquid phase is a physical demonstration of a high degree of nonrandomness; only one liquid phase can exist with an ideal mixture. Null[19] has pointed out that for systems with symmetrical activity coefficient curves, activity coefficients at infinite dilution in excess of about 7.4 will cause phase splitting. For the more common case of unsymmetrical activity coefficients, a higher value at one end must balance a value lower than 7.4 for the system to split into more than one liquid phase. The magnitude of a critical solution temperature gives some feel for the magnitude of activity coefficients. A critical temperature below process conditions suggests lower values. Conversely, a critical solution temperature on the order of 300°C would suggest almost total immiscibility at room temperature. The activity coefficient at room temperature in this case could be in the thousands. Reference 14 gives a convenient tabulation of critical solution temperatures.

C. First-Pass Screening of Extraction Solvents

How, then, can this information be used to screen extraction solvents? As a general rule, one can expect that if a solvent candidate exhibits a high degree of difference in critical solution temperatures between the solutes that must be separated, then it has a chance of doing the job. The spread can be used as a first-pass rating of various solvents, although there is the consideration of solvent capacity. Water is so polar that it exhibits a high separation factor for many systems. However, when dealing with hydrocarbons the solvency is so poor that an infinite amount is necessary. Thus, a balance must be achieved. The most common liquid extraction agents in the chemical industry are those used to separate various hydrocarbons. Aromatics are separated from paraffins, butadiene is separated from monoolefins, etc. Common solvents are sulfolane, *N*-methyl-2-pyrrolidone, dimethyl acetamide, dimethyl formamide, and acetonitrile. In the case of the last solvent, water is added to enhance the selectivity. So long as the concentration of water is not too high, the solvent maintains adequate carrying capacity.

D. Example

Critical solution temperatures listed in Reference 14 for dimethyl formamide are as follows.

	Critical solution temperature (°C)
n-Hexane	68
n-Pentane	63
n-Butane	64.6
1-Butene	−13.5
Isobutene	−26.7
1,3 Butadiene	<-50
sec-Butylbenzene	<-57

From these data one would be led to expect that as an extraction solvent, dimethyl formamide would be able to separate paraffins from olefins and diolefins from olefins. In fact, it is used commercially for extractive distillation of butadiene. Aromatics could also be recovered. However, the critical solution temperatures are probably too low for this solvent to be used successfully for liquid extraction of aromatics; solubility of the paraffins would be too high. For three polar solvents the following values are listed.

	Critical solution temperature (°C)	
	***n*-Heptane**	**1-Heptene**
Nitromethane	115	70
Acetonitrile	84.6	38
Carbitol	25	−25

For separation of *n*-heptane from 1-heptene by liquid extraction, nitromethane would be a feasible solvent, giving the lowest solubility with *n*-heptane. Acetonitrile might be possible, and would have greater solvency for the 1-heptene. The extraction temperature could be adjusted to optimize the extraction. However, carbitol could only work with expensive refrigeration.

Such conclusions are only tentative because more data are needed. The slopes of tie lines in the ternary system are required. The book on liquid extraction by Treybal[20] shows how to examine the tie lines and determine whether they point in the right direction to make the solvent useful. The value of this type of cursory screening is to reduce the field of possibilities. When the solubility and tie line data are found, as in the DECHEMA series, for example (see references in Chapter 4), those data are of course much more preferable. Nevertheless, critical solution temperatures may point to the advisability of taking extraction data on some system not included in that compilation. Other perspectives on the issue of screening extraction and extractive distillation solvents are given in References 26 and 27.

IV. INFINITE DILUTION ACTIVITY COEFFICIENTS

A. Significance of Infinite Dilution Region

When a molecule is completely surrounded by dissimilar molecules it usually exhibits its highest degree of nonideality. The measure of the escaping tendency of the molecule in that condition is the product of vapor pressure and activity coefficient. At infinite dilution the physical condition at the top and bottom of purification distillation towers is represented. (Inasmuch as deviations from ideality are usually positive, the separation is usually more difficult at the top than at the bottom of a tower.) It is often the case that most of the trays are needed to go from 95 to 99+% purity; however, data in that region have typically been the least accurate. As typically measured, the error in PTxy data goes to infinity as infinite dilution is approached. This uncertainty can result in a need for considerable overdesign of such columns. Fortunately, means have been developed to measure such values directly with just a few percent inaccuracy. Those techniques will be described in Chapter 7.

Extrapolation of activity coefficients from finite concentration to infinite dilution is difficult to do correctly. It is an exercise that is highly dependent on the form of the activity coefficient equation used. The older types of activity coefficient equations, such as the Van Laar, usually underpredict very significantly the value at infinite dilution. Seldom can graphical extrapolation be relied upon, unless the system is nearly ideal. If the activity coefficients at infinite dilution can be determined experimentally, however, the interpolation over the finite concentration range can be made without a large error. Thus, the author has a bias toward those data measurements that pin down the ends of the activity coefficient curve.

B. Infinite Dilution Region Case Study

Companies are happy to trumpet their successes but slow to discuss their blunders. Thus, it was unusual to find Monsanto's published account of a major blunder in the design of a fractionator.[21] The company needed to strip water from a solvent, probably butyrolactone. The design pressure at the top of the column was 112 mHg and the diameter at the top was 6.5 ft. There were 12 trays in the top section and 8 in the stripping section where the diameter increased to 10 ft. Bubble-cap trays were used.

The start-up was a disaster! The initial inventory of solvent disappeared in less than 2 days. Finally, at 4 times design reflux the column was marginally operable with solvent losses at 20 times design. Obviously, some quick backtracking was necessary to determine what went wrong and try to correct the problem. Foaming was eliminated as a problem, tray efficiencies were discovered to be not up to expectation, and most importantly a major problem was discovered with treatment of the VLE data. This is illustrated in Figure 10.

The last data point had been taken at roughly 75% water. This may have been due to use of weight fractions in determination of where to take the data point. The data had been extrapolated to infinite dilution. As the figure shows, even though a high slope was assumed at infinite dilution, the true slope was much higher. At infinite dilution the activity coefficient of the solvent in water was over 8, whereas a value of only 5.5 had been used. Differences of that magnitude are so large that they can affect the type of separation chosen, much less the number of equilibrium stages! It would appear the Van Laar or Margules equation was used to make the extrapolation. The Wilson equation would have given a more steeply sloped activity coefficient curve. In any event, additional data were taken after the fact, as shown in the figure. A few dollars saved on data measurements in the beginning had been lost within 1 hr during the start-up.

The conclusion of the paper gives a lesson which applies to virtually every separation that is considered. "The solution of these deficiencies proved surprisingly difficult even after the causes (VLE data and poor tray efficiency) were defined. The successful solution was the obvious one: change the separation involved."[21] Even though experimental phase equilibrium data may not be justified in the probing stages of a project, the true phase behavior of a system cannot be taken lightly. One must check for azeotropes and the presence of two liquid phases. If the infinite dilution activity coefficients can be found or predicted correctly it is much less likely that the recovery process will change as the project proceeds.

C. Prediction Methods

1. Regular Solution Theory

The earliest success in predicting the properties of mixtures from those of the pure components was based on "regular solution theory". Using "solubility parameters", Scatchard and Hildebrand devised simple equations for activity coefficients. The solubility parameters are obtained when the square root of the energy of vaporization into a vacuum

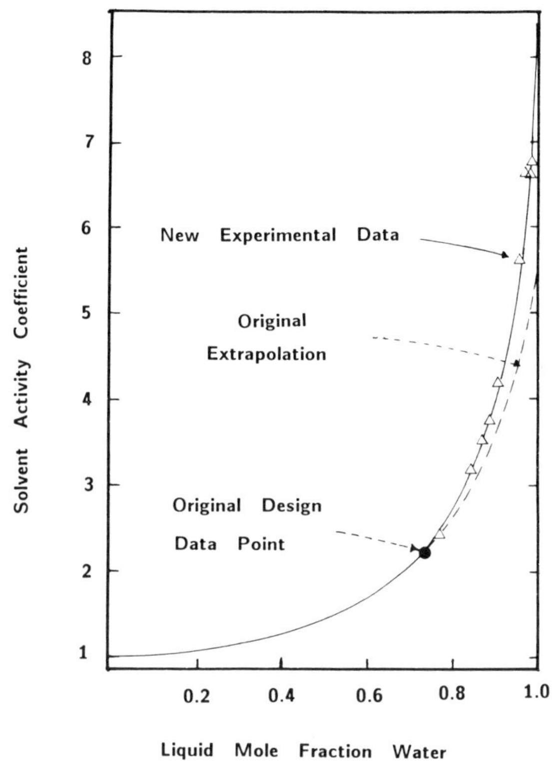

FIGURE 10. Activity coefficient of a modern solvent in water, comparing early data with the extrapolation made for design. Also shown are data measured after start-up, which indicate that the original design was in serious error because of erroneous extrapolation to infinite dilution. (Adapted from Martin, H. W., *Chem. Eng. Prog.*, 60(1), 50, 1964.)

is divided by the molar volume. The equation is not presented here because it is primarily of historical interest. Reference 22 gives a discussion in succinct terms and shows that the theory can be used to fit the benzene + *n*-heptane system. Research continues on extensions of the idea which may bear fruit in the future.

CRC Press[23] has published a useful handbook containing very extensive tables of solubility parameters. The use of solubility parameters is now widespread in the paint industry and can be used as an approximate guide in choosing solvents. However, this approach usually assumes that the excess entropy is zero. That assumption is virtually never valid, particularly when dealing with chemicals other than simple hydrocarbons. When applied to polar compounds the results can be worse than when ideal behavior is assumed. Extensions to the basic idea have brought better results. They are summarized in Reference 23.

2. *Null-Palmer Correlation*

The first significant attempts to expand regular solution theory to polar compounds are given in References 24 and 25. Neither of these correlations was applicable to strongly polar solutes or to strongly associating compounds. The extension to include those compounds was given in Reference 24. The effects included in the correlation were (1) the Scatchard-Hildebrand regular solution heat of mixing, (2) the polar energy and induced dipole effects described in References 24 and 25, (3) the entropy effects including association, and (4) free energy of broken polymer bonds as the associating component goes into solution.

Parameters, given in Reference 19, are based on infinite dilution activity coefficients published by Shell Development Company[28-30] and Gerster.[31]

The overall correlation was based on 845 data points for 84 compounds. This was not a group contribution approach, so while it gave some ability to predict new combinations between those 84 compounds it offered no ability to go outside that range. Today this method is almost obsolete. However, it may yet be of some use, since it addresses most of the solvents used in the chemical industry. These include *n*-methyl-2-pyrrolidone, γ-butyrolactone, β-chloropropionitrile, diethyl carbonate, diethylene glycol, dimethyl acetamide, dimethyl formamide, dimethyl sulfoxide, methyl cellosolve, sulfolane, and triethylene glycol.

3. ASOG Method

Not many years have passed since some of the most knowledgeable experts in thermodynamics were saying there never would be a respectably accurate method for prediction of mixture properties. That prognostication fell by the wayside almost overnight with demonstration of the power of analytical group contribution methods.

The theoretical breakthrough in group contributions was made by Wilson and Deal,[32] apparently the first to recognize that if a molecule were considered to be a mixture of groups, the activity coefficients of the individual groups could be plotted as a function of composition. Thus evolved the analytical solutions of groups (ASOG) method. The concept is quite simple, once the concept of ASOG is in place. Calculate a contribution to the activity coefficient based on size. That contribution is approximate, based on adding up the number of groups. Next, calculate interaction parameters for groups, based on the Wilson equation (see Chapter 5). Then add the two contributions. Since the number of potential mixtures of groups is finite, contrasted with mixtures of molecules which are essentially infinite in number, the problem is now tractable. All that is needed, in theory, is a table giving the two interaction parameters for each pair of groups.

For many years this method was kept as a proprietary correlation within Shell Development Company. Its first publication was obscure,[33] and the description of the method was deliberately vague. However, the method had promise because it gave not only the activity coefficients at infinite dilution, but also the composition dependence. The first popularized version was published in 1975.[34] The parameters developed for that paper were directed to acid and anhydride systems. An example will be given in Chapter 8.

Today, the best reference for ASOG parameters is a book by Kojima and Tochigi.[35] One of the problems with ASOG is a fairly large temperature dependence for the parameters.[36] The phase equilibrium section of the DIPPR *Manual for Predicting Chemical Process Design Data*[71] includes the parameters in a form suitable for extrapolation with respect to temperature. Specific application of the method to liquid-liquid systems also has been made.[37] The book by Kojima and Tochigi[35] gives a FORTRAN program for ASOG.

4. UNIFAC Method

Group contribution methods are usually extensions to groups of thermodynamic equations for molecules. After the UNIQUAC equation, described in Chapter 5, had been developed, it was a natural step to extend it into a group contribution method. In fact, that equation already included separate terms for interaction energy on the one hand, and size/shape effects on the other. The new development, called UNIFAC, was published by Fredenslund et al.[39] It also took advantage of the basic concept developed by Wilson and Deal. Several advantages of UNIFAC were immediately apparent. First, it took into account the size terms based on theoretical work by Bondi.[38] It also accounted for shape of the molecules. Further, the interaction parameters exhibited little temperature dependence, compared to ASOG. Finally, and best of all for the chemical industry, a large table of interaction parameters was presented. At that time, only a few interaction parameters for the ASOG method had been published.

The UNIFAC equation met an enormous unfilled need in the chemical industry. One year after publication, in January of 1977, there was a very important thermodynamics conference at Asilomar, Calif. It was attended by 135 engineers and scientists from around the world. The thermodynamics experts from most major chemical companies were present. It was discovered that within that year virtually every company had installed the UNIFAC correlation in its in-house process simulator packages. Today, all commercial simulators offer it. A thorough description and FORTRAN computer program are given in Reference 40. Although many other group contribution methods have been proposed, only ASOG and UNIFAC have gained general acceptance.

UNIFAC had 25 different interaction groups when initially published. Structural variants of the groups, such as CH, CH_2, and CH_3 are handled by the size and shape parameters; this helps to reduce the number of needed interaction parameters. A little over half of the group interactions were available initially. The limitation on filling in the triangular grid of interaction parameters has been the availability of experimental phase equilibrium data. As a result, a number of people and organizations have been trying since 1977 to fill in the grid by taking new data. This is a particularly worthy effort, and one which the chemical industry is supporting both financially and with encouragement. The U.S. National Science Foundation also has funneled money to professors through DIPPR. The express purpose was to fund measurements of phase equilibrium data which could be used to fill gaps in the group contribution table.

Most of the current UNIFAC research is carried on at the Instituttet fur Chemiteknik, Technical University of Denmark. Not only do the researchers measure new data and expand and revise the UNIFAC tables,[69] but also they are continually looking for new ways to improve the equation. Recent efforts have centered on finding modifications which allow simultaneous correlation with heat of mixing, and which also may predict LLE. The latter goal will be possible to achieve qualitatively. The problem with quantitative predictions of LLE is that the phase envelope is extremely sensitive to small variations in free energy. Thus, successes in that area will be modest. A separate table of parameters designed specifically for LLE predictions has been published.[70] It is hoped that the interaction parameter grid eventually will be filled in, even though new groups are added occasionally.

The role of UNIFAC in scoping designs is extremely important. Although it obviously requires computer computations, that is no longer a barrier; virtually every engineer has a terminal, a personal computer, or both. As a supplement to published phase equilibrium data, UNIFAC can be used to test many conceptual product recovery schemes. It is in fact very seldom that all of the data are available. Often it is desirable to test on paper the effect of adding entrainers or liquid extraction agents. In many but not all of these situations UNIFAC is a useful tool. It must be remembered that results for liquid extraction will be only approximate, but can be useful in flow sheet development. An example is the dehydration of a solvent by Monsanto; had they known the true activity coefficient of water in the solvent, they likely would have chosen a different unit operation such as extractive distillation.

UNIFAC has been used in a variety of applications beyond VLE and LLE. Two papers have covered the use of UNIFAC to predict solvent effects on chemical reaction rates.[47,48] With small modifications it has been used to describe the equilibria of polymer solutions.[49] It has even been used to predict the flash points of flammable mixtures.[50]

5. *Comparison of ASOG and UNIFAC*

A fairly thorough comparison of the ASOG and UNIFAC methods has been made, based on activity coefficients measured at infinite dilution.[54] The data included 2097 experimental values, with 129 solvents and 60 solutes. For the classes of solutes studied (paraffins, olefins, aromatics, naphthenes, and chlorides), the UNIFAC method using regular VLE parameters was clearly superior. In fact, for paraffins it was twice as good. For four pairs of chemicals such as pentane/1-pentene, the selectivity was predicted far more accurately by UNIFAC; thus it is believed to be significantly more accurate. However, this comparison did not include many compound classes, so it would be premature to overgeneralize. Sometimes the choice between prediction methods will be based on which method has the available parameters.

The size of the prediction errors was large in this comparison. Percentage errors with UNIFAC ranged from a mean of 24% with aromatics to 47% with chlorides. These mean errors ranged from 41 to 91% with ASOG. Obviously, individual errors were greater than the mean errors for some of the compounds. At infinite dilution the activity coefficients are usually at their maximum value, and some can have magnitudes into the hundreds. Thus, it might have been more meaningful to compare errors in the logarithms of the activity coefficients. Nevertheless, it is helpful to know that just because the computer can give many significant figures, the first one may be wrong in many cases. Improvement in both data and correlations is necessary to better this situation. On the other hand, the industry is grateful for these methods, because assumption of ideal mixing or use of the Scatchard-Hildebrand equation can lead to errors of orders of magnitude.

6. Screening for Potential Azeotropes with Infinite Dilution Activity Coefficients

The tables of azeotropes are not complete, and it may be necessary to use infinite dilution activity coefficients to predict whether one will be present. The approach is very simple. Assuming an ideal gas phase (fugacity coefficients could be included if desired), the criterion for a low-boiling azeotrope is found in Equation 12. For the less common occurrence of a high-boiling azeotrope the criterion is given in Equation 13.

$$\gamma_1^\infty > P_2^*/P_1^* > 1/\gamma_2^\infty \tag{12}$$

$$\gamma_1^\infty < P_2^*/P_1^* < 1/\gamma_2^\infty \tag{13}$$

7. Screening of Extraction Solvents

There are two classic articles[30,31] which illustrate the use of infinite dilution activity coefficients to choose liquid extraction solvents. Perhaps the most important liquid extraction process in the refining industry is aromatics extraction. Extraction is used, as opposed to extractive distillation, because it is necessary to separate aromatics having different volatilities from paraffins and cycloparaffins which likewise have varying volatilities.

The limiting separation factor for liquid extraction is given by the ratio of activity coefficients at infinite dilution for the two key components in the solvent. Using that criterion, consider a sampling of data presented by Deal and Derr.[30] The data points are given at 60°C.

	Activity coefficients		
Solvent	**Hexane**	**Benzene**	**Selectivity**
Water	18,500	1,800	10.2
Methyl ethyl ketone	3.15	1.15	2.7
n-Methyl-2-pyrrolidone	8.6	1.0	8.6
Acetonitrile	15.8	2.6	6.0
Phenol	12.0	2.5	4.8
Sulfolane	48	2.45	19.6

Although this represents just a small fraction of the data presented, it can be used to make several points. Water, it will be noted, gives a good separation factor of 10.2. Unfortunately, the activity coefficients are so high that the solvency for benzene is negligible. At that high activity coefficient value the solubility of benzene can be approximated as 0.055 mol %. Methyl ethyl ketone presents another problem. The activity coefficient with hexane is less than the critical value of 7.4 normally required to form two liquid phases. Thus, it cannot work at all as a liquid extraction agent in this system. The next three solvents are all potential candidates, although the volatility of acetonitrile poses a problem. Finally, it will be noted that sulfolane has the highest separation factor. At the same time, it has good solubility for benzene and a suitably low solubility for hexane. Thus, it should not come as too much of a surprise that sulfolane is widely used commercially.

In another study of extraction solvents for the benzene + *n*-hexane system, it was discovered that nitrile groups give good selectivity.[54] This is illustrated by the following selectivity factors.

Solvent	Selectivity
Dimethyl sulfoxide	22.7
Glutaronitrile[$(CH2)_3(CN)_2$]	21.1
Adiponitrile[$(CH2)_4(CN)_2$]	20.1
Pimelonitrile[$(CH2)_5(CN)_2$]	18.3
1,2 Ethanediol	12.8
Dimethyl formamide	12.5
Aniline	11.2

Solvency can be adjusted by increasing or decreasing operating temperature. It is almost universally true that solvency increases and selectivity decreases with increasing temperature. Solvency for hydrocarbons is sometimes deliberately decreased by adding water to the solvent.

8. Screening of Extractive Distillation Solvents

Extractive distillation has a number of advantages over liquid extraction. The equipment sometimes can be smaller, and relative densities of the different liquid phases are of no concern. On the other hand, the constituents to be separated must boil over a fairly narrow range. Separations by molecular type, as in aromatics extraction, are not possible over a range of carbon numbers. The limiting separation factor is calculated from the ratio of activity coefficients multiplied by the ratio of vapor pressures of the solutes.

$$\alpha_{ij} = \frac{\gamma_i P_i^*}{\gamma_j P_j^*} \tag{14}$$

If there is already a slight volatility difference favoring separation of two chemicals, it can be enhanced by the solvent. On the other hand, the vapor pressure differences may be in the wrong direction and negate the effect of the solvent. In the latter case, it may be advisable to choose a different type of solvent, or turn to liquid extraction.

It is of interest to note that Shell Development Co. licenses to others the sulfolane process for aromatics extraction. However, Shell also licenses to others a process based on a mixture of acetonitrile + water for extractive distillation recovery of butadiene from mixed C_4 streams. This illustrates the fact that no single solvent is best for all circumstances. It is sometimes the case that one of the components in a reactor effluent can serve quite nicely as the liquid extraction, extractive distillation, or azeotroping agent if its concentration is increased.

When the components to be separated differ significantly in electron donor capability (differing amount of unsaturation), it is possible to find a polar, electron-accepting solvent to separate them. For example, benzene forms a more ideal mixture with alcohols than does

Table 3
EFFECT OF SEPARATION METHOD ON FRACTIONATION REQUIREMENT

Method	Trays (n)	Solvent/ethanol	Fraction of ethanol heating value
Low-pressure distillation	60	—	0.433
Azeotropic distillation			
Pentane	19	3.0	0.368
Benzene	34	3.2	0.413
Diethyl ether	44	7.9	0.461
Extractive distillation			
Gasoline	19	3.0	0.346
Ethylene glycol	46	4.1	1.15

Data from Black, C., *Chem. Eng. Prog.*, September, 78, 1980.

hexane. Prausnitz and Anderson[68] have noted some useful generalizations about solvents: "The effectiveness of a solvent is related to its polarity (which should be large) and to its molecular size (which should be small). In cases where chemical effects are important or where the molar volumes of the hydrocarbons to be separated are only slightly different, selectivity also depends on the relative ability of the hydrocarbons in acting as electron donors and on the ability of the solvent to act as an electron acceptor in forming acid-base complexes."

9. *Azeotropic Distillation*

Azeotropic distillation is often considered in the same category as extractive distillation. Normally, it is used where water is the desired overhead product and a hydrocarbon immiscible with water is used to entrain it. One should look for higher activity coefficients of the entrained material in the entraining agent than in the material that comes out the bottom. It would be safe to assume that if the activity coefficient at infinite dilution for the entrained material in the solvent is not at least 10, the separation will not work.

10. *Comparison of Azeotropic and Extractive Distillation*

One cannot generalize that one method is better than another. An example is the recovery and dehydration of ethanol (Table 3). Low-pressure fractionation is incapable of yielding dry ethanol because of the azeotrope. Even recovery of the azeotrope requires a heat duty equal to 43% of the fuel value of ethanol. Using the Simulation Sciences process simulator, Black[16,55] has done a thorough comparison between three azeotropic and two extractive distillation methods for dehydration. His computations assumed a concentration of 2.6% ethanol in aqueous feed based on biomass conversion. If the product is destined for gasoline, extractive distillation with gasoline as the solvent is about equal to azeotropic distillation with *n*-pentane. Detailed computations using simulators are often necessary to make a judgment when several solvents or methods are considered.

V. APPLICATION OF THERMODYNAMICS TO MEMBRANE SYSTEMS

Thermodynamics has more applications than are commonly considered. Membrane separations are a prime example. Though membranes have never achieved much commercial success in separation of organic chemicals, this unit operation can serve as an interesting example of application of the techniques of phase equilibria.

There are three basic types of membrane separation for nonelectrolyte mixtures. Vapor permeation or gas diffusion is the most common. It is exemplified by Monsanto's Prism separator for recovery of hydrogen. There are two types of liquid permeation. Pervaporation occurs when there is a vapor phase on the permeate (product) side of the membrane. Typically, the vapor side is operated at low pressure.[41] A system called perstraction makes use of a liquid purge on the permeate side of the membrane.

Techniques for treatment of these three approaches have not been unified, even by authors taking data for different systems. Yet, when simple thermodynamic considerations are used, the results seem to converge on a common answer.

The rate of permeation of a liquid or gas, per unit time and unit area, is given by q_i. It is computed for vapor permeation by the equation

$$q_i = (\text{perm}_i/t) * \{(P_i)_{feed} - (P_i)_{permeate}\} \tag{15}$$

In this equation perm_i/t is the permeability coefficient divided by the thickness of the membrane. This quotient is analogous to a mass transfer coefficient. Further work showed that the partial pressures should really be replaced by fugacities.[42]

$$q_i = (\text{perm}_i/t) * \{(f_i)_{feed} - (f_i)_{permeate}\} \tag{16}$$

It is hypothesized that this equation is valid whether the system sees vapor or liquid. For a vapor, the fugacity is

$$f_i^v = y_i\, P\, \hat{\phi}_{i,P} \tag{17}$$

The fugacity coefficient at moderate pressures can be approximated by second viral coefficients as

$$\hat{\phi}_{i,P} = \exp[-BP/RT] \tag{18}$$

For liquids the fugacity is

$$f_i^L = x_i\, \gamma_i\, \phi_{i,P_i^*}\, P_i^* \tag{19}$$

The activity coefficient is that of the permeate in the mixture at the composition of the feed at a given point in the membrane unit. Obviously it will vary throughout the length of the unit, as will composition. Thus, a simulation of such a unit must include stepwise integration using changes in both composition and activity coefficients. On the permeate side the activity coefficient is that of the permeate in the sweep liquid. To maximize recovery the sweep liquid should be chemically similar to the permeate so that the activity coefficients will be close to unity. It also should have a volatility less than the permeate to facilitate recovery of permeate from dilute solutions.

The implication drawn from these equations is that the computed permeability coefficient should be approximately the same for all three types of permeation. Is that the case? The hypothesis appears to be qualitatively correct, as illustrated by Figures 11 and 12, which show permeability coefficients as a function of reciprocal temperature for methanol and ethyl acetate, respectively. The film was low-density polyethylene. Data by Bent[43] on liquid permeation (pervaporation) are in excellent agreement with data by two other authors[44,45] on vapor permeation. This excellent agreement will not always be found, particularly if there are flaws in the experimental data such as failure to eliminate completely the fugacity (back pressure) on the permeate side of the membrane. Further, there can be differences in swelling of the polymer film which actually change the permeability coefficient. Permeability

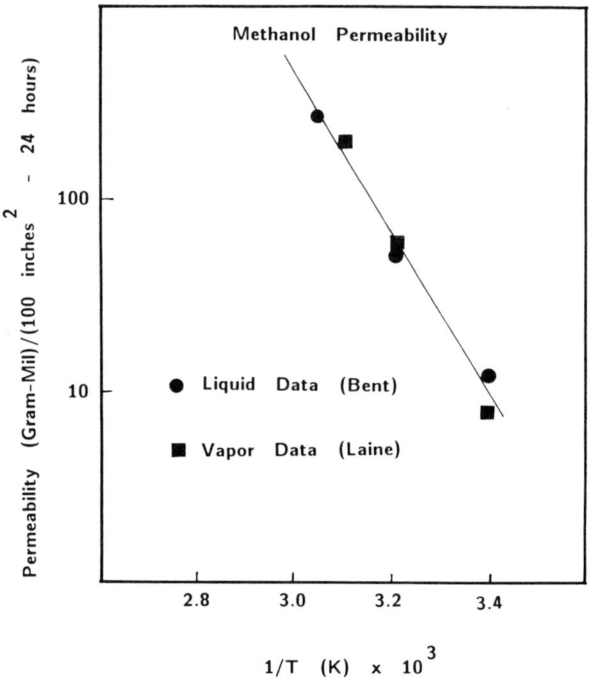

FIGURE 11. Permeability of both liquid and vapor forms of methanol through a low-density polyethylene film after data reduction by the fugacity method of this chapter.

coefficients are in fact a function of permeate solubility in the polymer. Nevertheless, these equations give a useful first-order estimate of the effect of switching from one type of permeation to another.

An additional implication can be drawn from these equations. While partial pressure differences are important for vapor permeation, the total pressure has no effect on the fugacity difference for perstraction. Typically, the temperature will be more important for liquid permeation because it raises the vapor pressure exponentially. The effect is decreased just slightly by the reduction in activity coefficients as temperature is raised.

VI. SUMMARY

Developing the flow sheet for a proposed new chemical process is a fundamental part of the process development. It requires application of creative skills, and almost always calls upon the phase equilibrium skills discussed in this chapter. A series of approaches have been examined which help to develop a qualitative feel for the chemicals being treated. This understanding, based on degrees of unsaturation, polarity, and oxygen content of the molecule, gives some hints as to how the product should be recovered and purified. Sometimes these skills must also be used to make the reaction proceed properly or to carry a reversible reaction to completion. The need for these skills diminishes as the process graduates from the probing to the evaluation stage and eventually to the development stage. Nevertheless, the methods for predicting activity coefficients remain useful for treatment of some byproducts or other binary mixtures where the separation factor is not crucial to the design.

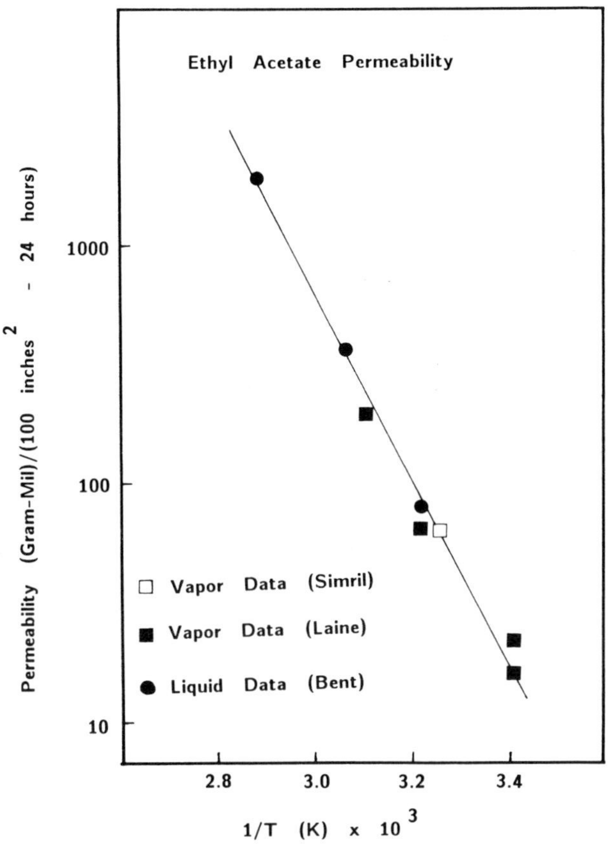

FIGURE 12. Permeability of both liquid and vapor forms of ethyl acetate through a low-density polyethylene film after data reduction by the fugacity method of this chapter.

REFERENCES

1. **Smith, L. A. and Huddleston, M. N.,** New MTBE design now commercial, *Hydrocarbon Process.*, March, 1982.
2. **Palmer, D. A. et al.,** Production of ortho-phthalic acid and its conversion and recovery of phthalic anhydride, U.S. Patent 4,215,053, 1980.
3. **Evans, H. D. and Sarno, D. H.,** Acetonitrile as an extractive distillation solvent, *Seventh World Petroleum Congress Proceedings,* Vol. 5, Elsevier, New York, 1967, 259.
4. **Redlich, O. and Kwong, J. N. S.,** On the thermodynamics of solutions. V. An equation of state. Fugacities of gaseous solutions, *Chem. Rev.*, 44, 233, 1949.
5. **Walas, S. M.,** *Phase Equilibria in Chemical Engineering,* Butterworths, Reading, Mass., 1985.
6. **Prausnitz, J. M.,** *Molecular Thermodynamics of Fluid-Phase Equilibria,* Prentice-Hall, Englewood Cliffs, N.J., 1969.
7. **Van Ness, H. C.,** *Classical Thermodynamics of Non-Electrolyte Solutions,* Pergamon Press, Elmsford, N.Y., 1964.
8. **Kyle, B. G.,** *Chemical and Process Thermodynamics,* Prentice-Hall, Englewood Cliffs, N.J., 1984.
9. **Acree, W. E., Jr.,** *Thermodynamic Properties of Nonelectrolyte Solutions,* Academic Press, Orlando, Fla., 1984.
10. **Daubert, T. E.,** *Chemical Engineering Thermodynamics,* McGraw-Hill, New York, 1985.
11. **Palmer, D. A. and Smith, B. D.,** Thermodynamic excess property measurements for acetonitrile-benzene-*n*-heptane system at 45°C, *J. Chem. Eng. Data,* 17, 71, 1972.

12. **Palmer, D. A.,** A practical synthesis of experimental and correlation techniques for modeling highly nonideal systems with many components, *Ind. Eng. Chem. Process Des. Dev.,* 23, 259, 1984.
13. **McClellan, A. L.,** *Tables of Experimental Dipole Moments,* W. H. Freeman, San Francisco, 1963.
14. **Francis, A. W.,** *Critical Solution Temperatures,* Advances in Chemistry Series, No. 31, American Chemical Society, Washington, D.C., 1961.
15. **Horsley, L. H.,** *Azeotropic Data, III,* Advances in Chemistry Series, No. 116, American Chemical Society, Washington, D.C., 1973.
16. **Black, C.,** Distillation modeling of ethanol recovery and dehydration processes for ethanol and gasohol, *Chem. Eng. Prog.,* September, 78, 1980.
17. **Sorensen, J. M., Magnussen, T., Rasmussen, P., and Fredenslund, A.,** Liquid-liquid equilibrium data: their retrieval, correlation and prediction. I. Retrieval, *Fluid Phase Equilibria,* 2, 297, 1979.
18. **Elgin, J. C. and Weinstock, J. J.,** Phase equilibrium at elevated pressures in ternary systems of ethylene and water with organic liquids. Salting out with a supercritical gas, *J. Chem. Eng. Data,* 4, 3, 1959.
19. **Null, H. R.,** *Phase Equilibrium in Process Design,* Wiley-Interscience, New York, 1970.
20. **Treybal, R. E.,** *Liquid Extraction,* McGraw-Hill, New York, 1963.
21. **Martin, H. W.,** Scale-up problems in a solvent-water fractionator, *Chem. Eng. Prog.,* 60(10), 50, 1964.
22. **Prausnitz, J. M.,** *Molecular Thermodynamics of Fluid-Phase Equilibria,* Prentice-Hall, Englewood Cliffs, N.J., 1969.
23. **Barton, A. F. M.,** *CRC Handbook of Solubility Parameters and Other Cohesion Parameters,* CRC Press, Boca Raton, Fla., 1983.
24. **Weimer, R. F. and Prausnitz, J. M.,** Screen extraction solvents this way, *Hydrocarbon Process.,* 44, 237, 1965.
25. **Helpinstill, J. G.and Van Wickle, M.,** Prediction of infinite dilution activity coefficients for polar-polar binary systems, *Ind. Eng. Chem. Process Des. Dev.,* 7, 213, 1968.
26. **Anderson, R. and Prausnitz, J. M.,** Selectivity in extractive distillation of hydrocarbons, *AIChE J.,* 7, 96, 1961.
27. **Tassios, D. P.,** Rapid screening of extractive distillation solvents, in *Extractive and Azeotropic Distillation,* Advances in Chemistry Series, No. 115, Tassios, D. P., Ed., American Chemical Society, Washington, D.C., 1972, 46.
28. **Pierotti, G. S., Deal, C. A., and Derr, E. L.,** Document #5782, American Documentation Institute, Washington, D.C., 1958.
29. **Black, C., Derr, E. L., and Papadopoulos, M. N.,** Systematic prediction of separation factors: screening estimates, *Ind. Eng. Chem.,* 55, 40, 1963.
30. **Deal, C. H. and Derr, E. L.,** Selectivity and solvency in aromatics recovery, *Ind. Eng. Chem. Process Des. Dev.,* 3(4), 394, 1964.
31. **Gerster, J. A.,** Selective solvents for separation of *n*-pentane from 1-pentene by extractive distillation, *J. Chem. Eng. Data,* 5, 423, 1960.
32. **Wilson, G. M. and Deal, C. H.,** Activity coefficients and molecular structure, *Ind. Eng. Chem. Fundam.,* 1, 20, 1962.
33. **Derr, E. L. and Deal, C. H.,** Int. Chem. Eng. Symp. Ser., 32, 3(40), 1969, in Int. Symp. on Distillation, Brighton, England, Proc., Institute of Chemical Engineers, London, 1969.
34. **Palmer, D. A.,** Predicting equilibrium relationships for maverick mixtures, *Chem. Eng.,* June 9, 80, 1975.
35. **Kojima, K. and Tochigi, K.,** *Prediction of Vapor-Liquid Equilibria by the ASOG Method,* Elsevier, London, 1979.
36. **Tochigi, K., Lu, B. C.-Y., Ochi, K., and Kojima, K.,** On the temperature dependence of ASOG parameters for VLE calculations, *AIChE J.,* 27, 1022, 1981.
37. **Tochigi, K., Hiraga, M., and Kojima, K.,** Prediction of liquid-liquid equilibria for ternary systems by the ASOG method, *J. Chem. Eng. Jpn.,* 13, 159, 1980.
38. **Bondi, A.,** *Physical Properties of Molecular Crystals, Liquids, and Glasses,* John Wiley & Sons, New York, 1968.
39. **Fredenslund, A., Jones, R. L., and Prausnitz, J. M.,** Group-contribution estimation of activity coefficients in nonideal liquid mixtures, *AIChE J.,* 21, 1086, 1975.
40. **Fredenslund, A., Gmehling, J., and Rasmussen, P.,** *Vapor-Liquid Equilibria Using UNIFAC,* Elsevier, London, 1977.
41. **Binning, R. C., Lee, R. J., Jennings, J. F., and Martin, E. C.,** Separation of liquid mixtures by permeation, *Ind. Eng. Chem.,* 53, 45, 1961.
42. **Stern, S. A., Mullhaupt, J. T., and Gareis, P. J.,** The effect of pressure on the permeation of gases and vapors through polyethylene. Usefulness of the corresponding states principle, *AIChE J.,* 15, 64, 1969.
43. **Bent, H. A.,** Permeability to liquids of polyethylene and irradiated polyethylene, *J. Polym. Sci.,* 24, 387, 1957.
44. **Laine, R. and Osburn, J. O.,** Permeability of polyethylene film to organic vapor, *J. Appl. Polym. Sci.,* 15, 327, 1971.

45. **Simril, V. L. and Hershberger, A.,** Permeability of polymeric films to organic vapors, *Mod. Plast.*, June, 97, 1950.
46. **Heidman, J. L., Tsonopoulos, C., Brady, C. J., and Wilson, G. M.,** High-temperature mutual solubilities of hydrocarbons and water, *AIChE J.*, 31, 376, 1985.
47. **Lo, H. S. and Paulaitis, M. E.,** *AIChE J.*, 27, 842, 1981.
48. **Gmehling, J. and Fellensiek, J.,** *Z. Phys. Chem. Neue Folge,* 122, 251, 1980.
49. **Oishi, T. and Prausnitz, J. M.,** Estimation of solvent activities in polymer solutions using a group contribution method, *Ind. Eng. Chem. Process Des. Dev.*, 17, 333, 1978.
50. **Gmehling, J. and Rasmussen, P.,** *Ind. Eng. Chem. Fundam.*, 21, 186 (Errata, 326), 1982.
51. **Tsonopoulos, C. and Wilson, G. M.,** High-temperature mutual solubilities of hydrocarbons and water. I. Benzene, cyclohexane and *n*-hexane, *AIChE J.*, 29, 990, 1983.
52. **Zudkevitch, D. and Joffe, J.,** Correlation and prediction of vapor-liquid equilibria with the Redlich-Kwong equation of state, *AIChE J.*, 16, 112, 1970.
53. **Tamir, A.,** Compilation and correlation of binary azeotropic data, *Fluid Phase Equilibria,* 5, 199, 1980/1981.
54. **Bastos, J. C., Soares, M. E., and Medina, A. G.,** Selection of solvents for extractive distillation. A data bank for activity coefficients at infinite dilution, *Ind. Eng. Chem. Process Des. Dev.*, 24, 420, 1985.
55. **Black, C.,** Simulation applied to special industrial problems, *Chem. Eng. Prog.*, December, 53, 1982.
56. **Wichterle, I.,** High pressure vapour-liquid equilibrium. I, *Fluid Phase Equilibria,* 1, 161, 1977.
57. **Wichterle, I.,** High pressure vapour-liquid equilibrium. II, *Fluid Phase Equilibria,* 1, 225, 1977/1978.
58. **Chementator,** High-purity isoamylene made by catalytic distillation, *Chem. Eng.*, August 19, 10, 1985.
59. **Adler, S. B.,** K-constants: water in hydrocarbons, *Hydrocarbon Process.*, April, 99, 1985.
60. **Anthony, R. G. and McKetta, J. J.,** Phase equilibrium in the ethylene-ethane-water system, *J. Chem. Eng. Data,* 12, 21, 1967.
61. **Brady, C. J., Cunningham, J. R., and Wilson, G. M.,** Water-Hydrocarbon Liquid-Liquid-Vapor Measurements, GPA RR-62, 1982.
62. **Gillespie, P. C. and Wilson, G. M.,** Vapor-Liquid and Liquid-Liquid Equilibria: H_2O–CH_4, H_2O–CO_2, H_2O–H_2S, H_2O–nC_5 and H_2O–CH_4–nC_5, GPKA RR-48, 1982.
63. **Griswold, J. and Kasch, J. E.,** Hydrocarbon-water solubilities at elevated temperatures and pressures, *Ind. Eng. Chem.*, 34, 804, 1942.
64. **Kobayashi, R. and Katz, D. L.,** Vapor-liquid equilibria for binary hydrocarbon-water systems, *Ind. Eng. Chem.*, 45, 51, 1953.
65. **Kudchadker, A. P. and McKetta, J. J.,** Solubility of benzene in water, *Hydrocarbon Process.*, 41(3), 191, 1962.
66. **Li, C. C. and McKetta, J. J.,** Vapor-liquid equilibrium in the propylene-water system, *J. Chem. Eng. Data,* 8, 271, 1963.
67. The Vapor-Liquid Equilibria of Binary Mixtures of C_6 Hydrocarbons with Water, API Rep. 3-68, Pennsylvania State University, University Park, 1968.
68. **Prausnitz, J. M. and Anderson, R.,** Thermodynamics of solvent selectivity in extractive distillation of hydrocarbons, *AIChE J.*, 7, 96, 1961.
69. **Gmehling, J., Rasmussen, P., and Fredenslund, A.,** Vapor-liquid equilibria by UNIFAC group contribution. Revision and extension. II, *Ind. Eng. Chem. Process Des. Dev.*, 21, 118, 1982.
70. **Magnussen, T., Rasmussen, P., and Fredenslund, A.,** UNIFAC parameter table for prediction of liquid-liquid equilibria, *Ind. Eng. Chem. Process Des. Dev.*, 20, 331, 1981.
71. **Danner, R. P. and Daubert, T. E.,** *Manual for Predicting Chemical Process Design Data,* American Institute of Chemical Engineers, New York, 1983.
72. **Terrill, D. L., Sylvestre, L. F., and Doherty, M. F.,** Separation of closely boiling mixtures by reactive distillation. I. Theory, *Ind. Eng. Chem. Process Des. Dev.*, 24, 1062, 1985.
73. **Cleary, W. and Doherty, M. F.,** Separation of closely boiling mixtures by reactive distillation. II. Experiments, *Ind. Eng. Chem. Process Des. Dev.*, 24, 1071, 1985.
74. **Null, H. R. and Palmer, D. A.,** Predicting phase equilibria, *Chem. Eng. Prog.*, 65(9), 47, 1969.

Section II: Process Assessments

Process assessment is the second major stage in carrying an idea to commercialization. Whatever name it is known by in different companies, this stage is characterized by a full-time laboratory effort. The more progressive companies require justification before undertaking this level of effort. Such justification usually involves a preliminary process design and economic analysis. It usually must be performed with minimal data from the laboratory on reactions, and usually with no new physical property data. The process engineer must leverage all the available information into a reasonably informed appraisal of the economic prospects for the new or improved process or product.

The first step in this evaluation involves work outside the scope of this book, namely, the assessment of experimental reaction data. There may be a modicum of material balance information as well. It often will be necessary to extrapolate from one type of reaction condition to another. For example, reactions run batch-wise may need to be converted for back-mixed reactor operation, requiring more residence time and perhaps multiple reactors. The conversion process is particularly difficult when the reaction involves a heterogeneous catalyst.

The second step involves development of the pure component data base. If the process is to be simulated by computer, the flow sheet simulator will require certain data such as critical properties, thermal properties, and vapor pressure. These are needed even for hand computations of the process. Choice of the flow sheet simulator as opposed to hand computations should depend on the probability of full development of the process, the number of options to be studied, availability of most of the pure compounds in the simulator data bank, and the engineer's familiarity with the simulator. It also may depend on simulation cost; is the simulator on an in-house computer or are royalties required?

Typically, new process development involves compounds as intermediates or by-products which are not well known; thus, more work is necessary to find their physical properties. Chapter 4 emphasizes the importance of using critically evaluated data, or of finding and critically evaluating the data. If this step of the evaluation is done properly and documented, it will not have to be repeated.

The third step of the evaluation usually involves locating accurate mixture phase equilibrium data. These data are very important because the separations section of a plant can represent up to 80% of the capital and operating cost. Poor data can result in design of the wrong unit operation or inoperable design of the correct operation. Chapter 4 examines not only how to find data, but also how to derive as much benefit from it as possible. This includes prediction of multicomponent information from binary data. Physical property data cannot be measured at every possible temperature, pressure, and composition. The power of thermodynamics is its ability to reduce the amount of data needed by giving fair representations of nature. This allows interpolation and extrapolation of the limited data at hand.

The necessary physical property data base for a new process is virtually never complete. Often, a majority of the data are missing. This is particularly true for mixtures and especially phase equilibria. Thus, the next step in the process assessment stage is to make judicious predictions. It is still too early to take physical property measurements in the laboratory, unless there is a crucial question to resolve. Thus, Chapter 5 examines the different prediction methods, emphasizing both pure component and mixture correlations. A wide range of problems is covered, including low- and high-pressure phase equilibria, gas solubility, electrolyte systems, polymer-solvent equilibria, solid-liquid equilibria, and properties of petroleum and synfuel fractions.

Chapter 4

A CRITICALLY EVALUATED DATA BASE FROM THE LITERATURE

TABLE OF CONTENTS

I. PURE COMPONENT DATA

A. Importance of Using Good Sources

It is unfortunate that the first pass at retrieval of physical property data is often totally disorganized and poorly done. Finding a desired physical property value in a prestigious handbook is not necessarily a guarantee that the number is accurate. Handbooks containing data are seldom compiled based on critical evaluation of physical property data. Therefore, the first rule is to use sources critically, because they are extremely variable in quality.

A large chemical plant was built in the mid-1970s using a heat of vaporization for the product that was in error by 20%! How could such a large error have been possible? A little detective work showed that the value used came from a standard handbook. The handbook value was based on differentiation of vapor pressure data. That is normally a fairly safe procedure. However, the vapor pressure points spanned only a few degrees! Had a more careful analysis been made, using all available vapor pressure data and perhaps remeasuring the vapor pressure, a more accurate value would have been discovered.

B. Critically Evaluated Data Sources

The first recourse should be to sources of data that are critically evaluated. This chapter lists those sources and indicates the organizations providing such data. Computer access to evaluated data is now a reality, and the means of using such programs will be indicated. Companies should not repeat critical evaluations that have been made already, but concentrate their efforts on data for compounds not previously subjected to such analysis.

National Bureau of Standards (NBS)

The NBS (Washington, D.C. 20234) is the only federal laboratory with the explicit mission of serving U.S. industry and science. The laboratories are located in Gaithersburg, Md. and Boulder, Colo. The NBS promotes interaction with industry by collaborating in cooperative research programs, responding to requests for assistance, and sponsoring many conferences. Some NBS research programs appear esoteric, but they are counterbalanced by genuine efforts to help the industry.

Within the NBS there are a number of efforts of special interest to people needing data. In recent work, the Center for Chemical Engineering correlated the existing quality data on water and steam. They used a new equation of state, the Harr-Gallagher-Kell equation. It has been accepted by the International Association for the Properties of Steam. Another computer correlation, known as TRAPP, predicts viscosity, density, and thermal conductivity of fluids. It is available from the NBS Boulder laboratories [(303) 497-3257]. The NBS Fluid Properties Data Center at the same location does experimental work of highest quality, literature evaluation, and equation-of-state development. A number of multiclient studies of industrial importance have been performed. That effort is under the direction of Dr. Neil A. Olien [(303) 499-1000].

The Office of Standard Reference Data (OSRD) is the conduit through which NBS data and quality information from many sources flow to the public. The OSRD makes computer programs and various publications available for reasonable fees. An example would be the NBS Chemical Thermodynamic Database. It contains the thermodynamic properties of 15,000 substances, with an effort to maintain thermodynamic consistency. The data base was originally developed by researchers such as D. Wagman and covers primarily inorganic compounds and organic compounds with no more than two carbon atoms. It is also available on-line through the Chemical Information System (CIS) [(800) CIS-USER]. Other data bases available include mass spectral data, of interest to chemists. A most important data base licensed by the OSRD is the Design Institute for Physical Property Data (DIPPR) data base which will be described later.

A useful contact in the OSRD is the manager of the Program on Industrial Process Data. At the time of this writing that role was ably filled by Dr. Howard White [(301) 975-2205]. He could usually point someone in the right direction when they needed a data source. His participation on international data committees gave him a broad perspective and window on publications coming from the Communist bloc countries. He could also help identify people within the NBS qualified to respond to specific questions. Electrolyte problems, for example, might be funneled to the Electrolyte Data Center. Other questions might be referred to the Chemical Thermodynamics Data Center or to the Equation of State Group. The NBS is willing to respond to many questions free of charge and, by the same token, is able to do work for a fee when the questions are really project-size in scope.

One of the most useful services of the OSRD has been its funding of data centers. This has permitted development of data evaluation expertise in a number of university settings. In 1984 there were 23 continuing data centers and 31 other data evaluation projects. Several of these have had special interest for the chemical industry. One center, now terminated, which provided a considerable amount of pure component and mixture data, was the Thermodynamics Research Laboratory (TRL), administered by Prof. B. D. Smith. He always strove for excellence in analysis of pure compound and mixture data. At this writing, part of the mixture work, on C_6 hydrocarbons and alcohols, remained to be published by the NBS.

Center for Information and Numerical Data Analysis and Synthesis (CINDAS)

Purdue University
2595 Yeager Road, West Lafayette, IN 47906
(317) 494-6300 and (317) 463-1581

CINDAS (originally TPRC) was established at Purdue University in 1957. It was originally directed by Y. S. Touloukian and now is under the direction of P. E. Liley. At present this center, which operates with some funding by the OSRD, concentrates on 14 thermophysical and 22 electronic properties. Information on 70 important fluids, with critical evaluation of their thermal conductivity, viscosity, and specific heat at constant pressure, has been published in hardbound volumes.

Every effort has been made by CINDAS to assess the data critically. A detailed description of the methodology used can be found in Reference 1. In this article the authors indicate that at that time they had already compiled more than 100,000 sets of experimental data. They attempt to show how this activity of critical evaluation, correlation, analysis, and synthesis " . . . brings order out of very chaotic experimental observations. While data evaluation and analysis deal mainly with the validity, reliability, and accuracy of existing experimental data, data synthesis is a process of generating a full field of new data based on limited fragmentary experimental information coupled with a knowledge of the theory."

CINDAS also makes information available on-line, including bibliographic information, and will accept telephone orders for copies of articles in its collection. A key to its holdings is in the useful bibliographic series it has issued (*Thermophysical Properties Research Literature Retrieval Guide 1900—1980*). The third volume is especially useful to the chemical industry, as it deals with organic compounds and polymeric materials. Properties covered include thermal conductivity, thermal diffusivity, specific heat, viscosity, and Prandtl number.

Thermodynamics Research Center (TRC)

Texas A & M University
College Station, TX 77843

Dr. K. N. Marsh, Executive Director
(409) 845-1451, (409) 845-5981, and (409) 845-4904

The TRC has a long history, with roots in the American Petroleum Institute Project 44. Voluminous tables of thermodynamic properties are now indexed. Some of the information now can be accessed on-line through the CIS [(800) CIS-USER]. The tables are far from complete, so the program has an option which helps determine availability of data on user-specified compounds. The CIS has published some in-depth studies on various chemicals, including phenol, cresols, benzene, xylenols, and furans.

More recently, the TRC has been performing a valuable service of publishing mixture phase equilibrium data in standard format. Experimental data will be measured increasingly at the TRC, including PVT, vapor-liquid equilibrium (VLE), and calorimetric. More detail on TRC capabilities is given in Chapter 6, Section III.

The TRC has recently purchased an extremely comprehensive collection of articles dealing with phase equilibria and pure component physical properties. The literature was accumulated at the TRL. Workers scanned every article ever published in journals which have been known to carry physical property information. Although transport properties were usually ignored, the articles represent a very important resource. It is hoped that they will be indexed by the TRC. Critical review of that compilation earlier led the TRL to issue to some industrial sponsors the correlation constants for approximately 1100 pure compounds. These included data on C1 to C9 hydrocarbons, C1 to C31 alcohols, C3 to C35 ketones, and C1 to C3 halogenated organics.

American Petroleum Institute (API)

1220 L N.W., Washington, DC 20005
Mr. Walter C. Retzsch
(202) 682-8153

The API pioneered the development of pure component and mixture data for the petroleum industry. The committees concerned with physical property data funded the manufacture, purification, and storage of pure hydrocarbon samples at Pennsylvania State University. They also funded pure component physical property measurements. Of particular value to the petroleum industry has been the ongoing publication of the API *Technical Databook — Petroleum Refining*. It contains a mixture of experimental data and prediction methods for pure hydrocarbons and petroleum fractions. The work is still funded by the petroleum industry, although at a reduced intensity.

Gas Processors Association (GPA)

1812 First Place, Tulsa, OK 74103
Mr. Carl Sutton, Executive Director
(918) 582-5112

The GPA Technical Committee has had a distinguished history of defining, funding, and contracting data projects. The work is aimed at components and mixtures of practical significance to the gas processing industry. The experimental studies have included both pure components and mixtures. Even difficult systems such as gas hydrates have been studied. Furthermore, some development of equations of state and computer programs has been funded. Methods used in this organization were used as a pattern in developing the procedures for DIPPR.

Design Institute for Physical Property Data (DIPPR)

American Institute of Chemical Engineers (AIChE)
345 East 47 Street, New York, NY 10017

1986 DIPPR Officers:

Dr. D. W. H. Roth, Chairman, Administrative Committee [(201) 455-5314]
Dr. D. A. Palmer, Chairman, Technical Committee [(312) 420-5056]
Dr. J. C. Forman, Executive Secretary, AIChE [(212) 705-7332]

DIPPR was created by thermodynamics experts from the chemical industry to meet industry needs. It was modeled after the successful cooperative organizations used by the gas processing and petroleum industries. These were the GPA and the API technical data committees. A new cooperative effort was found necessary because existing data efforts did not fill the comprehensive needs of the chemical industry. The AIChE offered an attractive organizational framework with its use of ''design institutes''. As a professional organization representing the largest core of potential users of the information, the AIChE also provided an excellent technical atmosphere in which the work could proceed and then be disseminated.

Projects within DIPPR are funded primarily by supporting companies, which now include most of the important chemical companies in the U.S. There are also foreign participants, i.e., engineering contractors, software companies, and government agencies. When companies vote on projects, they ballot with dollar commitments. The level of commitment can vary by a factor of eight, depending on the size of the company.

The NBS/OSRD provides money and technical support. It also has permitted the use of the prestigious logo of the National Standard Data Reference System on the *Manual for Predicting Chemical Process Design Data*[18] and the large handbook, *Aqueous Electrolyte Thermodynamics — Theory and Practice*. The logo likely will be used on other DIPPR publications as well. The Electrolyte Data Center of the NBS also has served as a DIPPR contractor.

The largest program in DIPPR has been the Data Compilation Project. Being carried on at Pennsylvania State University under the direction of Dr. Thomas E. Daubert (with earlier collaboration by Prof. Ronald P. Danner), this project has had support of a majority of the DIPPR participants. It involves the accumulation and critical evaluation of data on the 1000 compounds of most interest to the chemical industry. These involve all the commodity chemicals, useful solvents, and intermediates. The following pure compound constants are included.

Molecular weight	Melting point	Heat of fusion (mp)
Critical temperature	Critical pressure	Critical volume
Critical compressibility	Normal bp	Liquid molar volume
Triple pt. temperature	Triple pt. pressure	Standard net heat of comb.
IG heat of formation	IG Gibbs heat of formation	IG absolute entropy
Acentric factor	Radius of gyration	Solubility parameter
Dipole moment	van der Waals volume	van der Waals area
Refractive index	Flash point	Autoignition temperature
Lower flammability limit	Upper flammability limit	

The temperature-dependent properties included are the following.

- Density, solid, liquid, and vapor
- Heat capacity, solid, liquid, and vapor

- Viscosity, liquid and vapor
- Thermal conductivity, liquid and vapor
- Vapor pressure
- Heat of vaporization
- Second virial coefficients
- Surface tension

Where data are not available, predictions are used. However, there are designations in the compilation to alert the user to probable data accuracy and to flag those numbers based on prediction. In addition, an auxiliary computer program gives all the available raw data. It indicates the data points used, not used, and those rejected outright as erroneous. Of course the literature references are also given in both sources. The Pennsylvania State Data Compilation supplies accumulated expertise in data retrieval, organization, critical evaluation, and prediction.

In addition to the expertise of the researchers at Pennsylvania State University, one must also consider the role of the project steering committee which has had a very active role in guiding the project. Subcommittees review the results of the work done by Pennsylvania State in different classes of compounds, and recommend improvements or point out oversights. All these points of improvement are made to the satisfaction of the subcommittees before the compounds are included in the public version of the data bank. At least 1 year elapses subsequent to that determination before the compounds in question actually are released to the public. This gives incentive to the sponsors to continue to pay for the ongoing work.

At the time of the printing of this book, more than half of the targeted chemicals were in the public version of the data bank. There already has been completion of the first 1000 chemicals. Subsequent work will include review and upgrading of the previous evaluations. It is anticipated that the DIPPR Data Compilation will serve as the industry standard, facilitating work between engineering contractors and chemical company clients.

There are three means of accessing the DIPPR Data Compilation. The first is to purchase hard copy from the AIChE Publications Department. The hard copy tables are in SI units, and include correlation constants for the temperature-dependent properties. Computer tape is more practical for serious users, and is available from the OSRD. The tape includes all the raw data, quality codes, etc. not found in the printed tables. Finally, an on-line version is available from the Chemical Abstracts Service of the American Chemical Society. The computerized versions give choice of units and permit generation of tables for temperature-dependent properties.

The DIPPR *Manual for Predicting Chemical Process Design Data*[18] has already been published by the AIChE.[19] The principal author is Prof. Ronald P. Danner at Pennsylvania State University. At this writing a few chapters were not yet available to the public but would be within several years. The manual is analogous to the API *Technical Databook — Petroleum Refining* in that its focus is on predictions of data. The distinction between the two manuals is that the DIPPR manual includes the polar organic molecules that do not concern the petroleum industry. Within the DIPPR manual there are comparisons between different methods that have been published for prediction of various properties. For those particularly interested in certain properties, documentation giving more detail on the chapters is available from the AIChE Publications Department. In some cases, the documentation reports show analysis of predictive methods not even mentioned in the manual. The manual is divided into the following sections.

1. General data
2. Critical properties

3. Vapor pressure
4. Density
5. Thermal properties
6. Phase equilibria
7. Surface tension
8. Viscosity
9. Thermal conductivity
10. Diffusivity
11. Combustion
12. Environmental data

Another pure component project within DIPPR is data measurement. The objective is to fill important gaps identified in the Data Compilation. The first type of data taken under contract was vapor pressure, measured by Dr. Thomas E. Daubert. The second thrust was measurement of critical properties for compounds with low thermal stability, measured by Dr. Amyn S. Teja at Georgia Institute of Technology. In the future, no doubt, other types of data also will be measured and reported in publications of the AIChE.

Specialized projects also have been undertaken to solve difficult problems. One such project measured the effect of gas phase dimerization of organic acids. Thus, a full range of organic acids was studied, with both PVT measurements and heats of vaporization.

International Council of Scientific Unions Committee on Data for Science & Technology (CODATA)

CODATA secretariat
51 Boulevard de Montmorency, 75016 Paris, France
E. F. Westrum, Jr., Secretary General

CODATA is a European-based organization which sponsors conferences related to physical properties. It has published a number of items in the *CODATA Bulletin* in an attempt to set standards for publication of data. An example would be the 1978 *Guide for the Presentation in the Primary Literature of Physical Property Correlations and Estimation Procedures*. An example of a large data collection effort is the April 1984 *CODATA Directory of Data Sources for Science & Technology*, E. F. Westrum, Editor, Chapter 11, Chemical Thermodynamics, R. D. Freeman, Chapter Editor. It was used as a resource in compiling some of the information in this chapter.

The CODATA Task Group on Data for the Chemical Industry had plans to develop an industrial data base, but did not realize that dream. The need disappeared, of course, when DIPPR found a means to accomplish that goal and did so. In essence, CODATA has become a liaison between data groups of different types from around the world, a very useful function.

Fluid Properties Research Inc. (FPRI)

c/o Celanese Chemical Company, Inc.
P.O. Box 9077, Corpus Christi, TX 78469
Mr. J. David Chase, President
(512) 241-2343, ext. 4267

FPRI is a proprietary consortium of companies interested in transport properties. A primary interest is the role of transport properties and densities in design of heat exchangers. From its inception until August 1985, it was directed by Prof. Robert Maddox at Oklahoma State

University. The emphasis was on measurement of high-quality experimental data which were available only to members of the organization. Over 5000 proprietary data points were measured. Due to the retirement of Prof. Maddox, the FPRI operations and equipment were moved to the Georgia Institute of Technology in Atlanta. The principal investigator, Dr. Amyn S. Teja, is now soliciting new members.

With FPRI equipment it is possible to measure liquid densities within 0.25%, viscosities within 2.0%, and thermal conductivities within 5%. FPRI has done a large amount of literature data retrieval, as well as correlation work. Both pure component and mixture data of defined and undefined components have been covered.

ESDU International, Ltd.

Suite 200, 1495 Chain Bridge Road, McLean, VA 22101
Ms. Cynthia Albrecht, Sales Engineer
(312) 397-2277

ESDU is a multifaceted engineering organization which specializes in critical data analysis. One part of its mission includes data analysis for the chemical industry. At the time of this writing ESDU had an active sales force in the U.S. trying to market its series of data volumes. The collection is not computerized and there are apparently no plans to do so. ESDU publishes an index to its collection which is worth having in itself, since it gives critical properties, mp, and bp for over 1900 compounds. An example of one report which would be available to subscribers is #74032, Thermal Conductivity of Liquid Aromatic Compounds Containing Oxygen. This report gives references, correlating equations, critical and mp temperatures, and tables and graphs of the smoothed information for 20 compounds. The raw data are not included. The overall value of this set of data reports must be judged by each company with respect to its individual needs. An unfortunate aspect of this compilation is its lack of comprehensiveness. It appears that there is no set of compounds for which all relevant properties have been accumulated. Moreover, no temperature-dependent properties have been examined for most of the chemicals of interest. Vapor pressure and liquid thermal conductivity are the two properties most thoroughly covered. What this compilation enjoys in quality it unfortunately lacks in overall organization and strategy.

CSIRO-NPL Thermodata System

CSIRO Division of Mineral Chemistry
P.O. Box 124, Port Melbourne 3207, Australia
A. G. Turnbull, Director

This organization provides on-line access to heats and free energies of formation, plus heat capacities. Some 4000 substances are included. Data for over 1600 substances have been provided by the National Physical Laboratory in the U.K.

French Institute for Energy

3 rue Henri Heine, 75016 Paris, France
E. Maintreu, Director

This organization does literature searches and critical analysis for a fee. It systematically screens hundreds of journals for thermodynamic and transport properties, building internal files of evaluated data.

Anjlina Engineering Information Services

A-17, Dattaguru Society
Deonar Pada Road, Deonar, Bombay 400 088, India
S. A. Kudchadker, Director

This organization specializes in responding to inquiries for data on physical, thermodynamic, and transport properties, and claims to do a critical evaluation of the information retrieved. It actively cooperates with the TRC.

JUSE-AESOPP

Computation Center
Institute of the Union of Japanese Scientists and Engineers
5-10-11, Sendagaya, Shibuya-ku, Tokyo 151, Japan
Hiromi Shohno, Director

This organization serves as a data center, with guidance from the respected professor, Mitsuho Hirata. Both pure component and mixture data are available for about 2400 substances. Critical, thermodynamic, and transport properties are included. The center is open to all on a fee basis, and data can be retrieved by purchase or lease of a magnetic tape.

Physical Properties Data for Design in Chemical Industry

Prosynchem
Konstytucgi 11, 44-101 Gliwice, Poland
K. Cempiel, K. Klepacka, Directors

This data center concentrates on both pure component and phase equilibrium information. The products of the work are available in hard copy, data tape, and on-line. However, access is more difficult outside of Poland than within that country.

Institute for High Temperatures, Academy of Sciences

Department of Chemical Thermodynamics
Korovinskoye Chausse, Moscow 127412, U.S.S.R.
L. V. Gurvich, Director

This institute specializes in thermodynamic properties of the elements, in the manner of the NBS. An on-line data bank, "IVTANTERMO", can be accessed within the U.S.S.R.

C. Bibliography of Pure Component Data Sources

If the available critical evaluations do not provide the needed information, it will be necessary to dig further. Some help may be obtained from the organizations listed herein, but in many cases the user will have to rely on the literature. A bibliography of the more important references follows, although the list is not exhaustive. Whereas some of the references to the Soviet literature are difficult to obtain, they are often available through some of the data organizations previously mentioned. Often, the translation already will have been completed.

In retrieving physical properties from the literature it is important to recognize that all literature data are not created equal. Therefore, the following bibliography has been divided arbitrarily, according to standards of reliability. The categories used are

A. Data are critically selected and of a high degree of precision or accuracy. In some cases the uncertainty in the data is stated.

B. Data are generally good, but may contain estimated values which should be used with caution, depending on circumstances. Sources in this category are not quite as reliable as those in Category A.

C. This category comprises valuable collections of data which were not critically evaluated. Some contain unique data, estimated values, and collections of references. Scattered data may contain significant errors. Use with caution and verify any data used for design; consult the original references.

CATEGORY A BOOKS

1. **Armstrong, G. T. and Domalski, E. S.,** Thermodynamic Data for Industrial Incinerators, Rep. 10487, National Bureau of Standards, Washington, D.C., 1972.
2. **Chase, M. W. et al.,** in JANAF Thermochemical Tables, 2nd ed., NSRDS-NBS 37, National Bureau of Standards, Washington, D.C., 1971; supplements in *J. Phys. Chem. Ref. Data,* 3, 311, 1974; 4, 1, 1975; 7, 793, 1978; 11, 695, 1982.
3. **Cox, J. D. and Pilcher, G.,** *Thermochemistry of Organic and Organometallic Compounds,* Academic Press, Orlando, Fla., 1970.
4. **Daubert, T. E. and Danner, R. P.,** DIPPR Data Compilation, American Institute of Chemical Engineers, New York, in press (loose-leaf sheets).
5. **Dymond, J. H. and Smith, E. B.,** *Second Virial Coefficients of Pure Gases and Mixtures,* Oxford University Press, New York, 1980.
6. **Harr, L., Gallagher, J. S., and Kell, G. S.,** *The NBS/NRC Steam Tables,* Hemisphere Publishing, Washington, D.C., 1983.
7. **Jamieson, D. T., Irving, J. B., and Tudhope, L. S.,** *Liquid Thermal Conductivity — A Data Survey to 1973,* National Engineering Laboratory, Edinburgh, 1975.
8. **McClellan, A. L.,** *Tables of Experimental Dipole Moments,* W. H. Freeman, San Francisco, 1963.
9. **Parker, V. B., Wagman, D. D., and Garvin, D.,** Rep. #75-968, National Bureau of Standards, Washington, D.C., 1976.
10. **Pedley, J. B. and Rylance, J.,** *Sussex-NPL Computer Analyzed Thermochemical Data: Organic and Organometallic Compounds,* University of Sussex, Brighton, 1977.
11. *Physical Constants of Hydrocarbons C1 to C10,* ASTM Data Ser. Publ. DS 4 A, American Society for Testing and Materials, Philadelphia, 1971.
12. **Powell, R. W., Ho, C. Y., and Liley, P. E.,** Thermal Conductivity of Selected Materials, NSRDS-NBS No. 8, National Bureau of Standards, Washington, D.C., 1966.
13. Selected Values of Chemical Thermodynamic Properties, Circ. 461 (1947) and 500 (1952), National Bureau of Standards, Washington, D.C.
14. **Smith, B. D.,** *Component Physical Properties,* Vol. 1 to 4, Thermodynamics Research Laboratory Data Books, Washington University, St. Louis, 1984 (loose-leaf sheets).
15. **Starling, K. E.,** *Fluid Thermodynamic Properties for Light Petroleum Systems,* Gulf Publishing, Houston, 1973.
16. **Stull, D. R. and Prophet, H.,** in *JANAF Thermochemical Tables,* 2nd ed., NSRDS-NBS 37, National Bureau of Standards, Washington, D.C., 1971; supplements in *J. Phys. Chem. Ref. Data,* 3, 311, 1974; 4, 1, 1975; 7, 793, 1978.
17. **Stull, D. R., Westrum, E. F., and Sinke, G. C.,** *The Chemical Thermodynamics of Organic Compounds,* John Wiley & Sons, New York, 1969.
18. **Touloukian, Y. S., Liley, P. E., and Saxena, S. C.,** *Thermophysical Properties of Matter,* Vol. 3, Plenum Press, New York, 1970.
19. **Touloukian, Y. S. and Makita, T.,** *Thermophysical Properties of Matter,* Vol. 6, Plenum Press, New York, 1970; Vol. 6 (Suppl.), 1976.
20. **Touloukian, Y. S., Saxena, S. C., and Hestermans, P.,** *Thermophysical Properties of Matter,* Vol. 11, Plenum Press, New York, 1975.
21. **Wagman, D. B. et al.,** The NBS tables of chemical thermodynamic properties — selected values for inorganic and C1 and C2 organic substances in SI units, *J. Phys. Chem. Ref. Data,* 11 (Suppl. 2), 1982.
22. **Wilhoit, R. C. and Zwolinski, B. J.,** *Handbook of Vapor Pressures and Heats of Vaporization of Hydrocarbons and Related Compounds,* Thermodynamics Research Center, Texas A & M University, College Station, 1971.

23. **Zwolinski, B. J. and Wilhoit, R. C.,** Heats of formation and heats of combustion, in *American Institute of Physics Handbook,* Gray, D. E., Ed., McGraw-Hill, New York, 1972, chap. 41.
24. **Zwolinski, B. J. et al.,** *Comprehensive Index of API 44-TRC Selected Data on Thermodynamics and Spectroscopy,* Thermodynamics Research Center, Texas A & M University, College Station, 1974.
25. **Zwolinski, B. J. et al.,** *Tables of Selected Values of Properties of Chemical Compounds,* Thermodynamics Research Center Data Project, Texas A & M University, College Station, 1984 (loose-leaf sheets).

CATEGORY A ARTICLES

1. **Ambrose, D., Broderick, B. E., and Townsend, R.,** The critical temperatures and pressures of 30 organic compounds, *J. Appl. Chem. Biotechnol.,* 24, 359, 1974.
2. **Domalski, E. S.,** Selected values of heats of combustion and heats of formation of organic compounds containing the elements C, H, N, O, P, and S, *J. Phys. Chem. Ref. Data,* 1, 221, 1972.
3. **Domalski, E. S., Evans, W. H., and Hearing, E. D.,** Heat capacities and entropies of organic compounds in the condensed phase, *J. Phys. Chem. Ref. Data,* 13 (Suppl. 1), 1984.
4. **Haar, L. and Gallagher, J. S.,** Thermodynamic properties of ammonia, *J. Phys. Chem. Ref. Data,* 7(3), 635, 1978.
5. **Jasper, J. J.,** The surface tension of pure liquid compounds, *J. Phys. Chem. Ref. Data,* 1, 841, 1972.
6. **Kestin, J., Sengers, J. V., Kamgar-Parsi, B., and Levelt-Sengers, J. M. H.,** Thermophysical properties of fluid H_2O, *J. Phys. Chem. Ref. Data,* 13, 175, 1984.
7. **Kudchadker, A. P., Alani, G. H., and Zwolinski, B. J.,** The critical constants of organic substances, *Chem. Rev.,* 68, 659, 1968.
8. **Martin, J. F.,** The heat capacities of organic compounds, *Chemical Thermodynamics: A Specialist Periodical Report,* Vol. 1, The Chemical Society, London, 1973, 133.
9. **Matthews, J. F.,** The critical constants of inorganic substances, *Chem. Rev.,* 72, 71, 1972.
10. **Pilcher, G.,** Thermochemistry of chemical compounds, in *Thermochemistry and Thermodynamics, MTP International Review of Science, Physical Chemistry* (1st series), Vol. 10, Butterworths, London, 1972, 57.
11. **Pilcher, G.,** Thermochemistry of organometallic compounds containing metal-carbon linkages, in *Thermochemistry and Thermodynamics, MTP International Review of Science, Physical Chemistry* (2nd series), Vol. 10, Butterworths, London, 1972, 45.
12. **Wilhoit, R. C. and Zwolinski, B. J.,** Physical and thermodynamic properties of aliphatic alcohols, *J. Phys. Chem. Ref. Data,* 2 (Suppl. 1), 1973.

CATEGORY B REFERENCES

1. **Affens, W.,** Flammability properties of hydrocarbon fuels, *J. Chem. Eng. Data,* 11, 197, 1966.
2. **Ashcroft, S. J. and Mortimer, C. T.,** *Thermochemistry of Transition Metal Complexes,* Academic Press, Orlando, Fla., 1970.
3. *ASHRAE Thermodynamic Properties of Refrigerants,* American Society of Heating, Refrigerating and Air Conditioning Engineers, New York, 1969.
4. **Bolz, R. E. and Tuve, G. L., Eds.,** *CRC Handbook of Tables for Applied Engineering Science,* 2nd ed., CRC Press, Boca Raton, Fla., 1973.
5. **Boublik, T., Fried, V., and Hala, E.,** *The Vapor Pressure of Pure Substances,* Elsevier, New York, 1973.
6. **Braker, W. and Mossman, A. L.,** *Matheson Gas Data Book,* 6th ed., Matheson Gas Products, Secaucus, N.J., 1980.
7. **Burcat, A. E.,** Thermochemical data for combustion calculations, in *Combustion Chemistry,* Gardiner, W. C., Ed., Springer-Verlag, Basel, 1984, 455.
8. **Canjar, L. N. and Manning, F. S.,** *Thermodynamic Properties and Reduced Correlations for Gases,* Gulf Publishing, Houston, 1967.
9. **Cholinski, J., Szafranski, A., and Wyrzykowska-Stankiewica, D.,** *Second Virial Coefficients for Pure Organic Compounds and for Binary Organic Mixtures,* Polish Scientific publishers, Warsaw, 1982—1983.
10. **Christensen, J. J., Eatough, D. J., and Izatt, R. M.,** *Handbook of Metal Ligand Heats and Related Thermodynamic Properties,* 2nd ed., Marcel Dekker, New York, 1975.
11. **Coward, H. F. and Jones, G. W.,** Limits of Flammability of Gases and Vapors, Bull. 503, U.S. Bureau of Mines, Washington, D.C., 1952.
12. **Din, F., Ed.,** *Thermodynamic Functions of Gases,* Butterworths, London, Vol. 1 and 2, 1956; Vol. 3, 1961.
13. *Encyclopedia of Polymer Science and Engineering,* Vol. 1, 2nd ed., Wiley-Interscience, New York, 1985.
14. *ESDU Validated Engineering Data Index,* ESDU International, Ltd., London, 1983—1984.

15. **Handrick, G. R.,** Heats of combustion of organic compounds, *Ind. Eng. Chem.*, 48, 1366, 1956.
16. **Hilsenrath, J. et al.,** Tables of Thermal Properties of Gases, Circ. 564, National Bureau of Standards, Washington, D.C., 1955.
17. **Karapet'yants, M. K. and Karapet'yants, M. L.,** *Handbook of Thermodynamic Constants of Inorganic and Organic Compounds,* Humphrey Publishers, Ann Arbor, Mich., 1970.
18. **Keenan, J. H. et al.,** *Steam Tables, Thermodynamic Properties of Water Including Vapor, Liquid, and Solid Phases,* John Wiley & Sons, New York, 1966.
19. Matheson Company Unabridged Gas Data Book, The, Vol. 1 to 4, Matheson Gas Products, Secaucus, N.J., 1974.
20. **Ohe, S.,** *Computer Aided Data Book of Vapor Pressure,* Data Book Publishing, Tokyo, 1976.
21. **Rabinovich, V. A., Ed.,** *Thermodynamic Properties of Matter and Substances,* Moscow, Vol. 2, 1974; Vol. 4, 1975.
22. **Raznjevic, K.,** *Handbook of Thermodynamic Tables and Charts,* McGraw-Hill, New York, 1976.
23. **Rossini, F. D., Pitzer, K. S., Arnett, R. L., Brown, R. M., and Pimentel, G. C.,** *Selected Values of Physical and Thermodynamic Properties of Hydrocarbons and Related Compounds,* Carnegie Press, Pittsburgh, 1953.
24. **Sansonov, G. V.,** *Handbook of the Physicochemical Properties of the Elements,* Plenum Press, New York, 1968.
25. **Stull, D. R.,** Vapor pressure of pure substances — organic compounds, *Ind. Eng. Chem.*, 39, 517, 1947, corrections on p. 1684; Vapor pressure of pure substances — inorganic compounds, *Ind. Eng. Chem.*, 39, 540, 1947.
26. **Stull, D. R. and Sinke, G. C.,** *Thermodynamic Properties of the Elements,* American Chemical Society, Washington, D.C., 1956.
27. **Szafranski, A., Cholinski, J., and Wyrzykowska-Stankiewicz, A.,** *Solid-Liquid Equilibrium,* Polish Scientific Publishers, Warsaw, 1983—1984.
28. **Wichterle, I. and Linek, J.,** *Antoine Vapor Pressure Constants of Pure Compounds,* (in Czech), Academia, Prague, 1971.
29. **Yaws, C. L.,** *Physical Properties,* Chemical Engineering Division, McGraw-Hill, New York, 1977.
30. **Zabetakis, M.,** Flammability Characteristics of Combustible Gases and Vapors, Bull. 627, U.S. Bureau of Mines, Washington, D.C., 1965.

CATEGORY C REFERENCES

1. Aldrich Co. Chemicals Catalog.
2. **Dean, J. A., Ed.,** *Lange's Handbook of Chemistry,* 13th ed., McGraw-Hill, New York, 1985.
3. **Driesbach, R. R.,** *Physical Properties of Chemical Compounds,* Vol. 1 to 3, American Chemical Society, Washington, D.C., 1955, 1959, 1961.
4. **Gallant, R.,** *Physical Properties of Hydrocarbons,* Vol. 1 and 2, Gulf Publishing, Houston, 1970.
5. **Green, D. W., Ed.,** *Perry's Chemical Engineers' Handbook,* 6th ed., McGraw-Hill, New York, 1984.
6. **Horvath, A. L.,** *Physical Properties of Inorganic Compounds (in SI units),* Crane, Russak & Company, New York, 1973.
7. **Jordan, T. E.,** *Vapor Pressure of Organic Compounds,* Interscience, New York, 1954.
8. *Kirk-Othmer Encyclopedia of Chemical Technology,* John Wiley & Sons, New York, 1978—1984.
9. **Mellan, I.,** *Industrial Solvents Handbook,* 2nd ed., Noyes Data Corporation, Park Ridge, N.J., 1977.
10. **Nesmeyanov, A. N.,** *Vapor Pressure of Elements,* Academic Press, Orlando, Fla., 1963.
11. **Reid, R. C., Prausnitz, J. M., and Sherwood, T. K.,** *The Properties of Gases and Liquids,* McGraw-Hill, New York, 1973.
12. **Riddick, J. A. and Bunger, W. A.,** *Organic Solvents,* Wiley-Interscience, New York, 1970.
13. **Svebla, R. A.,** Estimated Viscosities and Thermal Conductivities of Gases at High Temperatures, NASA Tech. Rep. R-132, National Aeronautics and Space Administration, Washington, D.C., 1962.
14. **Timmermans, J.,** *Physico-Chemical Constants of Pure Compounds,* Vol. 1 and 2, Elsevier, New York, 1950, 1965.
15. **Utermark, W. and Schicke, W.,** *Melting Point Tables of Organic Compounds,* 2nd ed., Wiley-Interscience, New York, 1963.
16. **Vargaftik, N. B.,** *Handbook of the Thermophysical Properties of Gases and Liquids,* 2nd ed., Halsted Press, New York, 1975.
17. **Washburn, E. W., Ed.,** *International Critical Tables of Numerical Data, Physics, Chemistry, and Technology,* Vol. 1 to 7, McGraw-Hill, New York, 1926—1933.
18. **Weast, R. C., Ed.,** *CRC Handbook of Chemistry and Physics,* 64th ed., CRC Press, Boca Raton, Fla., 1983.
19. **Wendholz, M., Ed.,** *The Merck Index, An Encyclopedia of Chemicals, Drugs, and Biologicals,* annu. eds., Merck & Co., Rahway, N.J., 1983.

Primary Literature Retrieval

1. **Beilstein,** *Handbuch der Organischen Chemie*, Springer-Verlag, Berlin, 1918—present.
2. **Brielles, J., Dedit, A., Lallemand, M., LeNeindre, B., Lerov, Y., Vermese, J., and Vidal, D.,** Equation of state of gases at high pressures and low or moderate temperatures, *Experimental Thermodynamics*, Vol. 2, LeNeindre, B. and Vodar, B., Eds., Butterworths, London, 1975, 347.
3. **Chaney, J. F. et al., Eds.,** *Thermophysical Properties Research Literature Retrieval Guide 1900—1980*, Plenum Press, New York, 1982.
4. *Chemical Abstracts*, Chemical Abstracts Service, Ohio State University, Columbus, Ed., American Chemical Society, Washington, D.C., 1907—present.
5. **Freeman, R. D.,** CODATA directory of data sources for science & technology. Chemical thermodynamics, *CODATA Bull.*, 55, chap. 11, April 1984.
6. Publications of the GPA, Catalog of Research Reports, Gas Processors Association, Tulsa, Okla.
7. **Selover, T. B. and Klein, M., Eds.,** *Awareness of information sources, AIChE Symp. Ser.*, 80, 237, 1984.
8. **Tamir, A., Tamir, E., and Stephan, K.,** *Heats of Phase Change of Pure Components and Mixtures: A Literature Guide*, Elsevier, New York, 1983.
9. **Westrum, E. F., Jr., Ed.,** *Bulletin of Thermodynamics and Thermochemistry*, IUPAC Comm. on Thermodynamics and Thermochemistry Annu. University of Michigan, Ann Arbor, 1958—present.

D. Data Availability

Notwithstanding the hundreds of thousands of physical property data measurements reported, there remain enormous gaps in the literature. The experience of Pennsylvania State researchers in developing the DIPPR Data Compilation suggests where problems should be expected in finding experimental data. In this documentation of data gaps[2] only some of the physical properties were considered: 6 single-valued and 11 temperature-dependent values. The analysis was further restricted to the 193 largest-volume chemicals. The extent of the literature search could be considered thorough, not exhaustive. An attempt was made by Pennsylvania State to obtain data from current manufacturers of the chemicals, and to use product bulletins when they were the only sources of data.

The most surprising conclusion of this study was that there were no data at all for one third of the compound property combinations. (This did not include inappropriate properties such as vapor properties of sodium chloride.) One might have expected basic properties to be known for at least these commodity chemicals. The data base is best for paraffins, as might be expected. It is worst for esters and polyfunctional compounds, where approximately half the needed data are missing. The most serious omission is usually the lack of critical property data. Because of the importance of critical properties to corresponding-states prediction methods, DIPPR is attempting to rectify this problem experimentally. Surprisingly, liquid thermal conductivity data were unavailable for several of the largest-volume petrochemicals. In general, solid density data seldom have been reported as a function of temperature.

The following table gives the average number of missing properties (out of 17) per compound class.

Paraffins	1.2
Inorganics	1.9
Aromatics	2.7
Inorganic acids and bases	3.1
Halides	4.1
Olefins-acetylenes	4.4
Alcohols, phenols, and polyols	5.4
Aldehydes, ketones, and ethers	6.2
Nitrogen and sulfur content	7.3
Acids and anhydrides	7.5
Esters	8.6
Polyfunctional	8.9

The properties most often missing, in descending order, are

- Solid density
- Vapor viscosity
- Vapor thermal conductivity
- Critical specific volume
- Liquid thermal conductivity
- Critical pressure
- Solid heat capacity
- Solid heat capacity
- Critical temperature
- Liquid heat capacity
- Ideal gas heat capacity
- Heat of fusion
- Surface tension
- Liquid viscosity
- Liquid density
- Liquid vapor pressure
- Normal bp
- mp

E. Relative Importance of Physical and Thermodynamic Properties

The properties most needed for process design tend to be found at the bottom of the preceding list. However, the critical properties which are used as the basis for most predictive methods are measured less frequently than might be expected because of decomposition at high temperature.

Typically, the *vapor pressure* will be the most important property, since it affects separation efficiency. Together with the mixture properties discussed later, it affects not only the size but also the type of equipment used. Zudkevitch[27] did an economic study of the impact of a 1.5% bias in isopentane vapor pressures on deisopentanizer operation. This in fact represented the variance in values available to him. If his bias was optimistic, operating and capital charges were 10% too low, and the column would not meet design specifications. If the values were conservative, there would be 14% waste. There are many components for which the vapor pressures have much greater uncertainty than the 1.5% assumed.

Thermal properties follow in importance because they affect the energy inputs to the process on a continuing basis. Sometimes the accuracy requirements exceed data accuracy and correlation capabilities. For example, the GPA has set an objective of accuracy within 1 Btu/lb under all conditions.[28] However, the accuracy of correlations is now within 3 Btu/lb, still far from the objective. Around the critical point the errors are higher. Zudkevitch[27] has cited the example of a tower designed to separate ethylene from isobutane and hexenes. An error of 4 Btu/lb in the latent heat affected the internal vapor flow by 27%. The column operated inefficiently; the refrigeration/condenser system was inadequate for the actual column capacity. Vapor heat capacity is crucial to compressor design and performance.

Density is of considerable importance in liquid-liquid extraction (LLE) and critical importance in custody transfer. Thermal conductivity can be important, but usually has less effect on process costs than vapor pressures and thermal properties.

Each process will have its own special property needs, so generalizations can be misleading. As an example, consider a process to produce and isolate para-xylene from its isomers. High-temperature gas heat capacity, viscosity, and thermal conductivity are needed for the design of the gas phase isomerization of mixed xylenes. The product can be recovered by low-temperature crystallization. Thus, the same properties of the liquid, plus the heat of fusion, are required for design of the cold end of the process. The solid density, the property least often measured, is important.

F. Qualification of Pure Compound Data

Once the data have been accumulated, it is necessary to use consistency tests to be sure that the data are accurate. This is not simply a matter of curve fitting. It involves use of tools that just now are being developed. The DIPPR Data Compilation Committee has a subcommittee charged with development of these tools so that the Data Compilation can be continually improved. It is no less important, when data are used for design, to qualify raw data that come from the primary literature. The primary reference so far on this subject is by Chase.[3]

The first and most important process in data qualification is to make sure that the vapor pressure and heat of vaporization are thermodynamically consistent. As a first approximation, for systems far below the critical point, the following equations can be used.

$$\Delta H^v = A + B * T \tag{1}$$

$$\ln P^* = -A/RT + (B/R) \ln T + C \tag{2}$$

The relationship between these equations can be seen via the Clapeyron equation. Of course, more complicated heat of vaporization equations could be used, and density correction factors incorporated.

Thorough data qualification goes far beyond this simple approach. Chase has defined it in this manner: "Data qualification is the rating/testing of property data outside of the environment of their determination against external criteria. Ideally, the external criteria would include both property-oriented criteria and homologous series criteria; i.e., there are a multiplicity of criteria." He recommends qualifying basic data first, followed by the properties which vary with temperature. He would then qualify the data on a property basis followed by use of data on homologous series. It is amazing how quickly the plotting of data by computer helps to spot bad data! This effort should be done by computer because of common errors not easily spotted by hand such as misplacement of the decimal point. One useful method of plotting data on homologous series of isomers, while taking into account branching, is to use the compound critical temperature rather than carbon number. This was one of many approaches used in critical assessment and upgrading of the DIPPR Data Compilation on pure components.

G. Interactive Access to Data

Someone once said that mankind is drowning in information but thirsting for knowledge. The person searching for physical property information does not ultimately want correlation coefficients. He wants instead the actual value of a property under specific conditions, and in the units of the problem. To give some idea of the way in which the problem can get out of hand, consider the use of the DIPPR Data Compilation. When all 1000 components are completed the compilation will be roughly 3000 pages. It appears in SI units, fine for some but anathema to others. Thus, someone who needs a vapor thermal conductivity in British units must first solve a polynomial fitting equation at every temperature of interest, and then use a conversion factor. It is easy to see why the publication of very large volumes of data is becoming obsolete.

All data bases that hope to compete in the marketplace either will be available interactively, or will be part of simulation programs. For example, there are large data bases associated with both the Chemshare and Simulation Sciences simulation programs. The Aspen Project has used the data base from the appendix of the book by Reid et al.[4] Information on those ever-changing data bases is readily available from the respective companies, so the balance of this discussion emphasizes the various stand-alone computer programs that are available.

DIPPR Data Compilation

American Institute of Chemical Engineers
345 East 47 Street, New York, NY 10017
J. C. Forman, Executive Secretary
(212) 705-7332

A review of the interactive version of the DIPPR Data Compilation is given in Reference 22. A version with data base management capabilities will be on-line in 1987 through STN-Columbus, an affiliate of the Chemical Abstracts Service of the American Chemical Society (ACS). The command language structure used for other ACS data bases will be implemented here as well. A distinction is that this data base provides actual data, not just bibliography. For details contact

STN-Columbus, Attn. Brian P. Cannan
P.O. Box 2228, Columbus, OH 43202
(614) 421-3600

Computer tapes also can be leased for in-house use. This is handled by the NBS/OSRD, [(301) 921-2228].

Significant features of the DIPPR system are

- User-selected units
- Indexing by compound name, class, or formula
- Generation of tables, including temperature dependence
- Retrieval of correlation constants
- Estimates of probable uncertainty in the data
- Raw data, with specification of points used, not used, or rejected
- Literature references
- Generation of graphs of either raw or correlated data

The computer access to the compilation provides more information, such as the raw data, and is simpler to use than the hard copy version. Wherever relevant, missing data are filled in by prediction, and the use of predictive methods is flagged for the user. It is seen as preferable for the missing numbers to be filled in by experts than by someone less experienced. This system is not infallible; there will still be inaccurate or poorly predicted numbers for some chemicals. However, indications of higher probable error should flag those compounds so that experimental data can be planned when needed for final process design. User feedback and continuing review should reduce those problems within the limitations of available data. This system is recommended as the primary source of information, as it is virtually guaranteed to become the de facto standard in the U.S., due to the participation of most of the major chemical companies.

Physical Property Data Service (PPDS)

The Institution of Chemical Engineers
165—171 Railway Terrace, Rugby, Warwickshire CV21 3HQ, England
Dr. Beryl Edmunds, Manager
TELEX: 311780; telephone: Rugby (0788) 78214

The PPDS is operated by the U.K. National Engineering Laboratory, in association with the Institution of Chemical Engineers. Their data bank includes over 800 components, with most of the properties of interest in process design. The PPDS program operates more like simulator programs than does the DIPPR Data Compilation, in that it has mixing rules, equations of state, and the UNIFAC activity coefficient prediction method to give properties of mixtures. It is available interactively on a use basis, or the system may be leased. In the latter case there exists some possibility of linking the system to a process simulator. An interesting offshoot of that idea is that the National Engineering Lab in Glasgow has produced a piece of electronics which contains the PPDS package as ''firmware'' rather than software. It would be connected as any other peripheral to a computer system.

Several industrial users in the U.S. have reported good results with this package. This writer's own experience was not entirely positive, although the program did provide an interim source of information prior to the DIPPR data base. When DIPPR was formed, the PPDS package was already available, but the industrial experts agreed that it did not meet all their needs. It has no estimate of probable error, no access to the raw data, and no access to the primary literature on which it was based. PPDS claims that its compilation is critically evaluated. However, with very little access to the original information and no way of knowing whether a number is based on data or prediction, confidence in the final result suffers. Typically, this would require use of larger overdesign factors, with resulting capital cost penalties.

Chemical Information System (CIS)

Fein-Marquart Associates, Inc.
7215 York Road, Baltimore, MD 21212
Dr. A. E. Fein, President
(800) 247-8737 [(800) CIS-USER]
or (301) 321-8440

CIS is an interactive computer system which accesses information from several important data sources. The first is the NBS Technical Note 270 series. It has data on heat capacity plus free energy, entropy, and enthalpy of formation at 298 K for a large number of compounds. These are primarily inorganics, but C1 and C2 chemicals are also included. At present this data base contains about 14,000 entries. The same information is available in printed form.[5] The nature of the data suggests they would be easier to look up in the book than to find on the computer.

The second data base in CIS is that of the TRC at Texas A & M. As previously noted, the only practical way to use that data base is through an interactive system. It has the same properties as Technical Note 270, but emphasizes organic compounds. It also covers some additional properties such as critical properties, densities, compressibility factors, solubility of the solids in water, boiling and freezing points, and vapor pressure.

Finally, CIS has developed a simulation of physical properties for the gas processing industry which includes phase equilibria and thermal properties. The program uses the BACK equation developed by Kreglewski.[20]

Union Carbide Corporation

Union Carbide Corporation
P.O. Box 8361 South Charleston, WV 25303
Larry Orr, Product Manager
(304) 747-4814

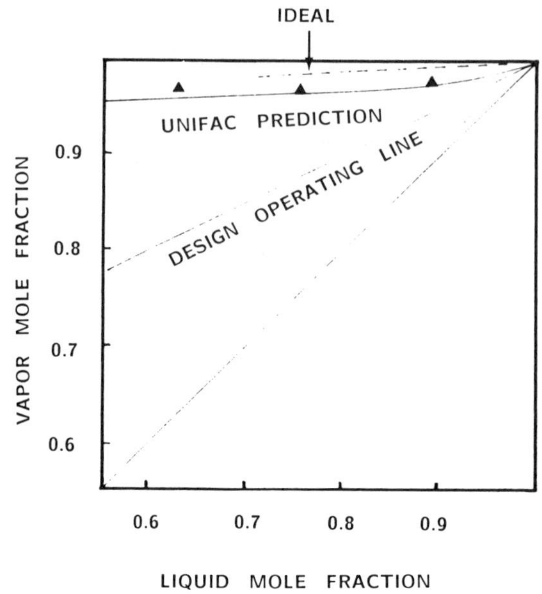

FIGURE 1. VLE of water + phenol, comparing actual behavior to the assumed ideal behavior. It is clear that the prediction using UNIFAC is not perfect but is a much better approximation to the experimental data. It indicates a ''pinch point'' at the top of the column.

Thermophysical property literature reference retrieval is offered by the Process Systems and Services Department of Union Carbide. The articles are flagged by a number of key words, including physical properties and components. Both pure component and mixture properties such as phase equilibria are included. Potential users of the CALLUP computer program either may purchase computer tape (and update copies) or use a telephone call-in service. The reference flags include information on correlations used.

Union Carbide routinely screens 53 journals and *Chemical Abstracts*. Brief comments added by the searchers are retrieved with the references. This may be a worthwhile service if a company generates a significant number of physical property search requests that require use of the primary literature.

II. MIXTURE EQUILIBRIUM DATA

A. Importance of Knowing the Phase Equilibrium

A phenol + water distillation column had to be designed. The engineer did not know how to find or estimate the vapor-liquid equilibrium (VLE), so he just assumed ideal behavior. The result of this miscalculation can be seen in the attached phase diagram (Figure 1). His operating line would have been fine if his assumption had been correct. However, the behavior of phenol in water is highly nonideal. The large activity coefficient for phenol at infinite dilution in water distorts the equilibrium curve in the water-rich region. If it were not for the large difference in vapor pressures, there would have been an azeotrope. As it turned out, the column operated as if there were an azeotrope. This can be seen on the phase diagram, where the operating line intersects the true equilibrium curve. When the tower was built, enormous unexpected phenol losses occurred. The column should have been designed with a much larger reflux ratio to circumvent the equilibrium problem, or a different separation should have been devised.

A methanol + water distillation column was designed for operation at 2 atm pressure.

The designer took pains to account for the shift in vapor pressures and to use activity coefficients to account for liquid phase nonideality. Unfortunately, on start-up the column did not meet specifications on the separation. What had happened? The problem in this case was more subtle. The vapor phase nonidealities had been ignored. Introduction of second virial coefficients to model the gas phase revealed that the column was indeed obeying the laws of nature.

A major chemical company was involved in a lawsuit with a vendor of high-efficiency distillation trays. Was the failure to achieve design separation a fault of the trays or was it due to faulty design data? Only accurate VLE data meeting thermodynamic consistency tests could reveal the truth.

Most of the effort and cost of development work on a new process revolve around the reaction. However, the separations section usually costs from 60 to 80% of the capital. It does not pay to scrimp on data related to the separations. A large distillation tower can use more energy in minutes than it costs to measure the VLE which so drastically affects the tower design. On large plants, savings of just 1% in energy can justify expenditure of $1 million for physical property data. Also, consider that many plants cost in excess of $100 million. It does not take much of a reduction in overdesign (contingency) factors to justify significant amounts of phase equilibrium data.

B. Sources of Phase Equilibrium Data

DECHEMA Data Series

The DECHEMA organization in West Germany has published an impressive set of volumes in English.[6-8] The work has been consistent, although not as carefully done as the work by the TRL. None of the thermodynamic parameters such as activity coefficients, fugacity coefficients, or excess free energies are tabulated. Fitted infinite dilution activity coefficients for the best activity coefficient method are listed. Parameters are given for various activity coefficient and vapor pressure equations used. Statistical evaluations are given for the degree of data fit by each of the activity coefficient equations. The degree of fit to the thermodynamically consistent equations gives some idea of consistency. There has been no area ratio test for thermodynamic consistency completed. (Section II.C in this chapter discusses thermodynamic consistency tests.) One of the DECHEMA volumes, dealing with volatile compounds,[8] lists equation of state interaction parameters instead of activity coefficient parameters.

One of the best features of the DECHEMA series is that it reprints, in common units, data which are extremely difficult to obtain from Eastern bloc countries. With the raw data the investigator can perform his or her own consistency tests. Thus, although the compilation is not the ultimate in critical evaluation, the chemical industry is very happy for it.

The Chemshare Corporation provides interactive access to all of the DECHEMA data developed at the University of Dortmund, as well as the data bank developed at Lyngby, Denmark. Together, they are claimed to contain over 1 million experimental data points for VLE, LLE and LLLE which have been reported in the literature. A data base management system assists in locating the desired types of information. The contact at Chemshare Corporation at the time of this writing was Dr. John Adams [(713) 627-8910].

Thermodynamics Research Center (TRC)

Under dynamic new leadership, the TRC has made itself an important disseminator of the first publication of VLE data. The format is uniform and gives substantially more thermodynamic information than the DECHEMA series. As long as the TRC sustains its International Data Series, this will continue to be a very important source of new phase equilibrium data.

Institute of Thermodynamics and Plant Design

Technische Universitat Berlin
Institut fur Thermodynamik und Anlagentechnik
TK7, Strasse des 17, Juni 135, Berlin 12, West Germany
H. Knapp, Director

This institute is highly regarded, and produces a significant amount of experimental data. Additionally, it has demonstrated considerable skill in development of thermodynamic correlations for mixture data. The institute is affiliated with the DECHEMA organization.

Solubility Data

Department of Chemistry
Emory University, Atlanta, GA 30322
H. Lawrence Clever, Director
(404) 329-6616

Few people have maintained such a sustained interest in the solubility of gases in liquids. Emory's efforts are coordinated with the Solubility Data Project of the International Union of Pure and Applied Chemists (IUPAC). Solubility Data has published a number of volumes dealing with solubility of a variety of gases in many liquids, including water and salt solutions. This organization responds to inquiries for gas solubility data, and provides copies of articles at reasonable cost.

Solubility Data Project

Department of Chemistry
Virginia Polytechnic Institute and State University
Blacksburg, VA 24061
Alan F. Clifford, Director
(703) 961-3591

This organization, also affiliated with IUPAC, is undertaking the extremely difficult project of updating the classic solubility volumes by Seidell. Given the unindexed and unevaluated state of those volumes, it will not be an easy task. Pending publication, the director is willing to respond to individual inquiries.

Thermodynamic Data for Technology

Institute for Physical Chemistry/PAN
M. Kasprzaka 44/52, 01-224 Warsaw, Poland
W. Zielenkiewicz, Director

This data center specializes in second virial coefficients and critical compilations of phase equilibrium data. Literature data have been recalculated in consistent units, correlated, and then published in a series of volumes. The principal author is A. Maczynski.

Design Institute for Physical Property Data (DIPPR)

Although DIPPR has not been involved in bibliographic work on VLE, it has been active in funding measurements. By January of 1986, almost 100 binary systems had been studied at two or three temperatures. These data were of two types. One type included data of industrial interest, typically systems that are toxic or otherwise difficult to measure. The other included binary systems measured to fill gaps in group contribution tables for correlations such as UNIFAC and Analytical Solutions of Groups (ASOG). The data are being published in regular Symposium Series volumes of the AIChE,[21] and should be added to any collection of phase equilibrium data. Companies anticipating a need for VLE data can have the data measured through DIPPR, gaining access to other data without spending more money.

American Chemical Society (ACS)

Four important sources of data have been published by the ACS.

1. **Francis, A. W.,** *Critical Solution Temperatures,* Advances in Chemistry Series No. 31, American Chemical Society, Washington, D.C., 1961.
2. **Furter, W. F., Ed.,** *Thermodynamic Behavior of Electrolytes in Mixed Solvents — II,* Advances in Chemistry Series No. 177, American Chemical Society, Washington, D.C., 1979.
3. **Horsley, L. H.,** *Azeotropic Data, III,* Advances in Chemistry Series No. 116, American Chemical Society, Washington, D.C., 1973.
4. **Linke, W. F. and Seidell, A.,** *Solubilities of Inorganic and Metal-Organic Compounds — A Compilation of Solubility Data from the Periodical Literature,* American Chemical Society, Washington, D.C., 1965. [Unevaluated and unindexed volumes].

On-line access to Chemical Abstracts is available through the DIALOG network.

Technical Database Services (TDS)

Wendy R. Stern
10 Columbus Circle, New York, NY 10019
(212) 245-0384

TDS has made partition coefficient information available on-line. This data base, called Log P, was developed at Pomona College.[23] The data give the common logarithm of the distribution of a solute between various solvents, such as octanol, and water. The data were typically measured at very low concentration of the solute. Over 12,000 organic compounds are included, with almost 300 solvents. Such data would be particularly useful in waste treatment areas of design work. They could also be useful in screening extraction solvents. The original thrust came from the pharmaceutical and biological systems areas.

Gas Processors Association (GPA)

This organization has sponsored a very large amount of research on mixture properties, primarily those of interest to the gas processing industry. Their catalog of publications can be obtained by writing to their headquarters, 1812 First Place, Tulsa, OK 74103.

Infinite Dilution Activity Coefficients

A compilation of measured infinite dilution activity coefficients has been developed in Portugal.[24] It includes 129 solvents, 60 solutes, and 1007 different mixtures. A total of 2029 experimental values are included; the effect of temperature was measured on many of these mixtures. This data base is a very valuable resource, and the authors have been most obliging in the past. The data base was too extensive for publication, so they have provided photocopies on request.

Other Sources

1. **Ambrose, D.,** *Vapor-Liquid Critical Properties,* NPL Rep. Chem. 107 (revised), National Physical Laboratory, Teddington, England, 1980. [Critical evaluation]
2. **Benson, M. S. and Zudkevitch, D., Eds.,** Experimental results from the Design Institute for Physical Property Data. I. Phase equilibria, *AIChE Symp. Ser.,* 81, 244, 1985. [Critically evaluated]
3. **Christensen, J. J., Hanks, R. W., and Izatt, R. M.,** *Handbook of Heats of Mixing,* John Wiley & Sons, New York, 1982. [Contains all available data with codes describing type of apparatus used]
4. **Chu, J. C., Getty, R. J., Brennecke, L. F., and Paul, R.,** *Distillation Equilibrium Data,* Van Nostrand Reinhold, New York, 1950. [Contains raw data not tested for thermodynamic consistency]
5. **Chu, J. C., Wang, S. L., Levy, S. L., and Paul, R.,** *Vapor-Liquid Equilibrium Data,* Edwards Brothers, Ann Arbor, Mich., 1956. [Contains raw data not tested for thermodynamic consistency]
6. **Hala, E., Wichterle, I., Polak, J., and Boublik, T.,** *Vapor-Liquid Equilibrium Data at Normal Pressures,* Pergamon Press, Oxford, 1968. [Data correlated with thermodynamically consistent equations]
7. **Hirata, M., Ohe, S., and Nagahama, K.,** *Computer Aided Data Book of Vapor-Liquid Equilibria,* Elsevier, New York, 1975. [Data tested via the thermodynamic consistency equations]
8. **Kehiaian, H. V., Ed.,** International Data Series (Ser. A: Selected Data on Nonreacting Species), Thermodynamics Research Center, Texas A & M University, College Station. [Continuing publication of high-quality vapor-liquid and heat of mixing data in standard formats]
9. **Linke, W. F.,** *Solubilities of Inorganic and Metal Organic Compounds,* Van Nostrand Reinhold, New York, 1958. [Unevaluated data]
10. **Maczynski, A., Maczynska, Z., et al.,** *Verified Vapor-Liquid Equilibrium Data,* Polish Scientific Publishers, Warsaw, 1976—1982. [Critically evaluated]
11. **Martell, A. E. and Smith, R. M.,** *Critical Stability Constants,* Vol. 1 to 5, Plenum Press, New York, 1974—1982. [Critically evaluated]
12. **Ogordnikov, S. K., Lesteva, T. M., and Kogan, V. B.,** Azeotropic Mixtures, A Handbook, *(in Russian),* Izdatel'stvo ''Khimiya'', Leningrad, 1971.
13. **Stephen, H. and Stephen, T.,** *Solubilities of Inorganic and Organic Compounds,* Macmillan, New York, 1963.
14. **Timmermans, J.,** *The Physico-Chemical Constants of Binary Systems,* Vol. 1 to 4, Wiley-Interscience, New York, 1959. [Unevaluated data]
15. **West, D. R. F.,** *Ternary Equilibrium Diagrams,* 2nd ed., Chapman & Hall, Methuen, N.Y., 1982.
16. **Wilhelm, E. and Battino, R.,** Thermodynamic functions of the solubilities of gases in liquids at 25°C, *Chem. Rev.,* 73, 1, 1973. [Critically selected data]
17. **Wilhelm, E., Battino, R., and Wilcock, R. J.,** Low-pressure solubility of gases in liquid water, *Chem. Rev.,* 77, 217, 1977.

Bibliographies

The more difficult approach to evaluation of VLE data is to find and evaluate the data yourself. However, the use of bibliographies can accelerate the effort. Be wary of data that are unevaluated; a large proportion of the reported VLE data are thermodynamically inconsistent, hence wrong. For the adventurous, here are the standard sources.

1. **Francis, A. W.,** *Liquid-Liquid Equilibriums,* Wiley-Interscience, New York, 1963.
2. **Goldberg, R. N.,** Compiled Thermodynamic Data Sources for Aqueous and Biochemical Systems: An Annotated Bibliography (1930—1983), NBS Spec. Publ. 685, U.S. Department of Commerce, Washington, D.C., December 1984.
3. **Himmbelblau, D. M., Brady, B. L., and McKetta, J. J.,** *Survey of Solubility Diagrams for Ternary and Quaternary Liquid Systems,* Spec. Publ. No. 30, Bureau of Engineering Research, University of Texas, Austin, 1959.
4. **Hiza, M. J., Kidnay, A. J., and Miller, R. C.,** *Equilibrium Properties of Fluid Mixtures: A Bibliography,* Plenum Press, New York, 1982.
5. **Oellrich, L., Plocker, U. J., and Knapp, H.,** *Vapor-Liquid Equilibrium,* Institute of Thermodynamics, Technische Hochschule, Berlin, West Germany, 1973.

6. **Westrum, I.,** *Bulletin of Thermodynamics and Thermochemistry,* IUPAC Comm. on Thermodynamics Annual, University of Michigan and Oklahoma State University. [More difficult to use but slightly more current than other bibliographies]
7. **Wichterle, I., Linek, J., and Hala, E.,** *Vapor-Liquid Equilibrium Data Bibliography,* Elsevier, New York, 1973; Suppl. I, 1976; Suppl. II, 1979; Suppl. III, 1983.
8. **Wisniak, J. and Tamir, A.,** *Liquid-Liquid Equilibrium and Extraction: A Literature Source Book,* Elsevier, Amsterdam, 1981.
9. **Wisniak, J. and Tamir, A.,** *Liquid-Liquid Equilibrium and Extraction: A Literature Source Book, Supplement 1,* Elsevier, Amsterdam, 1985.

C. Thermodynamic Consistency of VLE Data

VLE data are deceptively easy to measure but extraordinarily difficult to measure correctly. As a consequence, there is a very large body of incorrect literature data. The editors of the 6th edition of *Perry's Chemical Engineers' Handbook* wisely decided to delete the VLE data included in previous editions; a large portion seriously violated the rules of thermodynamic consistency.

Phase equilibrium data are usually the most critical to a process design, after definition of reactor output. Incorrect data can lead to underspecification of a separation, specification of an impossible separation, or use of the wrong unit operation to accomplish a desired task. Therefore, it is important to spend some time with the phase equilibrium data to be sure that the data are correct.

The phase equilibrium data are treated using fugacity coefficients to model the vapor phase. The liquid phase can be modeled with either activity coefficients (γ_i) or fugacity coefficients (ϕ_i). In the latter case, an equation of state (EOS) is used for both phases. It is somewhat more difficult to assess fugacities of the liquid phase, so the activity coefficient approach is preferred for data evaluation. In fact, this chapter emphasizes the activity coefficient approach. As improved EOS are developed, with the capability of handling highly polar and even associating chemicals, methods will be refined for testing of data using EOS. There will be a more thorough discussion of both activity coefficient equations and EOS in Chapter 5. The EOS method must be used for systems with large concentrations of a supercritical component in both phases, and for most high-pressure systems.

VLE data usually include measurements of the equilibrium temperatures and pressures, plus vapor and liquid compositions at that condition. Because of a fundamental relationship known as the Gibbs-Duhem equation (discussed in standard thermodynamics textbooks), the data measurements are redundant. This gives a means of assessing the data. Satisfaction of the Gibbs-Duhem equation is considered a necessary, although not sufficient, condition for accuracy. There are two means of testing the data, both involving a calculation of activity coefficients. That is done according to the following equation which uses the system pressure as the standard state.

$$\gamma_i = \frac{y_i P \, \hat{\phi}_{i,P}}{x_i P_i^* \, \phi_{i,P_i^*}} \exp\left[\frac{(P_i^* - P) \, V_i^L}{RT}\right] \tag{3}$$

Typically, the fugacity coefficients (ϕ_i) are computed from an EOS such as the virial or the Redlich-Kwong equation. A simple expression based on the virial equation is

$$\ell n \, \hat{\phi}_{i,P} = (2/V) \sum_j y_i \, B_{ij} - 1.0 - B/V \tag{4}$$

The first method, which has achieved considerable popularity, is to fit the activity coefficients vs. liquid composition to an activity coefficient equation which satisfies thermodynamic consistency. Such equations include the Van Laar, Wilson, NRTL, UNIQUAC, Margules, etc. The size of error in fitting the data gives some idea of inconsistency. Sometimes, however, the error is due to failure of the equation to model the system properly. Consider a set of data that essentially meets the criteria of thermodynamic consistency, benzene + cyclohexane at 333.14 K.[29] The standard deviation in fitting the total pressure of the system was 0.0057 for Margules, 0.0026 for Van Laar and NRTL, 0.0022 for UNIQUAC, and 0.0005 for the Wilson equation. The Wilson equation was also the best when fitting vapor compositions, activity coefficients, or the Gibbs excess free energy. Of course, the determination of which equation is ''best'' varies from one system to the next.

The second method is one that also can be strongly recommended. It has been in existence for some time and was apparently devised by Otto Redlich in the 1940s. For isothermal data, and ignoring the negligible contribution of excess volume, the Gibbs-Duhem equation can be integrated by the following equation.

$$\int_{x_1=0}^{x_1=1} \ell n(\gamma_1/\gamma_2) = 0 \tag{5}$$

For nonisothermal data, it is necessary to include a non-negligible heat of mixing term, so the test is somewhat less useful unless heat of mixing data are measured. The test for isothermal data is applied by plotting ln (γ_1/γ_2) vs. liquid mole fraction. The curve will slope from above the axis on the left-hand side to below the axis on the right. There is a certain amount of judgment involved in drawing the line if there is scatter in the data, and statisticians would not be happy with this approach. However, this gives the user a good feel for the nature of the data. If the extrapolation to the end points can be done with some certainty on such a plot, it is more reliable than extrapolating activity coefficient plots to infinite dilution by eye. The area above the axis on the left should equal the area below on the right. Interpretation of the results is really dependent on the degree of nonideality. It was thought for some years that area ratios below 0.8 should be rejected and that area ratios above 0.95 were excellent. However, in a very ideal system those criteria may be too stringent. Another method of interpreting these same data is to compute the absolute value of

$$\frac{\text{Area above x axis} - \text{Area below x axis}}{\text{Area above x axis} + \text{Area below x axis}}$$

The resulting quantity should usually be less than 0.02.

This approach is helpful in identifying individual data points that are in error. Systematic error, such as occurs through sampling or instrument calibration, will usually show up as a thermodynamic inconsistency. A valid criticism of this area ratio test is that it identifies problems in phase composition analysis but not in pressure measurement. However, most of the error is usually incurred in measurement of the vapor phase composition, and secondarily in measurement of the liquid phase composition. Temperature and pressure are usually measured within a much smaller margin of error. This is of course the basis for measuring PTx only, with calculation of the vapor mole fraction via the Gibbs-Duhem equation. In some cases, however, that procedure is not sufficiently sensitive to give accurate vapor compositions. There will be further discussion of this method in Chapter 7.

There are other tests that can be applied to the activity coefficient data. The vapor pressures reported by the investigator should be checked against critically evaluated vapor pressure data. Then a calculation should be made of the excess Gibbs free energy. This is done by the equation

$$G^E = RT \, \Sigma \, x_i \, \ell n \, \gamma_i \tag{6}$$

A plot of the excess Gibbs free energy is sensitive to pressure errors, so it complements the area ratio test. It should be possible to fit the data to a simple polynomial.

$$G^E = x_1x_2 \sum_{p=1}^{5} C_p(x_2 - x_1)^{p-1} \tag{7}$$

Alternatively, the data can be fitted to the original Gibbs free energy expression used in derivation of the activity coefficient equations. In that case, the parameters derived give the model parameters used for computation of the activity coefficients. A plot should also be made of G^E/x_1x_2. This plot gives an excellent idea of the scatter of the data.

If excess free energy data are available at more than one temperature and excess enthalpy data are also available, the data can be tested for thermodynamic consistency in another sense, using the Gibbs-Helmholtz equation.

$$(H^E)_{x_1} = \left[\frac{\partial(G^E/T)}{\partial(1/T)}\right]_{x_1} \tag{8}$$

Values of G^E/T at various compositions are plotted vs. inverse absolute temperature. The slope of the curve must equal the heat of mixing at that temperature and composition. Typically, this would be done at three compositions using G^E values obtained by fitting to a Redlich-Kister equation of the form

$$G^E/RT = x_1x_2[B + C(x_1 - x_2) + D(x_1 - x_2)^2 + \ldots] \tag{9}$$

Further information can be obtained if heat of mixing data are available at various temperatures. However, such nuances are usually left to full-time data evaluators.

D. Data Fitting

Fitting of activity coefficients to thermodynamically consistent equations can be done in many ways. Several investigators have recommended use of a statistical approach called "maximum likelihood".[25] It achieved a certain amount of credibility;[9] however, a recent publication shows that when errors are systematic, as they often are, the technique fails.[10] The author notes, "The M. L. principle offers a valuable tool in the limiting case of parameter estimation in a perfect model based on observations which are subject to random errors only. It does not, however, give any hope for estimation of reliable model parameters based on systematically erroneous data."

Even advocates of the M.L. principle acknowledge its limitations.[11] They note that imperfections in data cannot be overcome by "any degree of statistical sophistication". They believe, and probably correctly, that when there is an unacceptable inconsistency in the data, " . . . no method of data reduction which combines y and P data is acceptable. If one cannot regard the P's as reasonably accurate, the data set should be disregarded."

There are many possible means of regressing phase equilibria. Use of the ratio of activity coefficients $[\ln(\gamma_1/\gamma_2)_{exp} - \ln(\gamma_1/\gamma_2)_{calc}]^2$ is not good because it cancels out pressure effects and can give poor representation of the Gibbs excess free energy. The approach recommended when using PTx data is to use the difference in calculated and experimental total pressure (squared). This approach is also favored when otherwise accurate data are believed to have a high margin of error in the vapor compositions. The following equation is therefore relied upon quite often.

$$\text{Deviation} = \sum_i (P_{exp} - P_{calc})_i^2 \tag{10}$$

The use of differences in natural logarithms of the activity coefficients themselves has the advantage of giving most weight to the slope at the end points. On a Gibbs free energy curve, this has the effect of weighting the initial slopes more than the absolute value at the midpoint. The advantage of this approach is that it emphasizes the rectification and stripping sections of a distillation column in the composition regions where most of the trays are located. The disadvantage is that the experimental composition measurements are often in greatest error at the ends of the composition range. Where there does appear to be considerable scatter in data at the end points, it may be better to fit directly the values of the excess Gibbs free energy. Sometimes, however, the activity coefficients are measured accurately at infinite dilution by means to be described later. When the end of the activity coefficient curve is fixed in that manner, the direct use of natural logarithms of activity coefficients is preferred.

When correlation of the activity coefficients based on PTxy data gives an unacceptable fit, and this appears to be due to thermodynamic inconsistency, then the vapor phase compositions are usually suspect. In this case it is probably best to ignore the vapor compositions and use the total pressure as the objective function. In some circumstances this may give a set of usable data which otherwise might have been discarded. However, this should be done with considerable caution; erroneous results from such analyses have been published by competent investigators.

E. Prediction of Multicomponent Data from Binary Data

The power of thermodynamics is its potential for vastly reducing the amount of experimental data that must be taken to represent a system. A good illustration is prediction of multicomponent VLE from binary VLE. The ability to do this is now accepted, but before 1965 it could not be done accurately for nonideal systems. In that era, the algebraic expansions used for representation of activity coefficients needed ternary interaction parameters. It was the concept of local compositions, first embodied in the famous equation developed at the Massachusetts Institute of Technology by Wilson,[12] which opened the door. His approach to understanding interactions between two molecules may or may not be theoretically correct, but it has certainly solved a lot of problems. Recent molecular simulation work gives evidence that there are indeed regions of "local composition" in liquid mixtures.[26]

The Wilson equation was found to outperform completely any other two-parameter equation in fitting of alcohol-hydrocarbon VLE data. After making the first comprehensive test of the equation, Orye and Prausnitz[13] stated, "Whenever we compared the ability of the Wilson equation to fit data with that of the Van Laar equation or with that of the two-parameter Margules equation, we found that the Wilson equation was always as good as the other two and in many cases much better." When they applied the equation to prediction of ternary data from binary information, they were delighted to find that the results were excellent.

That study was followed by a more extensive review.[14] After examining 89 binary systems, the authors concluded that the Wilson equation was superior to the third-order Margules, a Van Laar, or a modified Van Laar equation. For ternary systems they examined 262 data points. With the Wilson equation giving by far the best results, they found that 92% of the predicted vapor compositions were within 2.0 mol % of the experimental data. Some 85% of the predicted temperatures were within 1.0°C. In general, the more ideal systems were more accurately predicted. One of the less satisfactory predictions was for the octane-ethylbenzene-Cellosolve system.

A disadvantage of the Wilson equation, its inability to predict liquid phase splitting, stimulated development of the NRTL equation. Renon and Prausnitz[15] introduced a "nonrandomness" third parameter, maintaining the local composition idea but using a different formalism. Tested against the Wilson on miscible ternary systems it was comparable. However, it was also applicable to partially miscible ternaries. Compared to the Wohl-type equations it did a much better job for ternary predictions. Although no ternary data were used to develop the NRTL parameters and the best ternary interaction constant was found for the Wohl equations, the NRTL equation gave the best result for six ternary systems compared to two for the Wohl equation. The average mole fraction deviation in vapor composition between the NRTL prediction and the experimental value was 0.4 mol %.

A separate study of the NRTL equation[16] concluded that the Wilson gave generally better predictions of ternary data than the NRTL, and concurred with the assessment of Holmes and Van Winkle[14] that the Wilson equation surpassed the results of the Margules form.

The largest body of work on such ternary predictions is included in the DECHEMA volumes.[6-8] From Reference 6, 44 ternary systems were chosen at random. The differences between experiment and prediction of the vapor mole fractions were then evaluated. The Wilson and UNIQUAC two-parameter equations and the NRTL three-parameter equation each had been fitted to binary VLE data before making the predictions, without inclusion of any ternary data. The averages for all these systems were 0.0125, 0.0123, and 0.0116 for the Wilson, UNIQUAC, and NRTL, respectively.

The Wilson and UNIQUAC do not give the same results for each system, but overall the results are about the same. The improved result for the NRTL is attributable to its additional fitting parameter. The multicomponent predictions will not be any better than the fit of binary data. In many cases the error band is almost due more to experimental error than to correlation problems. For example, Kwang-Chu Chao measured some very accurate data for his doctoral dissertation on the system ethyl acetate + benzene + cyclohexane. The Wilson equation fit was a remarkable 0.0011 average deviation from experimental, and the maximum deviation was only 0.0075! The three numbers quoted previously are interesting because they came from a range of very nonideal systems. Chemicals such as methylene chloride, the alcohols, water, ketones, carbon disulfide, acetonitrile, isoprene, nitromethane, phenol, aniline, and aromatics were included. Thus, given the probability of experimental error in the preceding numbers, the likely uncertainty in a predicted multicomponent point using one of these three equations is a mole fraction of 0.01. This of course assumes that accurate binary data for each of the pairs were available, tested for consistency, and analyzed properly. The accuracy of fit of the binary data also can be used as a guide. When there is any suspicion of formation of two liquid phases, the Wilson equation should be avoided.

F. Reduction of Liquid-Liquid Solubility Data to Activity Coefficient Parameters

Only one liquid phase can exist in an ideal mixture where the activity coefficients are unity. As nonidealities increase, close-boiling mixtures will form azeotropes. Further increases in the activity coefficients cause phase splitting. Null[17] has pointed out that for systems with symmetrical activity coefficient curves, activity coefficients at infinite dilution in excess of 7.39 will cause phase splitting. For the more common case of unsymmetrical activity coefficients, a higher value at one end must balance a value lower than 7.39 for the system to split into more than one liquid phase. Thus, this value serves as an approximate rule of thumb to suggest when more than one liquid phase is to be expected.

It is helpful to know when more than one liquid phase is present in a distillation column. A particularly difficult example of this was encountered during this author's pilot tests of azeotropic distillation of water and formic acid out of acetic acid. The entrainer was an aromatic hydrocarbon. There was a large temperature difference between the top and bottom of the column but, instead of being spread out across the column as one might expect, it happened almost entirely in a step change on one tray. Control was a nightmare because the crossover tray was moving up or down the column unpredictably in a range of 20 or more trays. The problem did not disappear when the commercial column was constructed. Presence of more than one liquid phase was suspected of being a contributor to the instability.

Entrainer may have built up in the downcomers and then dumped all at once without any acid. This could have resulted in very rapid temperature changes on the trays. Possible solutions to this type of problem would be use of dual-flow trays, which do not have downcomers, or use of a packed tower.

Not only can a separate liquid phase wreak havoc with column performance, but also failure to predict it can result in erroneous equilibrium computations. This is true because the assumption of a single phase when two are present will result in excessively high predictions of the equilibrium pressure. A simple computational test for two liquid phases can be made quickly without going through the more complex three-phase computations. Simply compute the activities (activity coefficient times mole fraction) of each of the components. Physically, that computed value cannot exceed unity. If it does, it is a sure sign that there is an extra liquid phase. This shortcut test will not give a false prediction of two liquid phases, but may miss two liquid phases when there is a narrow range of partial miscibility. Since it takes such a small amount of computational effort, this test should be included in distillation and equilibrium flash simulators.

Two liquid phases are in equilibrium when their activities are equal (at the same temperature). Therefore, activity coefficient parameters can be calculated by requiring

$$x_i' \gamma_i' = x_i'' \gamma_i'' \tag{11}$$

This equation is as applicable to multicomponent as to binary data, except that in the latter case there is no residual error for a two-parameter equation. These data are an excellent source of interaction parameters because of sensitivity of the free energy surface in the region of phase splitting. Therefore, it is usually found that with a good activity coefficient equation it is possible to predict VLE in the miscible range based on solubility data. However, predictions of the solubility envelope based on VLE data alone are seldom accurate.

III. SUMMARY

This chapter has emphasized the importance of using critically evaluated data in process designs. The best sources of such data have been organized into three different levels of reliability. Sources for physical property data access via computer also have been listed. Because the necessary information is sometimes available in the literature but not in critically evaluated data compilations, means of qualifying such data have been discussed. With particular emphasis on mixtures, different approaches to fitting of activity coefficient equations have been examined. The same principles will apply in general to equations of state, although their use for highly nonideal mixtures is just beginning; few equations of state have a form suitable for handling such systems.

If all the necessary data can be found from the literature and all are of high quality, the search will be essentially complete. This will not usually be the case. Often, the critical evaluation will demonstrate the nonavailability of good data. The predictions used to fill in those gaps may be adequate for preliminary design but experimental data will usually be required for final process design. In Chapter 5 the problems of filling gaps by prediction are addressed. In Chapters 6 and 7, the means of obtaining experimental data are examined.

REFERENCES

1. **Ho, C. Y. and Touloukian, Y. S.,** Critical evaluation, correlation, analysis, and synthesis of thermophysical, electrical, thermoelectric, and optical properties data, in *Proceedings of the Eighth Symposium on Thermophysical Properties — Volume II,* Sengers, J. V., Ed., American Society of Mechanical Engineers, New York, 1982, 419.
2. **Buck, E. and Frankl, E. M.,** Gaps in the pure component experimental physical property data base, *Chem. Eng. Prog.,* March, 82, 1984.
3. **Chase, J. D.,** The qualification of pure component physical property data, *Chem. Eng. Prog.,* April, 63, 1984.
4. **Reid, R. C., Prausnitz, J. M., and Sherwood, T. K.,** *The Properties of Gases and Liquids,* 3rd ed., McGraw-Hill, New York, 1977.
5. **Wagman, D. D. et al.,** The NBS Tables of Chemical Thermodynamic Properties, *J. Phys. Chem. Ref. Data,* 11 (Suppl. 2), 1982.
6. **Gmehling, J. et al.,** *Vapor-Liquid Equilibrium Data Collection,* Chemistry Data Series, DECHEMA, Frankfurt, 1977—present.
7. **Sorenson, J. M. and Arlt, W.,** *Liquid-Liquid Equilibrium Data Collection,* Vol. 1 to 3, DECHEMA, Frankfurt, 1979, 1980.
8. **Knapp, H., Doring, R., Oellrich, L., Plocker, U., and Prausnitz, J. M.,** *Vapor-Liquid Equilibria for Mixtures of Low Boiling Substances,* DECHEMA, Frankfurt, 1982.
9. **Renon, H.,** Qualities of models for evaluation, representation and prediction of fluid phase equilibrium data, *Fluid Phase Equil.,* 2, 101, 1978.
10. **Skjold-Jorgensen, S.,** On statistical principles in reduction of thermodynamic data, *Fluid Phase Equil.,* 14, 273, 1983.
11. **Van Ness, H. C., Pedersen, F., and Rasmussen, P.,** Vapor-liquid equilibrium: data reduction by maximum likelihood, *AIChE J.,* 24, 1055, 1978.
12. **Wilson, G. M.,** Vapor-liquid equilibrium — a new expression for the excess free energy of mixing, *J. Am. Chem. Soc.,* 86, 127, 1964.
13. **Orye, R. V. and Prausnitz, J. M.,** Multicomponent equilibria with the Wilson equation, *Ind. Eng. Chem.,* 57, 18, 1965.
14. **Holmes, M. J. and Van Winkle, M.,** Prediction of ternary vapor-liquid equilibria from binary data, *Ind. Eng. Chem.,* 62, 21, 1970.
15. **Renon, H. and Prausnitz, J. M.,** Local compositions in thermodynamic excess functions for liquid mixtures, *AIChE J.,* 14, 135, 1968.
16. **Abrams, D. S. and Prausnitz, J. M.,** Statistical thermodynamics of liquid mixtures. A new expression for the excess Gibbs energy of partly or completely miscible systems, *AIChE J.,* 21, 116, 1975.
17. **Null, H. R.,** *Phase Equilibrium in Process Design,* Wiley-Interscience, New York, 1970.
18. **Danner, R. P. and Daubert, T. E.,** *Manual for Predicting Chemical Process Design Data,* American Institute of Chemical Engineers, New York, 1983.
19. **Newman, S. A.,** AIChE's DIPPR Data Prediction Manual, *Chem. Eng. Prog.,* January, 16, 1985.
20. **Kreglewski, A.,** *Equilibrium Properties of Fluids and Fluid Mixtures,* Texas A & M University Press, College Station, 1984.
21. **Benson, M. S. and Zudkevitch, D., Eds.,** Experimental results from the Design Institute for Physical Property Data. I. Phase equilibria, *AIChE Symp. Ser.,* 81, 244, 1985.
22. **Feay, B. A., Daubert, T. E., Danner, R. P., and High, M. S.,** Interactive computer for DIPPR data, *Chem. Eng. Prog.,* August, 55, 1984.
23. **Leo, A., Hansch, C., and Elkins, D.,** Partition coefficients and their uses, *Chem. Rev.,* 71, 525, 1971.
24. **Bastos, J. C., Soares, M. E., and Medina, A. G.,** Selection of solvents for extractive distillation. A data bank for activity coefficients at infinite dilution, *Ind. Eng. Chem. Process Des. Dev.,* 24, 420, 1985.
25. **Anderson, T. F., Abrams, D. S., and Grens, E. A., II,** Evaluation of parameters for nonlinear thermodynamic models, *AIChE J.,* 24, 116, 1978.
26. **Nagata, I.,** Prediction of ternary phase equilibria from binary data, *Thermochim. Acta,* 56, 43, 1982.
27. **Zudkevitch, D.,** Design data — importance of accuracy, in *Encyclopedia of Chemical Processing and Design,* McKetta, J. J. and Cunningham, W. A., Eds., Marcel Dekker, New York, 1982.
28. **Albright, M. A.,** Design implications of the NGPA research programs, paper presented at 49th NGPA Annu. Convention, Denver, March 17—19, 1970.
29. **Boublik, T.,** *Collect. Czech. Chem. Commun.,* 28, 1771, 1963.

Chapter 5

FILLING DATA GAPS BY CORRELATION AND PREDICTION

TABLE OF CONTENTS

I. INTRODUCTION

A. The Role of Prediction Methods

Development of a critically evaluated physical property data base from the literature is one of the first steps in the engineering assessment of a proposed new or revised chemical process. The tools to assemble that data base were described in detail in Chapter 4. Usually, the matrix of data values thus obtained will be sparse. There is never an infinite amount of data, and the power of thermodynamics is its ability to interpolate and extrapolate limited information. Furthermore, thermodynamic equations sometimes require that the interpolations conform with physical reality. The thrust of this chapter is the use of prediction and correlation tools for both pure component and mixture phase equilibrium properties.

There are physical property prediction methods which are completely adequate for preliminary design purposes. When the components are not present in major amounts and are not expected to interfere with purification, it is often possible to go into final process design with estimates of their physical properties. However, when the physical properties are missing for the major components of a system, the initial thought is usually to have some data measured. That costly step may be premature when the project is still in the assessment phase. Unless a given piece of physical property data is absolutely crucial to the technical and economic viability of a process, measurements should be deferred to the development phase. A description of sources of contract data is presented in Chapter 6, and in Chapter 7 there is detailed information on relatively simple phase equilibrium experiments that can be carried out in-house.

B. Sources of Evaluated Prediction Methods

At present, there are three major compilations of pure component and mixture predictive methods. One source is the classic book by Reid, et al.[1] It includes evaluations by the original authors of their own new methods for each property. Emphasis on methods applicable to hydrocarbons is given in the American Petroleum Institute (API) *Technical Databook — Petroleum Refining.*[190] There is an emphasis on nonhydrocarbons in the *Manual for Predicting Chemical Process Design Data*[2] put out by the Design Institute for Physical Property Data (DIPPR) of the American Institute of Chemical Engineers (AIChE). The authors of both manuals, Prof. Ronald P. Danner and Dr. Thomas E. Daubert of Pennsylvania State University, used a large and carefully evaluated data base. Users also may wish to purchase from the AIChE the documentation report for each chapter of the data prediction manual. Each report describes all the methods evaluated and gives the specific numerical results for each. The comparison of methods on a common basis is important to get a true gauge of relative strengths of different methods.

One of the problems faced in comparing prediction methods is that the data base is often limited. Seldom have data been measured simply to fill gaps in the data base. More often, data are measured in universities on systems requiring inexpensive, nontoxic chemicals. Alternatively, data are measured by industry to fill specific process needs. Therefore, it is often the case that the predictive methods for pure components will work well for some classes of compounds and work poorly for others. For that reason, comments on deficiencies in the methods are included in the method discussions. One of the reasons for a decline in experimental work on physical properties is the false but widespread notion that virtually everything of interest can be predicted. Unfortunately, that misconception has resulted in serious deterioration of university-based experimental programs. A drastic cutback of such work at government laboratories also has been reported.[3] In development of the *Manual for Predicting Chemical Process Design Data,*[2] DIPPR contractors Danner and Daubert identified severe deficiencies in both the experimental data base and predictive techniques.

C. Organization of the Chapter

Chapter 5 contains a large number of prediction and correlation methods for both pure component and phase equilibrium properties. The discussions begin with pure component methods and then turn to guidelines for choice of phase equilibrium methods. The two methods are based on activity coefficients and equations of state (EOS). Within each section there is a critical review of the better methods in the literature. Those discussions are followed by special topics which include electrolyte systems, polymer-solvent equilibria, solid-liquid equilibria (SLE), and properties of petroleum fractions and syncrude. In the case of formaldehyde + water phase equilibrium, the correlation was developed by this author.

II. PURE COMPONENT PREDICTION METHODS

The search for literature data will usually lead to discovery of a number of significant gaps in the needed physical property data base. Some must be filled eventually with experimental data, while for others predictions may be adequate. However, in the process assessment phase, the gaps in pure component data usually will be filled with predictions. Confusing the issue is the publication of a great number of methods. As a source of reference, a single method is included here for each of the major properties. In some cases alternative methods are mentioned that can be used to handle situations when basic methods listed are not applicable.

The choice of methods was based on a single criterion. Each of the methods was the one used most often by Pennsylvania State staff in 1985 for filling gaps in the DIPPR Data Compilation. Their choice of methods was based on extensive experience with both the API *Technical Databook — Petroleum Refining*[190] and the DIPPR *Manual for Predicting Chemical Process Design Data.*[2] Their choices were also based on inputs from a large committee of the physical properties experts from major chemical companies, oil companies, and engineering contractors. For many of these properties there is an acceptable procedure for hydrocarbons in the API *Technical Databook — Petroleum Refining*. Therefore, the choice of procedures listed here emphasizes molecules that have atoms in addition to hydrogen and carbon. The procedures also will be applicable for hydrocarbons, although specialized methods for the hydrocarbons may be slightly more accurate.

A. Critical Properties

Critical properties are sometimes needed when process conditions are near the critical point. More often, they are needed because the principle of corresponding states predicts that "all properties related to intermolecular forces can be predicted based on the relationship between the critical properties of both model compounds and the components under study". For that reason, most useful prediction methods are based on critical properties. A high degree of accuracy is essential to these prediction methods. It has been pointed out that a 2% error in critical temperature can cause a 10% error in predicted thermal conductivity.[5]

Development of predictive methods has been hampered by a scarcity of critical property data. The only comprehensive measurement program has been in regard to hydrocarbons. The chemical families having the greatest paucity of data are anhydrides, aldehydes, acids, and nitriles. As new chemicals with several functional groups are developed or rise to importance, it is often found that there are no available physical property data. This problem is of increasing concern as more fine and specialty chemicals attain commercial importance. A good method is needed, therefore, for prediction of the critical properties of such compounds. That can be the entry point for prediction of many properties by the corresponding states approach.

The method of Lydersen[7] is easy to use, and even though it is more than 30 years old, it is still among the more accurate methods for the estimation of critical property data. The critical temperatures are calculated very easily by a group contribution method. Increments are totaled using the groups defined in Table 1[1] and then used in Equations 1, 2, and 3.

$$T_c = \frac{T_b}{(0.567 + \Sigma\,\Delta_T - (\Sigma\,\Delta_T)^2)} \tag{1}$$

$$P_c = \frac{MW}{(0.34 + \Sigma\,\Delta_P)^2} \tag{2}$$

$$V_c = 40 + \Sigma\,\Delta_v \tag{3}$$

Table 1
GROUP INTERACTIONS FOR LYDERSEN CRITICAL PROPERTY ESTIMATION METHOD

	$\Delta\tau$	$\Delta\rho$	$\Delta\nu$
Nonring increments			
$-CH_3$	0.020	0.227	55
$-CH_2$ (with one additional bond, ∣)	0.020	0.227	55
$-CH$ (with additional bonds above and below, ∣)	0.012	0.210	51
$-C-$ (with additional bonds above and below, ∣)	0.00	0.210	41
$=CH_2$	0.018	0.198	45
$=CH$ (with one additional bond, ∣)	0.018	0.198	45
$=C-$ (with one additional bond, ∣)	0.0	0.198	36
$=C=$	0.0	0.198	36
$\equiv CH$	0.005	0.153	(36)
$\equiv C-$	0.005	0.153	(36)
Ring increments			
$-CH_2-$	0.013	0.184	44.5
$-CH$ (with additional bonds above and below, ∣)	0.012	0.192	46
$-C-$ (with additional bonds above and below, ∣)	(−0.007)	(0.154)	(31)
$=CH$ (with one additional bond, ∣)	0.011	0.154	37
$=C-$ (with one additional bond, ∣)	0.011	0.154	36
$=C=$	0.011	0.154	36
Halogen increments			
$-F$	0.018	0.224	18
$-Cl$	0.017	0.320	49
$-Br$	0.010	(0.50)	(70)
$-I$	0.012	(0.83)	(95)
Oxygen increments			
$-OH$ (alcohols)	0.082	0.06	(18)
$-OH$ (phenols)	0.031	(−0.02)	(3)
$-O-$ (nonring)	0.021	0.16	20
$-O-$ (ring)	(0.014)	(0.12)	(8)
$-C=O$ (nonring) (with one additional bond, ∣)	0.040	0.29	60

Table 1 (continued)
GROUP INTERACTIONS FOR LYDERSEN CRITICAL PROPERTY ESTIMATION METHOD

	Δ_T	Δ_P	Δ_V
–C=O (ring) (with one free bond on C)	(0.033)	(0.2)	(50)
HC=O (aldehyde) (with one free bond on C)	0.048	0.33	73
–COOH (acid)	0.085	(0.4)	80
–COO– (ester)	0.047	0.47	80
=O (except for combinations above)	(0.02)	(0.12)	(11)
Nitrogen increments			
$-NH_2$	0.031	0.095	28
–NH (nonring) (with one free bond on N)	0.031	0.135	(37)
–NH (ring) (with one free bond on N)	(0.024)	(0.09)	(27)
–N– (nonring) (with one free bond on N)	0.014	0.17	(42)
–N– (ring) (with one free bond on N)	(0.007)	(0.13)	(32)
–CN	(0.060)	(0.36)	(80)
$-NO_2$	(0.055)	(0.42)	(78)
Sulfur increments			
–SH	0.015	0.27	55
–S– (nonring)	0.015	0.27	55
–S– (ring)	(0.008)	(0.24)	(45)
=S	(0.003)	(0.24)	(47)
Miscellaneous			
–Si– (with free bonds above and below)	0.03	(0.54)	
–B– (with free bond below)	(0.03)		

Note: There are no increments for hydrogen. All bonds shown as free are connected with atoms other than hydrogen. Values in parentheses are based upon too few experimental values to be reliable. From vapor-pressure measurements and a calculational technique similar to that of Fishtine,[8] it has been suggested that the >C<–H ring increment common to two condensed saturated rings be given the value of $\Delta_T = 0.064$.

where the units employed are Kelvins, atmospheres, and cubic centimeters per gram mole. The percentage errors in critical temperature and pressure predicted by the Lydersen method are given in Table 2.[2]

Another method, developed by Fishtine[8] and documented in the DIPPR *Manual for Predicting Chemical Process Design Data,*[2] is useful when there is a closely related compound for which there are already an accurate critical temperature and bp.

Table 2
LYDERSON METHOD: PERCENT DEVIATION FROM EXPERIMENTAL VALUES (AVERAGE OF ABSOLUTE VALUES)

	T_c	P_c
Miscellaneous halides	0.27	5.23
Amines	0.62	2.69
Phenols	0.66	7.03
Ethers and epoxides	0.70	3.43
Acids	0.80	11.97
Ketones	1.08	5.93
Mercaptans and sulfides	1.09	1.86
Esters	1.23	2.70
Chlorides	1.58	3.35
Fluorides	2.10	12.14
Nitriles	2.56	7.75
Aldehydes	2.56	4.50
Alcohols	2.93	5.64
Bromides	4.19	5.85
Anhydrides	5.40	6.06
Averages	1.53%	5.61%

A recommended method for critical volumes is given by Fedors.[9] This method involves summation of fewer groups than other methods and is more accurate. The equation is

$$V_c = 26.6 + v_i \tag{4}$$

The increments (v_i) are given in Table 3. The average error with this method is about 3%. Development of a more accurate method is limited by inaccuracies in the data base.

B. Heat, Entropy, and Free Energy of Formation

The Benson method is widely used because it applies to a very wide range of compounds, it is easy to use, and it yields predictions which in most cases approach experimental accuracy. It is being used in industry in the form of the CHETAH computer program discussed in Chapter 2. That computer program is getting technical support from both chemical industry experts and the National Bureau of Standards (NBS). Until now, updates for the data base have been infrequent, but recent changes within the American Society for Testing and Materials (ASTM) Committee E-27, which maintains the program, promise to improve this situation.

C. Vapor Pressure and Acentric Factor

Vapor pressure is extremely important for separation system design. Predictions of vapor pressure may be acceptable for the process assessment phase of a research program, depending on the other chemicals present and the difficulty of the separation. For key compounds, however, at least two experimental points are usually required. Therefore the problem is not one of prediction in the absence of data, but of interpolation and extrapolation. Unfortunately, there are relatively few vapor pressure data measured above atmospheric pressure (760 torr) or below 10 torr. Beware of downward extrapolation of vapor pressure

Table 3
GROUP CONTRIBUTIONS FOR FEDORS CRITICAL VOLUME PREDICTION METHOD

Atom	v_i (cc/mol)
C	34.426
H	9.172
O	20.291
O (alcohols)	18.000
N	48.855
N (amines)	47.422
F	22.242
Cl	52.801
Br	71.774
I	96.402
S	50.866

The structural feature corrections are as follows:

3-membered ring	−15.824
4-membered ring	−17.247
5-membered ring	−39.126
6-membered ring	−39.508
Double bond	5.028
Triple bond	0.7973
Each additional ring attached directly to another ring, i.e., biphenyl or naphthalene	35.524

data, particularly with the popular Antoine equation. Upward extrapolation of limited vapor pressure points is generally reliable using the Riedel equation.

This equation is used at Pennsylvania State for nonhydrocarbons such as oxygenates and halogenated compounds. It requires the critical pressure and temperature, plus the normal bp. The reduced temperature T_r at the bp is first used to calculate a function F(bp) from the following equation.

$$F(bp) = -35 + 36/T_r + 42 \ln T_r - T_r^6 \tag{5}$$

The Riedel factor is then computed from

$$\alpha_c = \{0.315\, F(bp) + \ln P_c\}/\{0.0838\, F(bp) - \ln T_r\} \tag{6}$$

The computed Riedel factor is used to calculate Q from

$$Q = 0.0838\,(3.758 - \alpha_c) \tag{7}$$

The parameters used in the vapor pressure equation are then given by

$$A = -35\,Q \tag{8}$$

$$B = -36\,Q \tag{9}$$

$$C = 42\,Q + \alpha_c \tag{10}$$

$$D = -Q \tag{11}$$

The vapor pressure equation[11] is given as a function of reduced pressure (P^*/P_c) as

$$\ln P_r = A - B/T_r + C \ln T_r + DT_r^6 \tag{12}$$

As a predictive equation this does well between the bp and the critical point. However, it is not particularly accurate for extrapolation below the bp. The equation can be used to fit vapor pressure data, and parameters for roughly 1000 of the most important industrial chemicals are available in the DIPPR Data Compilation, described in Chapter 4.

Downward extrapolation of vapor pressure is particularly important for high-boiling chemicals. Inasmuch as OSHA and other government regulations require the vapor pressure of nonvolatile chemicals at 25°C, accurate extrapolation is necessary. The best available method appears to have been developed from statistical mechanics.[6,10]

A new corresponding states equation has been developed at Pennsylvania State University which appears to be more accurate than the Riedel, the Halm and Stiel, or the Lee-Kesler methods. This is the Gupte-Daubert method,[12] which requires the critical point and the normal bp. Beyond that, the radius of gyration (available in the DIPPR Data Compilation) and the reduced dipole moment are needed. As published, the method is not truly general; a separate expression is needed for each different class of polar molecules. Those published include alcohols, ketones, amines, formates, and acetates.

One other equation, the Wagner, bears mentioning. An exceptional reference[13] includes correlation coefficients for 251 substances. They are applicable from the triple point to the critical point, and represent the data within an extremely small margin of error. Where the data were limited, as is often the case, the fit was constrained. The three constraining criteria are described. The constraints require isolation of correlation coefficients which reproduce physical phenomena oserved for all organic chemicals. These criteria also can be applied with other vapor pressure equations. The Wagner equation is as follows.

$$\ln P_r = [A(1 - T_r) + B(1 - T_r)^{1.5} + C(1 - T_r)^3 + D(1 - T_r)^6]/T_r \tag{13}$$

For hydrocarbon mixtures the Maxwell-Bonnell approach[14] is used. The reader should consult Procedure 5A1.13 in the API *Technical Databook — Petroleum Refining*[190] for details on implementation of the procedure.

The acentric factor is a parameter used in corresponding states correlations of various types to represent molecular complexity. For monatomic gases it is supposed to be zero. It represents the intermolecular forces which do not emanate from the center, but rather from the surface of the molecule. In essence, it correlates the shape of the vapor pressure curve. It is calculated from the common logarithm of the vapor pressure (in atmospheres) at a reduced temperature of 0.700. The resulting number plus 1.000 is the negative of the acentric factor.

D. Heats of Vaporization

Heats of vaporization have been reported for a large number of compounds, but for half of them the only data point is at the boiling temperature. Heats of vaporization are most conveniently determined from vapor pressure and gas density data by means of the Clapeyron equation. Calorimetrically measured "true" data are uncommon. Data obtained by means of the Clausius-Clapeyron equation are subject to increased error, and this error is amplified if low-quality vapor pressure data are used. If the investigators deriving a heat of vaporization from vapor pressure data are not known to be reliable, it is better to ignore the reported value and do one's own analysis of the vapor pressure data. For example, one handbook value for the heat of vaporization of maleic anhydride was found to be in error by 20%! The cause of the error was traced to a reported heat of vaporization that was determined by differentiation of vapor pressure data taken over just a few degrees.

The equation published in 1834 by Clapeyron[15] is still the best means of predicting the heat of vaporization. The equation, which rigorously relates the slope of the vapor pressure curve to the heat of vaporization, is

$$\frac{d\ \ell nP^*}{d(1/T)} = \frac{-\Delta H^v}{R\ \Delta Z} \tag{14}$$

In this equation the term ΔZ is the difference in compressibility factor between the gas (unity for an ideal gas) and the liquid (which can be very small). Thus, a useful assumption for systems at less than atmospheric pressure is that ΔZ is unity. For systems under pressure it usually will be less than unity, and it can be computed by EOS.

The heat of vaporization consistent with the Riedel equation is as follows.

$$\Delta H^v = RT_c\ \Delta Z(B + CT_r + 6DT_r^7) \tag{15}$$

An evaluation of this method for 179 polar and sometimes associating compounds gave quite good results.[16] Predictions were slightly better than with the Lee-Kesler method (described in Section V.B) and substantially better than with two other methods.

Computations in the vicinity of the critical point become difficult, so the Watson equation is sometimes used to interpolate in that regime.

$$\Delta H_2^v = \Delta H_1^v\{(1 - T_{2r})/(1 - T_{1r})\}^{0.38} \tag{16}$$

Another alternative is use of the Lee-Kesler equations, since those are noted for giving quite good enthalpy values.

Heats of vaporization of the organic acids and their mixtures are special cases, due to dimerization in the gas phase. Therefore, the gas phase nonideality must be handled with equilibrium constants. Data have been measured on many of the organic acids and are included in the DIPPR Data Compilation. In fact, a complete experimental project on organic acid PVT data and calorimetric heats of vaporization was carried out for DIPPR. A discussion of the theory and computations necessary for computing heats of vaporization of mixtures containing organic acids is given by Tamir and co-workers.[17]

E. Ideal Gas Heat Capacity

The Benson method appears to be the best approach for predicting ideal gas heat capacities. It is normally employed through use of one of the versions of the CHETAH computer program. Its high degree of accuracy is due to the fact that it goes beyond first-order group contributions which take into account only the groups themselves; rather, it takes into account the effect of neighboring groups on each of the groups in a molecule. Therefore, the Benson method results are sufficiently accurate for most purposes.

The correlation of known ideal gas heat capacities as a function of temperature for the DIPPR Data Compilation has been accomplished previously with an equation containing exponentials. It could not be easily integrated to give the ideal gas enthalpy. Therefore, a few DIPPR technical representatives have searched out improved equation forms. The equation form selected was developed from statistical mechanics by Aly and Lee.[229] and by Faheeka et al.[230]

$$C_p^* = B + C[(D/T)/\sinh(D/T)]^2 + E[(F/T)/\cosh(F/T)]^2 \quad (17)$$

$$H^* = A + BT + CT(D/T)\coth(D/T) - ET(F/T)\tanh(F/T) \quad (18)$$

It is anticipated that these correlations will be incorporated in future work on the DIPPR Data Compilation.

F. Liquid Heat Capacity

The data base for liquid heat capacity is deplorably shallow, considering the process design significance of this property. There are virtually no data above room temperature for non-hydrocarbons. The Chueh-Swanson method[18,19] is recommended for nonhydrocarbons. This procedure gives the liquid heat capacity at room temperature. The accuracy is generally within 2 to 3% and almost always within 5% at room temperature. Table 4 gives the group contribution parameters, which correspond to a liquid heat capacity at 20°C, for this method.

Methods are also included for extrapolation to near the critical point, using a fair amount of auxiliary data. The method is complicated and not totally reliable in extrapolation to high temperature. One of the problems in developing correlations for heat capacity temperature sensitivity is the small amount of reliable high-temperature liquid heat capacity information. To calculate high-temperature liquid enthalpies it is probably better to calculate the ideal gas heat capacity with the Benson method, estimate the enthalpy corrections for gas phase nonideality using an EOS, and then subtract the heat of vaporization.

G. Solid Heat Capacity

A rather old method is still used for solid heat capacity at 20°C. It is Kopp's rule,[20] which is simple. It involves a sum of the heat capacities of each of the elements times the number of appearances of each element in the molecule. The contributions are given in Table 5.

Kopp's rule is not elegant and probably not very accurate; better methods are certainly needed for this property. One potentially improved method will be included in a version of the CHETAH computer program not yet issued at the time of publication of this book.

H. van der Waals Volume and Area

These are derived properties which are used in the UNIQUAC and UNIFAC models to represent the volume and surface area of molecules. They are already computed for most compounds of interest and can be found in the DIPPR Data Compilation and in the volumes published by DECHEMA (referenced in Chapter 4). For other molecules the values can be computed using data given by Bondi.[21] Reference 22 explains the computations necessary to convert Bondi's parameters from molecular dimensions to the commonly employed van der Waals volume and area parameters.

I. Vapor Density

Deviations from the ideal gas laws can be computed with EOS, discussed later in this chapter. Up to moderate pressures (sometimes as high as 25 bar), the virial coefficient approach is the most accurate for vapor density. In spite of considerable work completed since the publication of the Tsonopoulos correlation[30] for second virial coefficients, it still appears to be the best approach for the gas phase. However it cannot be easily used to represent the liquid phase. It does require critical properties, acentric factor, and dipole moment. The correlation is presented later in this chapter.

J. Liquid Density

Some EOS yield reasonably accurate liquid density predictions; however, there are other specialized equations which are usually more accurate. The primary equation used by Penn-

Table 4
GROUP CONTRIBUTIONS FOR THE CHUEH-SWANSON LIQUID HEAT CAPACITY CORRELATION AT 20°C

Group	Value[a]	Group	Value[a]
Alkane		–C(H)=O	12.66
$-CH_3$	8.80	–C(=O)–OH	19.1
$-CH_2-$	7.26	$-CH_2OH$	17.5
>CH–	5.00	>CHOH	18.2
>C<	1.76	>COH	26.6
Olefin		–OH	10.7
$=CH_2$	5.20	$-ONO_2$	28.5
=CH–	5.10	Nitrogen	
=C<	3.80	H_2N-	14.0
Alkyne		>NH	10.5
–C≡H	5.90	>N–	7.5
–C≡	5.90	–N= (in a ring)	4.5
In a ring		–C≡N	13.9
>CH–	4.4	Sulfur	
>C= or >C<	2.9	–SH	10.7
–CH=	5.3	–S–	8.0
$-CH_2-$	6.2	Halogen	
Oxygen		–Cl (first or second on a carbon)	8.6
–O–	8.4	–Cl (third or fourth on a carbon)	6.0
>C=O	12.66	–Br	9.0
–C(=O)–O–	14.5	–F	4.0
		–I	8.6
		Hydrogen	
		H– (for formic acid, formates, hydrogen cyanide, etc.)	3.5

[a] Add 4.5 for any carbon group which fulfills the following criterion: a carbon group which is jointed by a single bond to a carbon group connected by a double or triple bond with a third carbon group. In some cases a carbon group fulfills the above criterion in more ways than one. In these cases 4.5 should be added each time the group fulfills the criterion. Exceptions to the above 4.5 addition rule: (1) No such extra 4.5 additions for $-CH_3$ groups. (2) For a $-CH_2-$ group fulfilling the 4.5 addition criterion add 2.5 instead of 4.5. However, when the $-CH_2-$ group fulfills the addition criterion in more ways than one, the addition should be 2.5 the first time and 4.5 for each subsequent addition. (3) No such extra addition for any carbon group in a ring.

Reprinted from Chueh, C. F. and Swanson, A. C., *Chem. Eng. Prog.*, 69(7), 83, 1973. With permission.

Table 5
GROUP CONTRIBUTIONS FOR KOPP'S RULE FOR SOLID HEAT CAPACITY

Elements	Cal/g mol) (K)
C	1.8
H	1.3
B	2.7
Si	3.8
O	4.0
F	5.0
P	5.4
S	5.4
All other	6.2

sylvania State, and one also popular in industry, is the Rackett equation;[222] this requires the critical temperature, pressure, and compressibility factor.

$$\frac{1}{\rho_s} = \left[\frac{RT_c}{MW\ P_c}\right] Z_c(1 + (1 - T_r)^{2/7}) \tag{19}$$

A second equation is used at Pennsylvania State when one or more liquid density values are available. It is the Spencer-Danner modification[223] of the Rackett equation. The modified equation includes an additional parameter obtained from the experimental liquid density data. This equation is as follows.

$$\frac{1}{\rho_s} = \left[\frac{RT_c}{MW\ P_c}\right] Z_{RA}^{(1+(1-T_r)^{2/7})} \tag{20}$$

The unknown parameter, Z_{RA}, is given in Table 6.

K. Solid Density

It has been discovered at Pennsylvania State that no good method is available to predict solid density. A reasonable estimate of the solid density is made simply by multiplying the liquid density at the triple point by a factor of 1.17. No information is readily available concerning how well it fits glasses, polymers, or other types of solids.

L. Vapor Viscosity

The Reichenberg method[23,24] gives a simple but accurate equation for extrapolation of one or more known viscosity data points.

$$\text{viscosity (micropoise)} = a^*T/[1 + 0.18(1 + 4/T_c)(T_R - 1)]^{1/3} \tag{21}$$

Evaluation of the equation simply requires solution for the constant, a*, using a known point. Practically all of the 79 organic compounds tested by Reichenberg gave maximum deviations (less than 4%) between his equation and the data. Therefore, he developed a generalized method to predict vapor viscosity in the total absence of viscosity data.

The group contributions for C_i, given in Table 7, give a* in Equation 22.

$$a^* = MW^{1/2}\ T_c/\sum_i n_iC_i \tag{22}$$

Table 6
VALUES OF Z_{RA} USED IN THE SPENCER-DANNER MODIFICATION OF THE RACKETT EQUATION FOR LIQUID DENSITIES

Compound name	Z_{RA}	Average percent error	Number of points
Paraffins and cycloparaffins			
Methane	0.2879	0.60	126
Ethane	0.2791	1.41	81
Propane	0.2763	0.32	141
n-Butane	0.2730	0.33	76
Isobutane	0.2750	0.30	105
2,2-Dimethylbutane	0.2733	0.10	15
2,3-Dimethylbutane	0.2704	0.22	22
n-Hexane	0.2637	0.49	97
2-Methylpentane	0.2672	0.10	25
3-Methylpentane	0.2690	0.03	26
n-Heptane	0.2610	0.46	81
n-Octane	0.2562	1.45	75
n-Nonane	0.2555	0.29	65
n-Decane	0.2528	0.79	50
n-Undecane	0.2500	0.12	34
n-Dodecane	0.2472	0.44	57
n-Tridecane	0.2471	0.621	37
n-Tetradecane	0.2270	1.05	32
n-Pentadecane	0.2420	0.162	19
n-Hexadecane	0.2386	0.138	16
Cyclopentane	0.2711	2.62	23
Isopentane	0.2617	8.56	39
Neopentane	0.2762	0.81	23
n-Pentane	0.2685	0.20	117
Cyclohexane	0.2729	0.14	27
Alkenes			
Ethylene	0.2807	0.41	32
Propadiene	0.2730	1.14	12
Propylene	0.2780	0.78	120
1,2-Butadiene	0.2686	1.07	13
1,3-Butadiene	0.2713	0.25	70
1-Butene	0.2734	0.41	55
cis-2-Butene	0.2705	0.16	23
trans-2-Butene	0.2721	0.19	18
Isobutylene	0.2727	0.49	87
Cyclopentadiene	0.2969	0.0	1
1-Pentene	0.2700	0.29	26
cis-2-Pentene	0.2687	0.07	3
trans-2-Pentene	0.2705	0.07	3
1-Hexane	0.2670	1.76	24
1-Heptene	0.2611	0.18	17
Dicyclopentadiene	0.2479	2.44	2
1-Octene	0.2565	0.00	5
1-Nonene	0.2533	0.00	3
Acetylenes			
Acetylene	0.2707	0.50	26
Alkylbenzenes and polyaromatics			
Benzene	0.2696	0.20	118
Toluene	0.2645	0.33	118
Styrene	0.2634	0.21	17
Ethylbenzene	0.2619	0.29	70
m-Xylene	0.2593	0.56	97

Table 6 (continued)
VALUES OF Z_{RA} USED IN THE SPENCER-DANNER MODIFICATION OF THE RACKETT EQUATION FOR LIQUID DENSITIES

Compound name	Z_{RA}	Average percent error	Number of points
o-Xylene	0.2627	0.64	57
p-Xylene	0.2590	0.46	96
Cumene	0.2616	0.38	67
Naphthalene	0.2611	0.62	30
Biphenyl	0.2744	0.43	6
n-Dodecylbenzene	0.2436	0.29	20
Aldehydes			
Acetaldehyde	0.2593	0.37	48
n-Propionaldehyde	0.2444	1.32	19
n-Butyraldehyde	0.2488	0.85	11
Isobutyraldehyde	0.2569	0.00	2
Alcohols and phenols			
Methanol	0.2332	2.43	46
Ethanol	0.2524	2.18	58
Isopropanol	0.2504	1.76	48
n-Propanol	0.2533	1.69	77
n-Butanol	0.2571	1.09	42
sec-Butanol	0.2545	1.01	32
tert-Butanol	0.2565	1.58	27
Isobutanol	0.2568	2.02	56
1-Pentanol	0.2586	0.92	34
Phenol	0.2767	0.95	48
Cyclohexanol	0.2433	0.55	11
1-Hexanol	0.2612	0.50	34
o-Cresol	0.2718	0.18	19
2-Ethyl-1-hexanol	0.2535	0.18	14
Ketones			
Acetone	0.2452	0.11	28
Diethyl ketone	0.2557	0.01	8
Cyclohexanone	0.2460	0.27	13
Methyl isobutyl ketone	0.2589	0.06	11
Polyols			
Ethylene glycol	0.2508	1.95	72
1,2-Propylene glycol	0.2633	0.38	11
1,3-Propylene glycol	0.2511	0.07	2
Glycerol	0.2534	2.89	21
1,4-Butanediol	0.2512	0.13	2
1,3-Butylene glycol	0.2597	0.32	6
Diethylene glycol	0.2520	1.31	20
1,5-Pentanediol	0.2507	0.56	5
1,6-Hexanediol	0.2511	0.53	2
Triethylene glycol	0.2421	4.54	37
Sorbitol	0.2833	0.74	8
Tetraethylene glycol	0.2444	0.11	3
Esters			
Methyl formate	0.2580	0.61	47
Ethyl formate	0.2587	0.36	54
Methyl acetate	0.2553	0.53	51
Methyl acrylate	0.2554	0.20	6
Vinyl acetate	0.2578	0.10	16
Ethyl acetate	0.2538	0.47	57
Methyl propionate	0.2562	0.56	57
n-Propyl formate	0.2592	0.39	56

Table 6 (continued)
VALUES OF Z_{RA} USED IN THE SPENCER-DANNER MODIFICATION OF THE RACKETT EQUATION FOR LIQUID DENSITIES

Compound name	Z_{RA}	Average percent error	Number of points
Ethyl acrylate	0.2583	0.00	1
Methyl methacrylate	0.2535	0.07	10
n-Butyl formate	0.2740	0.25	12
Ethyl propionate	0.2546	0.29	57
n-Propyl acetate	0.2554	2.22	28
Ethyl methacrylate	0.2571	0.0	1
n-Butyl acetate	0.2628	0.28	30
Isobutyl acetate	0.2643	0.00	2
n-Butyl acrylate	0.2426	0.021	6
Dimethyl terephthalate	0.2430	0.56	3
Dioctyl phthalate	0.2424	0.10	2
Anhydrides			
Acetic anhydride	0.2712	1.38	24
Maleic anhydride	0.2520	0.00	1
Phthalic anhydride	0.2615	0.23	7
Ethers and Epoxides			
Ethylene oxide	0.2593	0.08	25
Dimethyl ether	0.2740	0.82	19
1,2-Propylene oxide	0.2493	0.15	4
1,4-Dioxane	0.2689	1.58	20
Diethyl ether	0.2643	0.40	23
Methyl *t*-butyl ether	0.2714	0.07	3
Isopropyl ether	0.2698	0.27	5
Di-*n*-butyl ether	0.2575	3.45	18
Chlorides and other halides			
Chloroform	0.2750	0.27	15
Dichloromethane	0.2618	0.205	27
Carbon tetrachloride	0.2721	0.20	34
Dichlorodifluoromethane	0.2762	0.33	59
Trichlorofluoromethane	0.2756	0.11	21
Methyl bromide	0.2306	0.05	2
Methyl chloride	0.2679	0.31	55
Methyl iodide	0.2667	0.03	5
Perchloroethylene	0.2669	0.53	16
Trichloroethylene	0.2697	0.14	7
Chloroacetyl chloride	0.2543	1.56	2
Vinyl chloride	0.2754	0.44	21
Ethylene dibromide	0.2664	0.09	9
1,2-Dichloroethane	0.2655	0.34	34
Ethyl chloride	0.2639	2.40	27
Ethyl iodide	0.2856	0.40	32
Mercaptans and sulfides			
Dimethyl sulfide	0.2674	0.75	37
Sulfolane	0.2517	0.36	7
Nitrogen compounds — amines			
Methylamine	0.2597	0.22	13
Acetonitrile	0.2010	0.25	7
Dimethylamine	0.2641	0.31	7
Ethylamine	0.2639	0.10	13
Monoethanolamine	0.1987	0.84	15
Acrylonitrile	0.1956	0.30	6
n-Propylamine	0.2644	0.37	11
Trimethylamine	0.2711	0.30	11

Table 6 (continued)
VALUES OF Z_{RA} USED IN THE SPENCER-DANNER MODIFICATION OF THE RACKETT EQUATION FOR LIQUID DENSITIES

Compound name	Z_{RA}	Average percent error	Number of points
n-Butylamine	0.2666	0.40	18
Diethanolamine	0.2045	0.59	19
Aniline	0.2617	0.28	20
Triethanolamine	0.2009	0.47	8
Hydrogen cyanide	0.1915	0.02	11
Polyfunctional compounds			
Phosgene	0.2791	0.21	10
Acetyl chloride	0.2691	0.05	5
Chloroacetic acid	0.2325	0.12	4
Dimethyl formamide	0.2232	0.11	13
Monochlorobenzene	0.2650	0.34	32
Nitrobenzene	0.2473	0.02	30
α-Epichlorohydrin	0.2475	1.09	14
Glycolic acid	0.3152	0.00	1
Furfural	0.2584	0.05	4
ϵ-Caprolactam	0.2536	0.07	2
Carbonyl sulfide	0.2779	0.20	7
Acids			
Formic acid	0.2226	3.11	20
Acetic acid	0.2242	2.3	46
Acrylic acid	0.2447	0.01	2
Propionic acid	0.2486	0.37	24
n-Butyric acid	0.2743	0.28	35
Adipic acid	0.2293	0.74	3
Inorganic acids			
Nitric acid	0.2185	1.31	14
Hydrogen chloride	0.2657	1.50	21
Hydrogen fluoride	0.1473	0.72	21
Hydrogen iodide	0.2882	0.05	8
Sulfuric acid	0.1710	0.48	18
Phosphoric acid	0.1610	0.51	21
Inorganic bases			
Sodium hydroxide	0.1526	0.10	4
Ammonia	0.2466	0.39	29
Other inorganics			
Bromine	0.2945	1.89	21
Carbon monoxide	0.2900	0.84	24
Carbon dioxide	0.2729	0.70	11
Carbon disulfide	0.2852	0.12	6
Hydrogen (normal)	0.3139	6.70	25
Chlorine	0.2769	0.53	32
Water	0.2374	3.29	58
Sodium chloride	0.1331	0.34	5
Hydrogen sulfide	0.2818	1.73	25
Nitric oxide	0.2652	4.66	11
Nitrogen	0.2893	0.25	21
Nitrous oxide	0.2748	2.42	44
Oxygen	0.2890	0.27	26
Sulfur dioxide	0.2667	1.13	32

Reprinted from Danner, R. P. and Daubert, T. E., *Manual for Predicting Chemical Process Design Data,* American Institute of Chemical Engineers, New York, 1983.

Table 7
VALUES OF THE GROUP CONTRIBUTIONS C_i FOR THE ESTIMATION OF a* IN THE REICHENBERG GAS VISCOSITY METHOD

Group	Contribution C_i (μP)
$-CH_3$	9.04
$>CH_2$ (nonring)	6.47
$>CH-$ (nonring)	2.67
$>C<$ (nonring)	−1.53
$=CH_2$	7.68
$=CH-$ (nonring)	5.53
$>C=$ (nonring)	1.78
$\equiv CH$	7.41
$\equiv C-$ (nonring)	5.24
$>CH_2$ (ring)	6.91
$>CH-$ (ring)	1.16
$>C<$ (ring)	0.23
$=CH-$ (ring)	5.90
$>C=$ (ring)	3.59
−F	4.46
−Cl	10.06
−Br	12.83
−OH (alcohols)	7.96
$>O$ (nonring)	3.59
$>C=O$ (nonring)	12.02
−CHO (aldehydes)	14.02
−COOH (acids)	18.65
−COO− (esters) or HCOO (formates)	13.41
$-NH_2$	9.71
$>NH$ (nonring)	3.68
=N− (ring)	4.97
−CN	18.13
$>S$ (ring)	8.86

Reprinted from Reichenberg, D., *AIChE J.*, 21, 181, 1975. With permission.

That quantity in turn is used in Equation 23 to calculate the vapor viscosity in micropoise. In tests made on a variety of compounds, the average error using the predictive form of this equation was about 2%.[25]

$$\eta = \frac{a^* T_r}{[1 + 0.36\, T_r(T_r - 1)]^{1/6}} \tag{23}$$

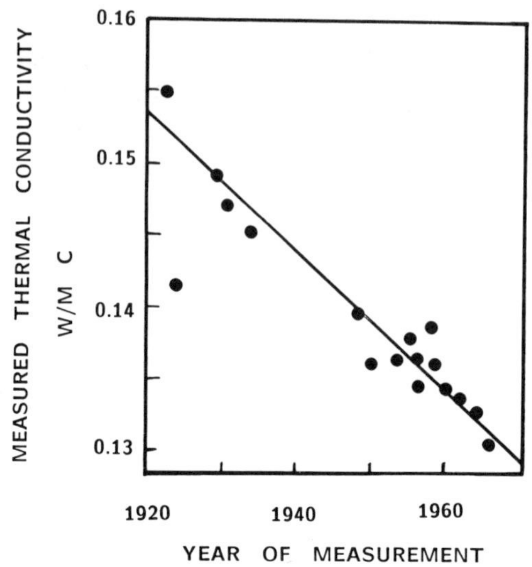

FIGURE 1. Correlation of the measured thermal conductivity of toluene vs. the year of measurement. This illustrates the continuing improvement in apparatus, with reduction in extraneous convection of heat as a function of time. (Adapted from Mallan, G. M., Michaellan, M. S., and Lockhart, F. J., *J. Chem. Eng. Data,* 17, 412, 1972.)

M. Liquid Viscosity

The Van Velzen et al. method[26] is used for liquid viscosity. The method is complex. It does not work well for the first member of each homologous series, and even for higher homologs the errors can be substantial. Viscosity is seldom a crucial physical property, so it needs to be taken into account during the process assessment only when it is high. In the development phase, most viscosities can be measured easily.

N. Vapor Thermal Conductivity

The Roy-Thodos method is used to predict vapor thermal conductivity. It requires critical temperature and pressure, plus molecular structure. A description of the fairly involved computation procedures and group contributions used for this method is given in the *Manual for Predicting Chemical Process Design Data,*[2] and beginning at page 481 of Reference 25. The data base for vapor thermal conductivities is extremely sparse for nonhydrocarbons.

O. Liquid Thermal Conductivity

Thermal conductivity data for liquids are notoriously poor. Take, for example, the representation in Figure 1 of the thermal conductivity of toluene vs. year of measurement.[224] It is clear that early experimental techniques were inadequate, with convection giving unrealistically high data measurements. However, measurement techniques have been improving with time. There is a similar published correlation for *n*-pentane. Data on liquids are more plentiful than for vapors, but there are gaps and significant measurement error for some chemicals. There are very few high-pressure data points available.

Two methods are employed at Pennsylvania State for liquid thermal conductivity predictions. Whenever applicable, the Baroncini method[27] is used.

When there is one experimental data point, it can be extrapolated with respect to temperature using the following equation.

$$\lambda = \frac{A(1 - T_r)^{0.38}}{T_r^{1/6}} \tag{24}$$

The data point is used to solve for A. Reference 27 gives the value of A for 144 organic liquids.

The molecular weight, bp, and critical temperature are used in a somewhat more general expression to calculate the thermal conductivity for general groups of paraffins, organic acids, alcohols, ketones, esters, ethers, aromatics, cycloparaffins, olefins, and refrigerant compounds. The general equation is

$$\lambda = \frac{A(1 - T_r)^{0.38}\, T_b^{6/5}}{MW^n\, T_c^{1/6}\, T_r^{1/6}} \tag{25}$$

The values of A and n for the groups of compounds are

	A	n
Paraffins	0.00350	$^1/_2$
Organic acids	0.00319	$^1/_2$
Alcohols	0.00339	$^1/_2$
Ketones	0.00383	$^1/_2$
Esters	0.0415	1.0
Ethers	0.0385	1.0
Aromatics	0.0346	1.0
Cycloparaffins	0.0310	1.0
Olefins	0.0361	1.0

Modified equations are given in Reference 27 for organic halide-type refrigerants. Overall, average errors in development of these generalized correlations were between 2 and 5 %. This is very good considering the data uncertainties. It would be helpful to have this simple generalized correlation extended to other compound classes.

The modified Missenard method[28,29] is used for other compound classes. Given the bp, molecular structure, and heat capacity and density at 0°C, the thermal conductivity can be calculated at the same temperature.

P. Surface Tension

Little attention has been given to surface tension, so the best method is apparently one published in 1924.[31] The equation is only approximate, but gives a reasonable temperature depencence for nonassociating chemicals. The equation is

$$\sigma = 0.657\, T_c^{1/3}\, P_c^{2/3}(1 - T/T_c)^{1.20} \tag{26}$$

The calculated result is in dynes per centimeter. This equation is a simplification of the approach used at Pennsylvania State. Small errors introduced by this simplification should have little effect, however, on most designs. Where the surface tension is critical it should be measured.

III. GUIDELINES FOR USING ACTIVITY COEFFICIENTS VS. THE EOS METHOD IN PHASE EQUILIBRIUM PREDICTIONS

A. Overview

It is well known to students of thermodynamics that there are two fundamentally distinct methods of representing phase equilibrium data. These are the activity coefficient approach and the EOS approach. They are routinely used to correlate or predict vapor-liquid equilibrium (VLE). Liquid-liquid-vapor equilibrium (LLVE) also can be predicted with some of the methods. With some modification, solid phases likewise can be predicted. Monsanto Co.

personnel once reported informally their successful representation of six different phases simultaneously. Thus, the methods that will be described have considerable potential for representing complex phase equilibria.

The two approaches are both thermodynamically correct, and even interrelated, yet the results can be quite different. In practice it is known that there are many situations where one method is preferred over the other. An attempt has been made to develop a set of heuristics, for inclusion in an expert system computer program, which would guide the user to the best EOS or activity coefficient equation.[32] That is a reasonable approach, and the recommendations are generally sound; however, somewhat more understanding and information are preferred. From the information provided in this book the reader can formulate a personal set of rules with which to make judgments.

B. EOS Approach

The criterion for phase equilibrium is that the chemical potential, and thus the fugacity, of the components in coexisting phases be equal. (It is assumed that the reader has been introduced to the concept of fugacities and fugacity coeffiecients in prior thermodynamics courses.) Thus, through calculation of volumetric properties and derived fugacities for the components in the coexisting phases, it is possible to compute the composition, density, and either temperature or pressure of the phases. The use of such equations is described in some detail by Chao and Greenkorn.[33] A variety of equations are available. The most popular are cubic in molar volume because of computational efficiency. A review of such equations has been presented by Abbott.[34]

EOS have the advantage of being able, in theory, to represent most properties of interest. Since they are not perfect, most are tuned for emphasis on one or another property while the others are less accurately represented. Starling[35] and co-workers attempted to remedy this situation by using simultaneous regression of a number of properties in development of a modification to the BWR equation, now known as the Benedict-Webb-Rubin-Starling (BWRS). Unfortunately, what such equations offer in capability is often applicable for only limited components or conditions. For example, the BWRS equation finds little use outside the fields of light-hydrocarbon processing and carbon dioxide flooding of oil fields. EOS are also very demanding of computation time, requiring many iterations just to determine one set of phase equilibrium conditions. Furthermore, they often fail to converge due to multiple roots. This problem requires skillful programming to avoid frequent program failures.

In general, EOS are chosen when the systems are close to the critical region. Thus, ethylene plants are usually simulated by EOS. However, butadiene extraction systems are usually simulated by the activity coefficient method, illustrated in Chapter 8. Systems that are essentially subcritical but which have one or more supercritical components can be handled by either approach.

C. Activity Coefficient Approach

In this approach the gas phase is again treated by an EOS. The simple virial EOS often is used because there are available a considerable number of experimental second virial coefficients.[36] There are also several good correlations; the Tsonopolous correlation is recommended.[30] Typically, this approach is not extended to pressures in excess of 25 bar. More general EOS such as the Redlich-Kwong and Peng-Robinson also are used; these latter equations have the advantage of being applicable to higher pressure.

The liquid phase is represented by first deriving an expression for the excess Gibbs free energy of mixing. That is differentiated appropriately to give activity coefficients; they are unity for ideal solutions. Activity coefficients vary greatly with composition and temperature but are relatively unaffected by pressure. However, to the extent that pressure affects the temperature, they will change. Increasing temperature usually lowers the activity coefficients, making the liquid phase more ideal.

The expression for the equilibrium ratio, based on equal fugacity in both phases, is

$$K_i = \frac{y_i}{x_i} = \frac{\gamma_i P_i^* \phi_{i,P_i^*}}{P\hat{\phi}_{i,p}} \exp\left(\frac{(P - P_i^*)\, V_i^L}{RT}\right)$$

where T is the absolute temperature, R is the gas constant, P_i^* is the vapor pressure of component i, P is the system pressure, $\hat{\phi}_i$ is the mixture fugacity coefficient, $\phi_{i,Pi*}$ is the pure component fugacity coefficient, γ_i is the activity coefficient with standard state of the system temperature and pressure, V_i^L is the liquid molar volume of component i, x_i is the liquid mole fraction, and y_i is the vapor mole fraction.

An advantage of activity coefficient equations is their ability to represent very high degrees of nonideality. Thus, many of these equations can predict partial miscibility and tie lines.

Computations by the activity coefficient method are much faster than by the EOS approach, and fewer convergence difficulties are encountered. The method usually will be chosen where there is a possibility of more than one liquid phase. There also will be a tendency to choose it for polar molecules. With these few guidelines there will still be some systems that pose serious problems. Those systems are typified by having some components in close proximity to their critical temperatures, and particularly when the mixture contains polar and/or associating components. These problems are being addressed by ongoing research.

A modified activity coefficient approach, the Chao-Seader method,[37] was developed many years ago to serve the needs of the petroleum industry. Solubility parameters were used to predict the liquid phase activity coefficients. The Grayson-Streed method[38] is a modified version of the Chao-Seader equation, and was designed to handle hydrogen-hydrocarbon systems. These two specific methods are mentioned briefly here because of their extremely widespread historical use in the petroleum industry. A fairly extensive test was done on them in unpublished work by S. J. Kramer of the Amoco Oil Co. He compared their prediction accuracy against a modern variant on the Redlich-Kwong EOS[39] developed by Soave.[40] The methods were compared using 105 binary systems, each containing many data points.

For hydrocarbon/hydrocarbon systems the Soave equation was the best method three quarters of the time. In 16 systems of hydrogen and a hydrocarbon, 10 were best represented by the Soave equation; the others, by Grayson-Streed. This result is perhaps surprising, given the fact that the Grayson-Streed method was developed specifically to handle such systems. Chao-Seader was a clear loser in such systems. The Soave equation was best for all the carbon monoxide systems. The same comment, with just a couple of exceptions, can be made for carbon dioxide and nitrogen mixtures with hydrocarbons. It should be noted, however, that the nitrogen data were not well represented by any of these equations.

Hydrogen sulfide systems, common in the petroleum industry, are not well represented by any of these three methods. Mean errors in relative volatility varied from 7 to 20%. Likewise, these methods fail for the ammonia + water system. It is essentially impossible for these equations to handle a polar liquid, such as methanol. The best prediction for the carbon dioxide/methanol system was given by Chao-Seader, with over 300% error. Another particularly bad prediction was for hexane/isopropanol with a 90% error. By contrast, a fit of that system with the Wilson equation gave only a 10% error.

These comparisons suggest that the widely used Chao-Seader and Grayson-Streed equations should be abandoned by the chemical industry. They were never claimed to be applicable to polar compounds, and their application in such circumstances is usually worse than using Raoult's law. Even for hydrocarbon systems, it is usually better to use the Soave EOS than the Chao-Seader or Grayson-Streed method. Presumably, the Peng Robinson equation[41] would give results similar to those of Soave. Newer EOS are needed to handle polar compounds, although some improvement might be realized through fitting of the binary interaction parameter in the Soave equation.

These conclusions reflect only on the Chao-Seader and Grayson-Streed methods, which have activity coefficients predicted by regular solution theory. It is not an indictment of the activity coefficient approach in general. Until an EOS is proven to be accurate for polar compounds, the activity coefficient approach should be used when such molecules are present. High-pressure systems with polar compounds and some constituents near their critical points are best handled with EOS developed for such purposes. However, testing is needed, as there is no *a priori* assurance of accuracy.

IV. CORRELATION OF LOW-PRESSURE PHASE EQUILIBRIA WITH ACTIVITY COEFFICIENTS

A. Role of Excess Functions

Thermodynamics textbooks usually have extensive discussion of the use of excess functions. Prausnitz[42] defines them as follows: "Excess functions are thermodynamic properties of solutions which are in excess of those of an ideal (or ideal dilute) solution at the same conditions of temperature, pressure, and composition." Thus, in an ideal solution all excess functions are zero. For example, the excess enthalpy (the heat of mixing) is zero for an ideal mixture. For real mixtures, components mixed together adiabatically at the same temperature will cause a change in temperature. The heat needed to maintain isothermal conditions is the heat of mixing. When heat must be added, the heat of mixing is positive. When it is negative, heat must be removed to maintain isothermal conditions. Such data are usually measured with isothermal titration calorimeters having no vapor space above the liquid.

Excess volumes, of more theoretical than practical interest, are easily measured. Extremely accurate density measurements can give the true values; however, it is usually more accurate to use a dilatometer. Figure 2 shows how this is done. Two liquids are inserted into the arms of the dilatometer, using flexible syringe needles. The liquids are separated by mercury, and the quantities are determined by weighing the syringes before and after injection. Then a capillary tube is inserted and mercury rises to a certain point in the tube. The position on the tube is noted after the entire assembly comes to equilibrium in a constant temperature bath. The dilatometer is then rocked back and forth, forcing all the test liquids into a single arm. After equilibrium has been reached again in the bath, the height of mercury in the capillary is once again measured. It is clear from the change in height that there is indeed a volume change.

Unfortunately, an entropy meter has not yet been invented. If it had been, the Gibbs free energy could be calculated from the expression

$$G^E = H^E - T\,S^E \tag{27}$$

The best that can be done in the absence of a direct measurement for excess entropy is to measure the excess Gibbs free energy indirectly. The method most analogous to the measurements for excess enthalpy and volume is measurement of equilibrium pressure vs. composition at constant temperature. If Raoult's law holds, the partial pressure will be a linear function of composition, varying from one pure component vapor pressure to the vapor pressure of the second component. In fact, the true pressure curve usually deviates substantially from ideality. Those nonidealities can be caused by both the gas and liquid phases. However, if there is no chemical association in the gas phase, as with organic acids or hydrogen fluoride, and if the pressures are not too high, it is possible to calculate the Gibbs free energy vs. composition. The details are given in Chapter 7.

There are limitations to the previous approach, referred to as the Ptx method. The chemicals must be of very high purity, the data of very high accuracy, the modeling of gas phase

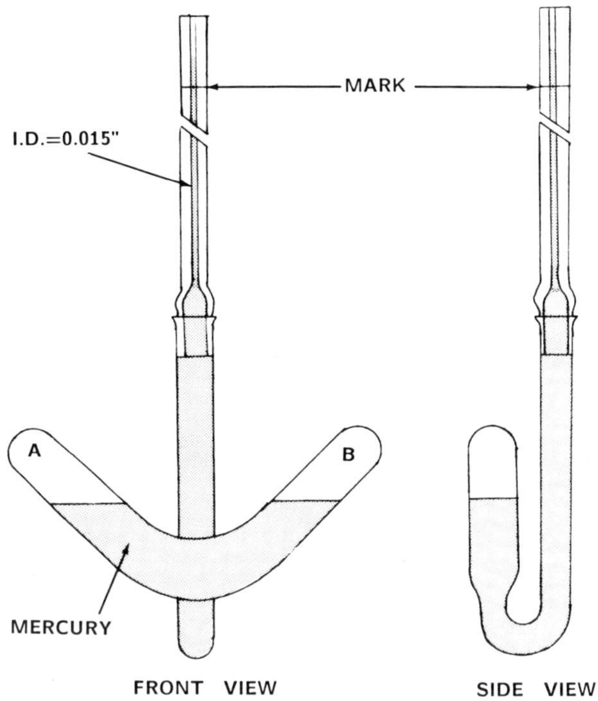

CAPILLARY CELL

FIGURE 2. Cross-sections of a dilatometer used to measure excess volumes of mixing. It is shown before components A and B are mixed. The apparatus is immersed in a constant temperature bath, with the capillary tube protruding out the top. Changes in height of the capillary due to subsequent mixing of the chemicals are measured by a cathetometer in reference to the mark on the capillary.

nonidealities very accurate, and the data in the infinite dilution range especially good. Furthermore, there cannot be any slow decomposition of the chemicals or the results will be in error. These conditions cannot always be met, so the more traditional approach to VLE continues to be used as well. This includes measurement of the vapor and liquid phase compositions, together with pressure and temperature. Usually, the data are taken at constant temperature to facilitate data evaluation. Addition of the vapor compositions gives redundant information, so it is possible to test the data using the Gibbs-Duhem equation. The raw data are reduced to activity coefficients, partial molar quantities related to the Gibbs free energy by

$$G^E = RT \sum_i x_i \ln \gamma_i \tag{28}$$

The activity coefficients themselves are calculated by the equation

$$\gamma_i = \frac{y_i P \hat{\phi}_{i,P}}{x_i P_i^* \phi_{i,P_i^*}} \exp\left[\frac{(P_i^* - P) V_i^L}{RT}\right] \tag{29}$$

The symbols in Equation 29 were defined previously in this chapter.

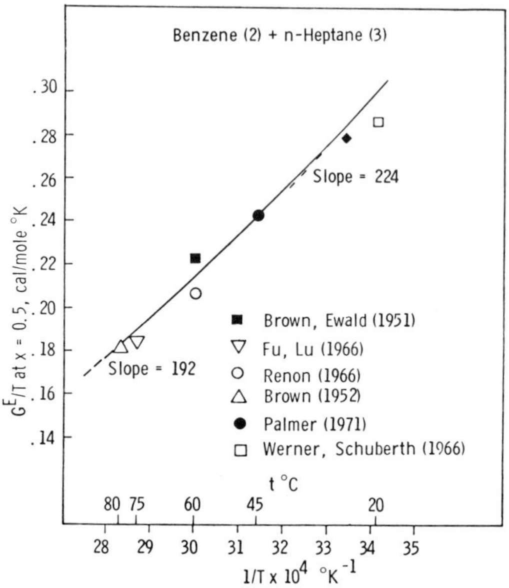

FIGURE 3. Application of the Gibbs-Helmholtz test to excess free energy data on the benzene + *n*-heptane binary system. The slope of the line is based on excess enthalpy data. The composition used was an equimolar mixture, but other mixtures also could have been chosen.

There are relationships between the different excess properties. For example, at a given composition the effect of temperature on G^E is related to the excess enthalpy by

$$(H^E)_{x_1} = \left[\frac{\partial(G^E/T)}{\partial(1/T)}\right]_{x_1} \tag{30}$$

This gives rise to the Gibbs-Helmholtz test, wherein data measured at different temperatures are plotted as G^E/T vs. $1/T$. The slope of the curve should be equal to H^E. This is illustrated for the benzene + *n*-heptane system in Figure 3. The slope of the line is adjusted over the temperature range to be consistent with measured excess enthalpies. There is quite good agreement among these sets of data, with the exception of the data by Werner and Schuberth.

Similarly, the partial molar heat of mixing can be used to predict the effect of temperature on activity coefficients. Given the proper apparatus and technique, it is not difficult to obtain the partial molar heat of mixing. For example, automatic titration isothermal calorimeters generate over 100 data points for a binary mixture. Polynomial equations can be used to fit heat of mixing curves and they can be differentiated. The end points are always pinned down (at zero). The equation to be used is

$$\left[\frac{\partial \ln \gamma_i}{\partial T}\right]_{P,x} = \frac{H_{\text{pure } i} - \overline{H}_i}{RT^2} = \frac{-\overline{H}_i^E}{RT^2} \tag{31}$$

The derivative of the excess Gibbs free energy with respect to pressure is equal to the excess volume. The small values of excess volume determined experimentally show that the effect of pressure on the activity coefficient must be quite small up to reasonable pressures. At high pressures the uncertainty in the fugacity coefficients will become sufficiently great that there will be a need to shift to an EOS approach in any case.

Since the excess Gibbs energy must go to zero when one component is present in 100% concentration, the activity coefficient must approach unity tangentially for that component. Usually, although not always, the activity coefficient reaches a maximum when the constituent it represents is at infinite dilution. It is also useful to remember that in a binary system, a maximum in one activity coefficient must be balanced by a minimum in the other at the same composition.

Activity coefficients can be derived from literature experimental data and tested for thermodynamic consistency as described in Chapter 4. The next problem is correlation of those values with respect to composition and temperature.

B. Activity Coefficient Equations

There is no need to review the entire history of activity coefficient equations. However, the primary equations in use today are presented, with some commentary, in this section. It should be remembered that all such equations are derived by differentiation of an expression for the variation of Gibbs excess free energy with respect to composition.

1. Algebraic Expressions

An equation which assumes symmetrical activity coefficient curves is $G^E = Ax_1x_2$. It is unsatisfactory for most purposes. Two-parameter equations were therefore developed, going back to the turn of the century. They were simply algebraic expansions. Illustrative of these equations were the Wohl expansions and the derived or related equations of Margules, Van Laar, Redlich-Kister, and others. The Van Laar equation was derived from the van der Waals EOS. Perhaps demonstrating the inadequacies of that equation, it only works when the parameters are allowed to vary to fit the data. The Van Laar equation is as follows.

$$\ln \gamma_i = \frac{B_{12}}{[1 + (B_{12}x_1/B_{21}x_2)]^2} \tag{32}$$

$$\ln \gamma_2 = \frac{B_{21}}{[1 + (B_{21}x_2/B_{12}x_1)]^2} \tag{33}$$

In this equation the parameters are equal to the logarithm of the activity coefficients at infinite dilution. However, the parameters can also be determined by regression analysis of activity coefficients across the composition range.

The two-parameter, three-suffix Margules equation is

$$\ln \gamma_1 = x_2^2[A_{12} + 2x_1(A_{21} - A_{12})] \tag{34}$$

$$\ln \gamma_2 = x_1^2[A_{21} + 2x_2(A_{12} - A_{21})] \tag{35}$$

Extensive tests of the Margules and Van Laar equations have been made in the published DECHEMA series. They were compared to the more modern, local composition equations. Out of a total of 3563 binary systems, Walas[43] computed that the Margules was the best equation 20% of the time while the Van Laar was the best only 13% of the time. The Margules handled best those systems with aromatics and systems with mixtures of alcohols, ketones, and ethers. An advantage of the Margules equation is its ability to predict activity coefficient curves that pass through a maximum. However, such systems are rare. When the Van Laar equation is used to fit mixtures with a highly polar constituent, it underpredicts by a substantial amount the activity coefficients at infinite dilution.

In spite of the fact that these totally empirical equations can fit single binary systems reasonably well, the equations have fallen out of favor in industry for two principal reasons.

First, the parameters are very temperature-dependent. For example, the parameters in the Van Laar equation are actually the logarithm of the activity coefficients at infinite dilution, which are highly temperature-dependent. Second, industry usually has to deal with multicomponent systems. Although some individual distillation columns may treat only binary mixtures, most streams in chemical plants are more complex. The algebraic equations typically require multicomponent interaction parameters which are almost never available. Null[44] and Chien and Null[45] developed a format for use of the Van Laar equation without multicomponent parameters. This modification avoids problems with the Van Laar equation when the parameters have opposite signs. It is claimed that the multicomponent mixing rule permits the use of different activity coefficient equations simultaneously. However, this approach has not been tested thoroughly, and the ten ternary systems that were studied seemed to be fit more reliably with the Wilson equation. Given the lack of theoretical justification of the Chien-Null equation, it cannot be relied upon. In fact, Walas[43] gives some significant evidence that the modification by Null is invalid.

2. The Wilson Equation

A concept which has essentially revolutionized the field of solution thermodynamics was embodied in the equation developed by Wilson.[46] The equation was built around the idea of nonrandomness in liquid mixtures. Assuming nonrandomness, the composition environment around a molecule of A must be different from the average composition environment around molecule B. These local compositions are computed and used in the Wilson equation. The local compositions are defined in terms of interaction energies in a Boltzmann distribution. This equation form does a reasonable job of making the parameters temperature-independent. However, over a large temperature range it is usually necessary to make them temperature-dependent.

The Wilson equation does an exceptional job of fitting alcohol + hydrocarbon mixtures, which are fit so poorly by the algebraic expressions. In fact, for all the mixtures covered by the DECHEMA evaluation of VLE data, the Wilson equation fit the data best for 30% of the mixtures. It is small wonder that this one equation has spawned hundreds of journal articles, evaluating it, extending it, modifying it, and explaining it.

The most impressive aspect of the Wilson equation is that from binary parameters alone it is possible to compute multicomponent properties. In fact, it is now clearly demonstrated that for miscible systems the multicomponent predictions are quite good. This has been of tremendous help to the chemical industry. In fact, elimination of the multicomponent interaction parameters alone would have been enough to establish the equation without the added benefit of improved fitting of binary data.

The Wilson equation is as follows

$$\ln \gamma_1 = -\ln(x_1 + \Lambda_{12}x_2) + x_2\left\{\frac{\Lambda_{12}}{x_1 + \Lambda_{12}x_2} - \frac{\Lambda_{21}}{\Lambda_{21}x_1 + x_2}\right\} \tag{36}$$

$$\ln \gamma_2 = -\ln(x_2 + \Lambda_{21}x_1) - x_1\left\{\frac{\Lambda_{12}}{x_1 + \Lambda_{12}x_2} - \frac{\Lambda_{21}}{\Lambda_{21}x_1 + x_2}\right\}$$

where

$$\Lambda_{12} = \frac{V_2^L}{V_1^L}\exp\left\{-\frac{(\lambda_{12} - \lambda_{11})}{RT}\right\} = \frac{V_2^L}{V_1^L}\exp\left\{-\frac{A_{12}}{RT}\right\}$$

$$\Lambda_{21} = \frac{V_1^L}{V_2^L}\exp\left\{-\frac{(\lambda_{12} - \lambda_{22})}{RT}\right\} = \frac{V_1^L}{V_2^L}\exp\left\{-\frac{A_{21}}{RT}\right\} \tag{37}$$

The multicomponent form is

$$\ln \gamma_i = -\ln\left(\sum_j x_j\Lambda_{ij}\right) + 1 - \sum_k \frac{x_k\Lambda_{ki}}{\sum_j x_j\Lambda_{kj}}$$

$$\Lambda_{ij} = \frac{V_j^L}{V_i^L}\exp - \left[\frac{(\lambda_{ij} - \lambda_{ii})}{RT}\right] = \frac{V_j^L}{V_i^L}\exp\left[-\frac{A_{ij}}{RT}\right]$$

$$\Lambda_{ii} = \Lambda_{jj} = \Lambda_{kk} = 1 \qquad (38)$$

The Wilson equation parameters can be found only by iteration, but convergence is rapid. The simple equations for the parameters at infinite dilution are

$$\ln \Lambda_{12} + \Lambda_{21} = 1 - \ln \gamma_1^\infty$$

$$\Lambda_{12} + \ln \Lambda_{21} = 1 - \ln \gamma_2^\infty \qquad (39)$$

When the activity coefficients at infinite dilution are less than unity, two or three unique solutions are possible.

The temperature dependence of the activity coefficients is related to the heat of mixing. In vacuum distillation there is usually a significant temperature difference between the top and bottom of the column. Heat of mixing effects are therefore relatively more important under vacuum conditions than in pressure distillation. It has been shown that unless the Wilson parameters are made functions of temperature, preferably quadratic, the results will be in significant error for many systems.[47] Without use of temperature-dependent parameters, the heat of mixing predicted can fall far short of the actual value. These considerations were examined in case studies of the design of a vacuum tower for separation of caprolactam from cyclohexanol. The following conclusions[48] were reached:

1. The heat of mixing is positive and much larger than that predicted from the Wilson equation.
2. The Wilson equation with temperature-independent parameters resulted in a design giving either off-specification product or operation at only 71% of design.
3. The most reliable design was obtained with the Redlich-Kister equation with temperature-dependent parameters.

Attempts to develop one-parameter forms of the Wilson equation have met with mixed success. The approach has some merit when the components are completely miscible and there is only one infinite dilution activity coefficient available. As proposed by Hankinson et al.[49] and Tassios,[50] the Wilson pure component energy parameters are replaced as follows with the heat of vaporization.

$$\lambda_{ii} = RT - \Delta H_i^v \qquad (40)$$

For component 1 the value is established at the atmospheric bp temperature of component 2 and the system pressure. The pressure effect is usually a negligible correction, and given the other uncertainties can probably be ignored. The number of systems studied was too few to arrive at general conclusions concerning expected accuracy. Work done subsequently suggested that use of UNIFAC is preferable to use of a one-parameter Wilson equation.[51]

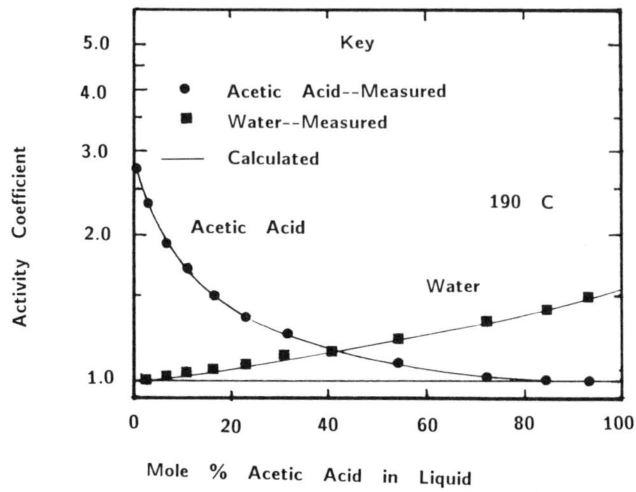

FIGURE 4. Correlation of activity coefficients for water + acetic acid at 189°C. Data measured by Wiltec for AIChE/DIPPR and correlated by a modified Wilson equation. (Adapted from Freeman, J. R. and Wilson, G. M., *AIChE Symp. Ser.*, 81(244), 14, 1985.)

3. A Modified Wilson Equation

The most serious drawback of the Wilson equation is that it cannot predict phase splitting. Its form is such that it takes into account entropic contributions to the Gibbs free energy but not enthalpic contributions. Therefore, a number of researchers have suggested modifications to the Wilson equation. The modification suggested by Wilson himself is as follows.

$$\ln \gamma_1 = -c\left[\ln(x_1 + \lambda_{12}x_2) + \frac{x_1}{x_1 + \lambda_{12}x_2} + \frac{x_2\lambda_{21}}{x_1 + \lambda_{21}x_1} - 1\right] \tag{41}$$

The interaction parameters in this equation were correlated with respect to temperature by an expression of the form $\lambda = \exp(A + B/T + C/T^2)$. This equation has considerable power, and has been tested on the acetic acid + water system. That is a particularly difficult test because other equations such as NRTL and UNIQUAC (discussed subsequently) cannot represent properly the composition dependence in that system. A particularly encouraging aspect of this work by Wilson was that the activity coefficients were fitted well over a very wide temperature range. In fact, the equation extrapolated extraordinarily well to predict data at subatmospheric pressures. This is illustrated by Figures 4, 5, and 6. Further testing of this equation will be necessary before it is recommended for general use. However, it represents an interesting alternative when other equations appear inadequate.

4. The NRTL Equation

An early attempt to use the concept of local compositions and still predict partial miscibility led to the NRTL equation.[53,54] It also uses the energy interaction parameters but has a different form and furthermore uses a third, "nonrandomness" parameter, α_{ij}. The binary form of the NRTL equation is

$$\ln \gamma_1 = x_2^2\left\{\tau_{21}\left(\frac{G_{21}}{x_1 + x_2G_{21}}\right)^2 + \frac{\tau_{12}G_{12}}{(x_2 + x_1G_{12})^2}\right\}$$

$$\ln \gamma_2 = x_1^2\left\{\tau_{12}\left(\frac{G_{12}}{x_2 + x_1G_{12}}\right)^2 + \frac{\tau_{21}G_{21}}{(x_1 + x_2G_{21})^2}\right\}$$

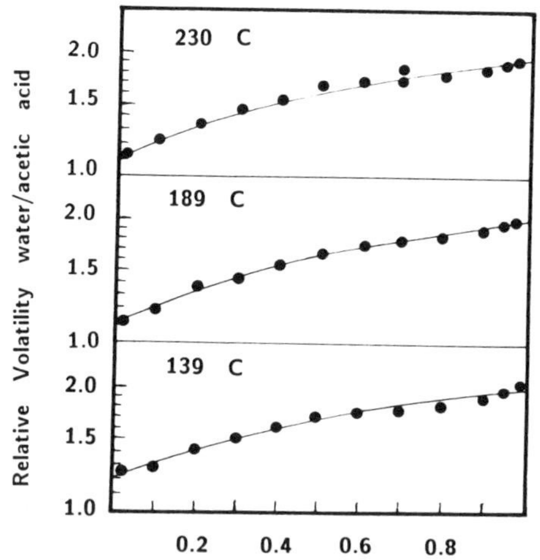

FIGURE 5. Correlation of the relative volatility in the water + acetic acid binary system as a function of temperature and weight fraction acetic acid. Modified Wilson equation used. Data measured by Wiltec for AIChE/DIPPR. (Adapted from Freeman, J. R. and Wilson, G. M., *AIChE Symp. Ser.*, 81(244), 14, 1985.)

where

$$G_{12} = \exp(-\alpha_{12}\,\tau_{12})$$

$$G_{21} = \exp(-\alpha_{21}\,\tau_{21})$$

$$\alpha_{12} = \alpha_{21}$$

$$\tau_{12} = \left(\frac{g_{12} - g_{22}}{RT}\right) = \frac{A_{12}}{RT}$$

$$\tau_{21} = \left(\frac{g_{12} - g_{11}}{RT}\right) = \frac{A_{21}}{RT} \qquad (42)$$

The adjustable parameters in this equation are A_{12}, A_{21}, and α. The multicomponent form of the NRTL equation is

$$\ln \gamma_i = \frac{\sum_j \tau_{ji}G_{ji}x_j}{\sum_k G_{ki}x_k} + \sum_j \frac{x_j G_{ij}}{\sum_k G_{kj}x_k}\left[\tau_{ij} - \frac{\sum_n x_n \tau_{nj} G_{nj}}{\sum_k G_{kj}x_k}\right] \qquad (43)$$

In spite of having three adjustable parameters, the NRTL was the best of five fitting equations for only 23% of the binary systems in the DECHEMA evaluation. For its time, it was a significant advance and found considerable use. Marina and Tassios[55] found that simply by setting the nonrandomness parameter to -1 they were able to use it effectively as a two-parameter equation. The NRTL equation is subject to multiple roots over a wider range than the Wilson equation. Reference 43 gives a number of graphs showing the ranges where problems can be expected. Even activity coefficients greater than unity are not immune from the problem when the nonrandomness parameter is set to -1. The NRTL equation is used less than would otherwise be the case because of other developments.

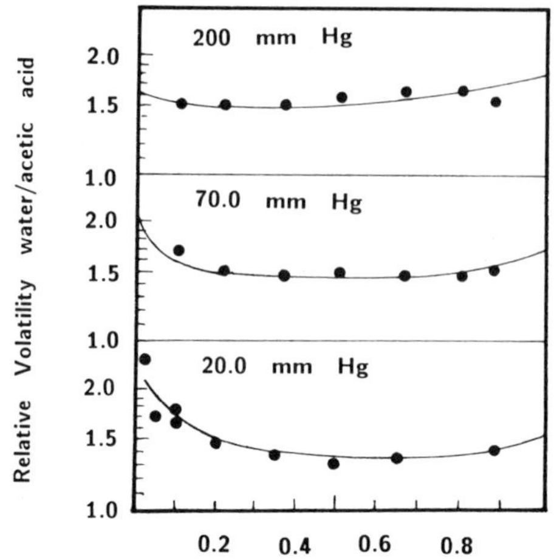

FIGURE 6. Prediction of the relative volatility of the water + acetic acid binary system as a function of pressure. This illustrates an extraordinary extrapolation from data taken above atmospheric pressure to conditions existing at high vacuum. Data measured by D. Othmer et al. (Adapted from Freeman, J. R. and Wilson, G. M., *AIChE Symp. Ser.*, 81(244), 14, 1985.)

5. The UNIQUAC Equation

Abrams and Prausnitz[22] developed the UNIQUAC equation, attempting to improve on the NRTL method. They developed a true two-parameter equation that could handle partial miscibility. It takes advantage of information on both the size and shape of molecules to calculate a ''combinatorial'' contribution to the activity coefficient. Then a contribution due to energetics, the ''residual'' part, is added. This is analogous to the entropy and enthalpy contributions to the Gibbs energy, but those terms cannot be factored out. The best source for the r and q size and shape factors for molecules is the DECHEMA collection. When those factors cannot be found, they often can be calculated from crystallographic data using the book by Bondi[21] and the instructions in Reference 22.

The binary activity coefficient equation for component 1 is as follows

$$\ln \gamma_1^C = \ln \frac{\varphi_1}{x_1} + \frac{z}{2} q_1 \ln \frac{\vartheta_1}{\varphi_1} + \varphi_2\left(l_1 - \frac{r_1}{r_2} l_2\right)$$

$$\ln \gamma_1^R = -q_1 \ln(\vartheta_1 + \vartheta_2\tau_{21}) + \vartheta_2 q_2\left(\frac{\tau_{21}}{\vartheta_1 + \vartheta_2\tau_{21}} - \frac{\tau_{12}}{\vartheta_1\tau_{12} + \vartheta_2}\right) \tag{44}$$

$$\ln \gamma_1 = \ln \gamma_1^C + \ln \gamma_1^R$$

where

$$l_i = \frac{z}{2}(r_i - q_i) - (r_i - 1)$$

$$z = 10$$

$$\vartheta_i = \frac{q_i x_i}{\sum_j q_j x_j} \quad \text{area fraction of component i}$$

$$\varphi_i = \frac{r_i x_i}{\sum_j r_j x_j} \quad \text{volume fraction of component i}$$

$$\tau_{ji} = \exp\left(-\frac{u_{ji} - u_{ii}}{RT}\right) \tag{45}$$

The multicomponent form of the full UNIQUAC equation is as follows.

$$\ln \gamma_i^C = \ln \frac{\varphi_i}{x_i} + \frac{z}{2} q_i \ln \frac{\vartheta_i}{\varphi_i} + l_i - \frac{\varphi_i}{x_i} \sum_j x_j l_j$$

$$\ln \gamma_i^R = q_i \left[1 - \ln\left(\sum_{j=1}^{m} \vartheta_j \tau_{ji}\right) - \sum_{j=1}^{m} \frac{\vartheta_j \tau_{ij}}{\sum_{k=1}^{m} \vartheta_k \tau_{kj}}\right]$$

$$z = 10$$

and finally,

$$\ln \gamma_i = \ln \gamma_i^C + \ln \gamma_i^R \tag{46}$$

The unknown parameters in this equation share the characteristic of other local composition equations in demonstrating less temperature dependence than the algebraic-type equations. However, one must watch for multiple solutions for equation parameters. As with the Wilson equation, the parameters exhibit a certain amount of intercorrelation.

The author's experience in application of the UNIQUAC equation confirms that of DECHEMA. The UNIQUAC equation is superior for many systems but not for all. It shares the general features of the Wilson equation, but it also can be applied to partially miscible mixtures. Another feature of the UNIQUAC equation is its use as the basis for the very popular UNIFAC equation. It fits almost exactly the composition dependence of activity coefficients predicted by UNIFAC. That linkage has given UNIQUAC a fairly secure place in industrial process simulators. A general discussion of the use of UNIQUAC for representing ternary and quaternary liquid-liquid equilibrium (LLE) data was made by Fredenslund et al.[56]

Tassios (see Krumins et al.[51]) developed a one-parameter form of the UNIQUAC equation by substitution of the heat of vaporization for the configurational free energy of the molecule:

$$g_{ii} = \frac{2(\Delta H^v - RT)}{z\, q_i} \tag{47}$$

where Z = 10.

In a study of many systems this model appeared to be better than the one-parameter Wilson model mentioned previously. However, it is probably better to use UNIFAC, wherever possible, as opposed to this one-parameter equation.

6. Other Activity Coefficient Equations

Each individual company uses activity coefficient equations of its own development or of some historical significance. Those researchers who feel insulted that their equations have not been mentioned may be consoled by the fact that the author has not even mentioned his own.[57] To gain widespread acceptance it is necessary that equations be tested thoroughly, preferably by a third party.

7. Recommendations on Choice of Activity Coefficient Equations

When use of an activity coefficient equation approach is preferred over the EOS approach, the Wilson equation should be used wherever possible. When more than one liquid phase is likely to be encountered, the UNIQUAC equation should be used. However, a careful and critical examination of the fit to binary data should be made. If it does not seem to work well over the entire composition range, examination of the applicability of an extended Wilson equation would be in order. The NRTL equation also might be tried, but with some care because of the possibility of multiple roots. For those rare cases where the activity coefficients go through a maximum, it may be necessary to use the Margules equation. However, it may be preferable in those situations to absorb some error by use of the Wilson equation in order to preserve reasonable predictions of multicomponent VLE.

C. Correlation of LLE with Activity Coefficients

In process simulators it is necessary to represent all forms of phase behavior with thermodynamically consistent equations. The author has used both EOS and activity coefficient equations to correlate and predict LLE data. It is preferable to use activity coefficients for LLE whenever the system falls within the scope of that approach. This judgment may change in the future as EOS are further improved in scope of application.

The old empirical correlation methods of Hand[227] or Othmer-Tobias[228] are obsolete. The replacement of empiricism by thermodynamics has significant advantages. First, various types of binary data can be used to give predictions of multicomponent information. Second, the tie lines and vapor phase composition can be calculated simultaneously for the liquid phase.

1. Calculation Methods

It may not be immediately apparent how to go about predicting multiple liquid phases using activity coefficients. The approach used by Null[58] is an interactive method which makes the valid assumption that the activities in both phases are equal. A Newton-Raphson procedure is used to find the fraction going into each phase. If the computations converge on zero or unity there is presumably only one phase. Reference 34 includes sample FORTRAN programs for both two- and three-liquid phase systems. These techniques are easily applied but are not always reliable. They are particularly difficult to use as the number of components increases beyond three or four. More advanced mathematical techniques are presented in Reference 59.

2. Suitable Activity Coefficient Equations

Prediction of phase splitting cannot be handled by all activity coefficient equations. The Wilson and the Flory-Huggins equations, based entirely on entropic contributions to the free energy, cannot predict two liquid phases. Algebraic expansions such as the Wohl equations require ternary interaction parameters; regardless of the approach taken to fitting the data the results are unsatisfactory without such parameters. The same is generally true for the Redlich-Kister and the Van Laar equations.

The NRTL and UNIQUAC equations are particularly suitable for correlation of LLVE. Both employ the concept of local compositions, and are now the most accepted. Neither is perfect and both eventually may be replaced. Nevertheless, they are sufficiently powerful to give reasonable predictions of multicomponent phase envelopes from binary data. Naturally, the predictions are enhanced by inclusion of some ternary information. The UNIQUAC equation is preferred by many since it has only two parameters per binary pair, compared to three for NRTL. If the third parameter in the NRTL equation is fixed, the UNIQUAC equation fits the data better. A review of a wide variety of equations, including the most significant aspects of previous work in this area, is given in Reference 60.

If ternary systems are considered to be representative of the general case of multicomponent systems, there is a fairly large body of data on the correlation capability of the UNIQUAC equation. The approach used by Sorensen et al.[60] is to fix the UNIQUAC parameters with binary solubility data. For Type I systems (one partially miscible pair) this leaves four parameters which can be fitted to the other tie lines. In case of a Type II system (two partially miscible pairs) there are only two parameters allowed to vary to match the phase envelope. The match is outstanding in 17 examples given. This would be the appropriate approach when trying to model a liquid extraction system. However, for systems with a vapor-liquid separation this likely would not be adequate.

The thermodynamic prediction of three liquid phases in special ternary systems is difficult. In theory, this is just an extension of work on two-liquid phase systems. Null[58] gives a FORTRAN computer program that converged on an answer for the nitromethane + lauryl alcohol + ethylene glycol mixture. The prediction gave only a fair representation of reality. This may be due to inadequacies in the data base used; more likely, it is a problem with the correlation used (Null's version of the Van Laar). A more recent paper tested the same system again.[61] The computations indicated that the UNIQUAC equation gave a slightly better prediction than the Van Laar, but there was still room for considerable improvement. Both correlations gave a good representation of the aniline + heptane + water system.

3. Fitting of Tie Lines

A quite reasonable prediction of the solubility envelope should be obtainable by fitting solubility data for partially miscible binary pairs and VLE data for miscible pairs. An example of this approach is given for the acetonitrile + benzene + *n*-heptane system. The correlation work done on this system was completed prior to development of the UNIQUAC equation. It was based on a new local composition equation with entropy-weighting functions. This particular approach has not been pursued in detail or widely tested, so it is not recommended for general use. However, its similarity to the other local composition equations makes the fitting results interesting. For the aforementioned ternary system, illustrated below, excess free energy, enthalpy, and volume data were measured for all three binary mixtures. On the partially miscible mixture, VLE data gave G^E within the region of miscibility, and solubility data were measured as well. LLVE were measured for the ternary region where more than one liquid phase was present. In addition, VLE data were measured for other parts of the ternary regime. This rather comprehensive set of data provided a good basis for testing different approaches. With each approach to data reduction, the root mean squared deviation (RMSD) was computed in calories for all of the 51 data points. Those data points included points at the ends of the tie lines for the binodal curve (region of two liquid phases). It should be noted that the equation had two adjustable parameters per binary, plus one temperature derivative which allowed inclusion of heat of mixing data.

The most successful approach was simultaneous correlation of excess free energy and enthalpy for each of the miscible binaries, plus use of solubility data for the acetonitrile + *n*-heptane binary; this gave a RMSD for all the ternary data of 8.9 cal. The worst result evolved from derivation of all six interaction parameters from tie line data; that gave an unacceptable RMSD of 82.1 cal. This suggests why regression of ternary tie lines is not recommended when the object is correlation of the VLE. When excess free energy and enthalpy data were substituted for solubility data on the partially miscible binary, the RMSD climbed from 8.9 to 12.7 cal. This is not bad, but it does illustrate the power of accurate solubility data. For comparison, the NRTL equation was fitted to all of the binary excess free energy data. The RMSD doubled to 25.7 cal in spite of using a total of nine variable parameters.[57]

In summary, the fitting of tie lines to derive the parameters for miscible binaries is risky. Since there is a range of parameter values which will give a close fit for most binary data,

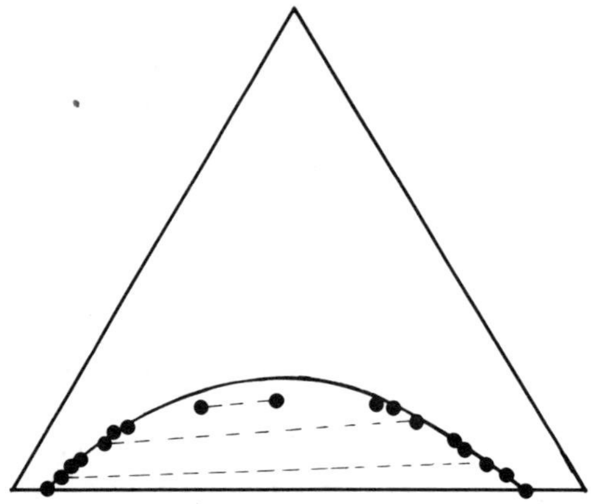

FIGURE 7. Prediction of LLE for the acetonitrile + benzene + *n*-heptane system using the Nagata modification of the UNIQUAC equation. Data by Palmer and Smith at 45°C.

addition of some ternary data such as tie lines may be beneficial. In the absence of ternary data, good binary VLE data and solubility data form the best basis for ternary predictions.

4. Nagata Modification of UNIQUAC

A significant improvement on the UNIQUAC equation for ternary liquid-liquid predictions was presented by Nagata.[62,63] The Nagata UNIQUAC uses an additional value of q, called q*. He used q* values of 0.99 for methanol, 0.92 for ethanol, 0.89 for isopropanol, 0.96 for water, 1.29 for nitromethane, and $q^* = q^{0.2}$ for other components. His revised equation is as follows.

$$\ln \gamma_1 = \ln \frac{\phi_1}{x_1} + 1 - \frac{\phi_1}{x_1} - \left(\frac{Z}{2}\right) q_1\left(\ln \frac{\phi_1}{\nu_1} + 1 - \frac{\phi_1}{\nu_1}\right) - q_1^* \ln(\nu_1 + \nu_2\tau_{21})$$
$$+ q_1\nu_2\left[\left(\frac{q_1^*}{q_1}\right)\left(\frac{\tau_{12}}{\nu_1 + \nu_2\tau_{21}} - 1\right) - \left(\frac{q_2^*}{q_2}\right)\left(\frac{\tau_{12}}{\nu_2 + \nu_1\tau_{12}} - 1\right)\right] \quad (48)$$

The effect of switching to this form of the UNIQUAC equation is to reduce dramatically the typical overprediction of the plait point. The acetonitrile + benzene + *n*-heptane system previously referred to was predicted much better than had been previously accomplished using two different approaches. The prediction is reprinted here as Figure 7. In fact, prediction of 51 ternary data points in this system was accomplished with a reduction of almost 50% in deviation between calculated and experimental values! The average deviation in mole fraction was 0.0043 with Nagata's revised UNIQUAC equation.

The Nagata equation also has been applied to alcohol mixtures with three phases.[64] However, it needs more testing before it can be recommended without reservation. This method represents another step forward in the quest for an equation that does a good job of representing nature.

5. Corrected Activity Coefficients

A very different approach to modification of the UNIQUAC and other activity coefficient equations for liquid-liquid predictions has been presented by Cha and Prausnitz.[4] It recognizes the problems mentioned previously in predicting LLE from VLE. To circumvent those problems, the authors multiply the excess Gibbs free energy expression by a correction factor which degenerates to unity for any of the binary mixtures. It is actually composed of two coefficients which are determined from the multicomponent tie lines.

This method requires that the model binary parameters first be determined. They are obtained from VLE for the miscible binary pairs and from solubility data for the partially miscible pairs. Multicomponent tie lines are then used to determine parameters in the following equation:

$$C = \exp[\alpha(x_1x_2x_3)^{\beta}] \tag{49}$$

The constant C is then used as a multiplier to the activity coefficient correlated without the ternary information, γ_i^0, according to the following equation:

$$\ln \gamma_i = C \ln \gamma_i^0 + C\beta \left(\frac{1}{x_i} - 3\right) (\ln C)(g^E/RT)_{123}^0 \tag{50}$$

Cha and Prausnitz found an essentially perfect fit to the tie line data for the acetonitrile + benzene + *n*-heptane system previously mentioned. Furthermore, they discovered that the small correction factor had very little effect on the calculated vapor phase mole fractions in the miscible ternary region. More experience is needed to determine whether this approach will be useful. It is quite clear that it is a somewhat general approach which corrects for inadequacies in the activity coefficient equation. It cannot, however, compensate for models that are grossly inadequate. Improvements to the activity coefficient equations should eliminate the need for this type of empirical correction.

D. Vapor Imperfections Modeled with Second Virial Coefficients

The use of activity coefficients almost always requires modeling of the vapor phase. Any good EOS can be used for that purpose. However, for systems below 25 atm pressure it is often easiest to use the virial EOS; therefore, the virial equation is emphasized in this section on use of activity coefficients. It cannot be used by itself to compute phase equilibria, since it cannot compute liquid phase fugacity coefficients accurately.

1. Applicability of the Virial EOS

The virial EOS is the simplest method for handling gas phase nonidealities at low pressure. It diverges as the pressure increases, so it generally would not be used above about 25 atm pressure. This specification is higher than normally accepted, but is substantiated by recent research. Third virial coefficients are virtually never used in industry, but use of second virial coefficients is widespread. Thus, the compressibility factor (Z) is given by a simple equation truncated after the second virial coefficient.

$$Z = PV/RT = 1 + BP/RT \tag{51}$$

In a mixture of components, the virial coefficient is calculated from

$$B = \sum_i \sum_j y_i y_j B_{ij} \tag{52}$$

The fugacity coefficients are calculated from

$$\ell n\, \hat{\phi}_{i,P} = (2/V) \sum_j y_i B_{ij} - 1.0 - B/V \tag{53}$$

2. Second Virial Coefficient Data Sources

An exhaustive critical compilation of virial coefficients is given by Dymond and Smith.[65] Second virial coefficients are also included in the DIPPR Data Compilation. In the latter reference, missing coefficients have been filled in by prediction methods.

3. Corresponding States Correlations

Quite often the necessary data are not available, so correlations are required. The Pitzer-Curl equation[66] is useful for hydrocarbons, correlating the virial coefficient as a function of critical properties and the acentric factor.[67] As a predictive equation it is lacking when applied to polar compounds, although the acentric factor can be used as a simple fitting parameter. It was the first of a series of corresponding-states equations for the virial coefficient. An attempt was made by O'Connell and Prausnitz[68] to improve on the Pitzer-Curl equation. However, this author tested the O'Connell-Prausnitz correlation on acetonitrile virial coefficients measured by Prausnitz. The results indicated that the equation has the wrong temperature dependence for such polar compounds. The Nothnagel et al. correlation[69] also has serious deficiencies, pointed out in later studies.[70] One of its problems is the need for a third parameter for polar compounds which can be obtained only from experimental data.

4. Hayden-O'Connell Method

The Hayden-O'Connell method[71] is used extensively by DECHEMA, and it has the advantage of handling organic acid dimerization. This corresponding-states method uses the dipole moment directly for polar nonassociating compounds. It has a total of four parameters. The average RMSD observed in DECHEMA tests was 40.7 cc/g mol. To put that in perspective, an error in cubic centimeters per mole of 300/P (atm) gives an error of 1% in the fugacity coefficient. Thus, at a pressure of 7 atm the average error in fugacity would be 1%. That level of uncertainty may or may not be acceptable, depending on the situation. This correlation gives a very poor prediction for pyridine. In cross-coefficient prediction there are large errors in polar-polar interactions. At Air Products and Chemicals Inc. this correlation was extended in terms of the chemical theory to handle mixtures of amines with methanol.[72] The authors correlated the virial coefficients in terms of the chemical equilibrium constants according to

$$B_{12} = -0.5\ RT\ K_{12} \tag{54}$$

The Tarakad-Danner correlation[70] has reasonably good accuracy and the advantage of treating polar and association effects together. Only one characteristic parameter instead of two, is used for polar effects. However, its determination requires experimental virial coefficients at a reduced temperature of 0.6; therefore, the method has not won great acceptance.

5. Tsonopoulos Correlation

The Tsonopoulos method,[30,73,74] developed at Exxon Corp., has been widely used. The correlation is as follows.

$$\frac{BP_c}{RT_c} = f^0 + \omega f^1 + a/T_r^6 - b/T_r^8$$

$$f^0 = 0.1445 - \frac{0.330}{T_r} - \frac{0.1385}{T_r^2} - \frac{0.0121}{T_r^3} - \frac{0.000607}{T_r^8}$$

$$f^1 = 0.0637 + \frac{0.331}{T_r^2} - \frac{0.423}{T_r^3} - \frac{0.008}{T_r^8} \tag{55}$$

$a = -2.140 \times 10^{-4} \mu_r - 4.308 \times 10^{-21} \mu_r^8$ where $\mu_r = 10^5 \mu_p^2 P_c/T_c^2$ and where b is reserved for hydrogen-bonding species.

This correlation is widely accepted and utilized in industry. Because of some component-specific parameters for polar compounds, there are certain components for which it is inapplicable. In comparison with an accurate experimental study of cross-virial coefficients, the Tsonopoulos method gave outstanding predictions of the data.[75] For the carbon dioxide/methane system the prediction was −64.6 cc/mol, differing by only 0.2 from the experimental value. For carbon dioxide/ethane the experimental value was − 122 and the prediction was − 127 cc/mol. The appropriate mixing rules are given by Tsonopoulos.[30,73,74]

6. Group Contribution Methods

A group contribution method is given in Reference 76. This uses the concept of second-order groups developed by Benson; thus, the effects of nearest neighbors on groups are included. The advantage of inclusion of those nearest neighbors is an ostensible improvement in accuracy. The disadvantage is a large expansion of possible groups and consequent lack of generality. There are 77 groups listed with appropriate parameters, representing virtually all second virial coefficient data measured before 1980. Carbon, oxygen, nitrogen, halogens, and sulfur are included in the groups listed. The virial coefficient is computed as a simple expansion in reduced temperature using the additive coefficients. This approach eliminates a number of problems with other approaches, giving a much more general prediction method. When the necessary groups are available, this method should find considerable use.

7. Virial Coefficients: Recommendations

The preceding discussion is really relevant only when using the activity coefficient approach. In that context, the importance of accuracy increases as the pressure goes up. In other words, vacuum operation K values will be affected by the vapor phase nonidealities only in the case of chemical association. At 2 atm pressure, the vapor phase nonidealities are significant enough under many circumstances to affect the design of a distillation column.

Which is the best correlation? There is no clear-cut answer. In some cases the "best" correlation may be the only one capable of making the prediction. The Hayden-O'Connell, Tsonopoulos, and Tarakad-Danner equations are all relatively good. However, caution should be exercised when dealing with mixtures of polar compounds and particularly with chemicals containing sulfur or halogens. These conclusions are not unexpected and are essentially the same as for more general EOS. Overall, there is now an excellent basis for prediction of the virial coefficient and gas phase fugacity of common chemicals.

E. VLE in Dimerizing Systems

Organic acids form such strong dimers in the gas phase that the VLE cannot possibly be modeled without taking them into account. In fact, VLE data look thermodynamically inconsistent if an ideal gas phase is assumed. The organic acids cannot be ignored, as is done by some authors, because these acids have considerable importance in the chemical industry. Acetic acid is used in manufacture of acetate fibers and is the solvent used in production of terephthalic acid, the main building block for polyester. Further, it is such a good solvent that it is used as a reaction medium in many other situations. For this reason, DIPPR established a special project just to measure its PVT properties and VLE in mixtures with water up to and including reaction pressures. Hydrogen fluoride forms hexamers which can be treated with the same approach.

1. Chemical Models

Only chemical models can represent the gas phase for aliphatic organic acids. Judgment is reserved on aromatic acids pending completion of data measurements by DIPPR. An equilibrium constant for the dimerization is defined as

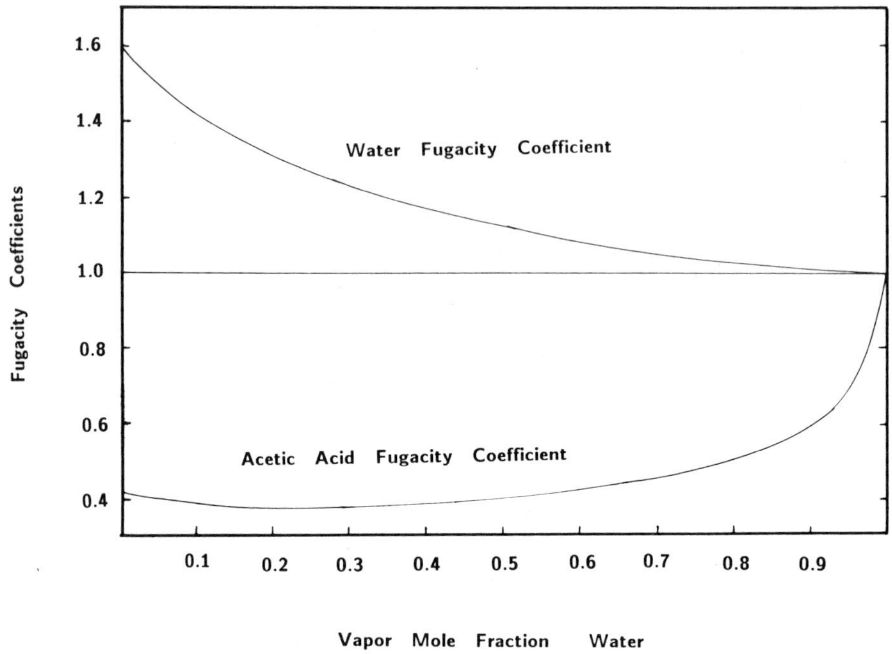

FIGURE 8. Mixture fugacity coefficients for the water + acetic acid binary vapor phase mixture at 1 atm pressure. Note that the low fugacity coefficient for acetic acid at high acetic acid levels is matched by a correspondingly high fugacity coefficient for water.

$$\ell n\ K_A = A + B/T + C/T^2 \tag{56}$$

The monomer vapor pressure is then computed from

$$P^*_{AM} = \frac{-1 + \sqrt{1 + (4/K_A)\ P^*_A}}{2/K_A} \tag{57}$$

The value of K_A for acetic acid, developed for DIPPR by Wiltec, is

$$\ln K\ (\text{mmHg}^{-1}) = -16.637 + 1457.7/T + 1132000/T^2 \tag{58}$$

2. Mixed Acid Systems

Figure 8 shows the fugacity coefficients for the acetic acid + water system as a function of composition at 1 atm pressure. The low values for acetic acid cause the values for water to exceed unity.

For mixed acid systems the available data are scanty. It has been suggested by Prausnitz[42] that the following combining rule be used:

$$K_{AB} = 2\sqrt{K_A K_B} \tag{59}$$

This mixing rule seemed to work adequately for the acetic acid + propionic acid mixture. However, when the author applied it to the formic acid + acetic acid mixture, the data[77] were not in agreement with prediction. A reanalysis will be in order once DIPPR reevaluates the dimerization of formic acid and measures additional mixture PVT data. In the meantime, the proposed mixing rule should be used with caution.

3. Mixed Acids and Other Chemicals

How can one use this information simply when there is a mixture of two organic acids and one or more nonassociating components? This is done by computing the monomer vapor pressure for each acid using the appropriate equilibrium constants. Then the partial pressure of each monomer is calculated from

$$x_A \, P^*_{AM} \, \gamma_A \tag{60}$$

The partial pressure of each dimer is computed from

$$\frac{(x_A P^*_{AM} \, \gamma_A)^2}{K_A} \tag{61}$$

Finally, the partial pressure of the cross-dimer is computed from

$$\frac{x_A x_B \, P^*_{AM} \, P^*_{BM} \, \gamma_A \gamma_B}{K_{AB}} \tag{62}$$

These partial pressures and those of the nondimerizing components are added together to give a computed pseudopressure somewhat higher than the bubble point pressure. Finally, the mole fractions of nondimerizing components in the vapor are computed from their predicted partial pressures divided by the computed pseudopressure. The vapor mole fraction of an acid is given by its monomer partial pressure plus its cross-dimer partial pressure plus two times its dimer partial pressure, all divided by the pseudopressure. The author has used the preceding approach with some success to model complex organic acid systems of industrial importance.

F. Gas Phase Properties of Hydrogen Fluoride

Hydrogen fluoride is a highly toxic chemical which is extremely corrosive in many environments, yet it has considerable importance in the chemical industry. It is used as a raw material, and even as a separating agent. For example, in combination with boron trifluoride it is used to separate and even to isomerize meta-xylene. However, like the aliphatic acids it has a very nonideal vapor phase.

There is strong evidence that hydrogen fluoride vapors exist as monomers and as hexamers. Evidently, the hexamer form is shaped like a benzene ring and has considerable thermodynamic stability. The equilibrium constant K for the reaction

$$6HF \leftrightarrow (HF)_6 \tag{63}$$

$$K = \frac{P_{(HF)6}}{P^6_{HF}} \tag{64}$$

is given by the following equation:

$$\log K = 8910/T - 43.65 \tag{65}$$

where temperature is in degrees Kelvin and the equilibrium constant has units of ($mmHg^{-5}$).[226]

The vapor pressure of the system includes the vapor pressure of the monomer plus the vapor pressure of the hexamer. At some temperatures, the partial pressure of the hexamer exceeds that of the monomer, because of the large amount of hexamer formation. There is even evidence that a large number of hexamers form in the liquid as well as the vapor phase. Details of the utilization of these data are given in Reference 226.

G. VLE of Formaldehyde + Water Systems

As amazing as it may seem, industry for the most part has failed in its attempts to simulate formaldehyde + water phase equilibrium. This cavalier attitude with respect to the basic phase equilibrium of the formaldehyde + water system is surprising in view of the commercial importance of formaldehyde. It is one of the largest-volume petrochemicals. Formaldehyde is very reactive, so it is involved in many types of chemistry. Virtually the entire plywood and particleboard industry uses formaldehyde-based adhesives. Formaldehyde is one of the raw materials of polyurethane foam, now used extensively for exterior insulating board on houses. Urea-formaldehyde foams are now obsolete because of the emissions problem. Formaldehyde emissions are important because of suspicions that the chemical is a carcinogen. Therefore, vapor partial pressures are technically important. Yet the vapor pressure per se is not usually important; formaldehyde is sold and used either in aqueous solution or as an oligomer called "paraformaldehyde". The VLE of the aqueous solution is of particular technical interest.

For these reasons, the first published workable solution to the VLE of formaldehyde + water mixtures is presented here. It can be extended as necessary to other components. The problem with formaldehyde is that it adds on to the formaldehyde + water complex to form polymers of varying molecular weight. Because of the effect of temperature on the equilibrium constants for these reactions, there is a very large effect of temperature on the VLE. At low pressures, water is more volatile than formaldehyde, in spite of the fact that the vapor pressure of pure formaldehyde is much higher than that of water. At high pressures, the volatility of small concentrations of formaldehyde exceeds that of water.

Having promised not to include any derivations in this book, the author will mail the derivation of the following equations to those who are particularly interested. In summary of what was done, equilibrium constants having a temperature dependence were developed for the reaction of formaldehyde with water, and of the resulting complex with another mole of formaldehyde. There are similarities here to a development by Brandani et al.;[78] however, their coefficients did not reproduce the data. The approach has been made simpler by eliminating use of the unsymmetric convention and ignoring the insignificant gas phase dimerization.

In this new approach, the partial pressure of formaldehyde is assumed to be contributed entirely by the formaldehyde monomer. By using the convergence properties of geometric series, the equilibrium constants beyond K_2 are not needed. The determined constants are

$$K_1 = \exp(-5.9017 + 4054.1/T)$$

$$K_2 = \exp(-5.6905 + 4363.6/T)$$

$$\text{fugacity of formaldehyde (atm)} = \exp(19.2201 - 2674.2/T) \tag{66}$$

The first equilibrium constant relates to the formation of a complex between formaldehyde and water, known as methylene glycol. The second equilibrium constant relates to the addition of another formaldehyde molecule to the methylene glycol. Longer chains also are developed by addition of more formaldehyde molecules, but two equilibrium constants and an empirically derived formaldehyde fugacity are sufficient to model the system. It also should be noted that in this formulation the physical activity coefficients are set equal to unity.

The utilization of this correlation is illustrated in Figure 9 up to 65 mol % formaldehyde for temperatures of 90 and 120°C. Figure 10 does the same for temperatures of 60, 97, and 130°C. Finally, Figure 11 compares data with prediction at a constant pressure of 3800 mmHg. It is evident from these figures that the data have been rather well correlated over a wide range of composition and temperature using just four parameters plus a fugacity expression for formaldehyde. For future reference it is of interest to plot the correlated

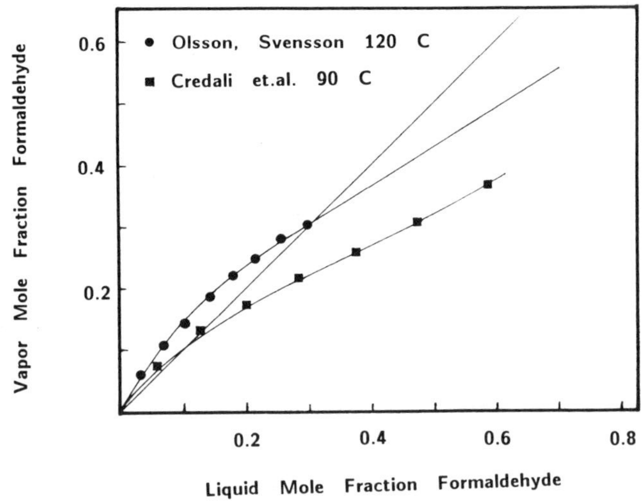

FIGURE 9. Correlation of the VLE of formaldehyde + water at 90 and 120°C.[79-83]

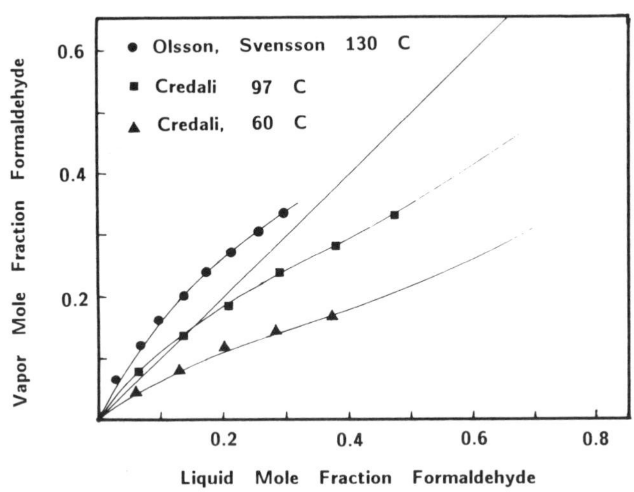

FIGURE 10. Correlation of the VLE of formaldehyde + water at 60, 97, and 130°C.[79-83]

azeotrope as a function of temperature in Figure 12 and pressure in Figure 13. Below 60°C the azeotrope disappears, permitting concentration of formaldehyde under vacuum.

H. Heat of Mixing Conversion to VLE

Various researchers have examined the possibility of predicting VLE from excess enthalpy (heat of mixing) data. This is an appealing prospect because these data can be obtained readily. Even high pressures, reacting systems, and high temperatures can be handled in flow calorimeters. Variation of the flow rates of the different components allows rapid generation of entire binary curves in the absence of a vapor space. The Thermochemical Institute at Brigham Young University has pioneered some of these techniques. Furthermore, considerable amounts of calorimetric data have been previously measured and are compiled in Reference 84. The large data base has made possible the development of group contribution

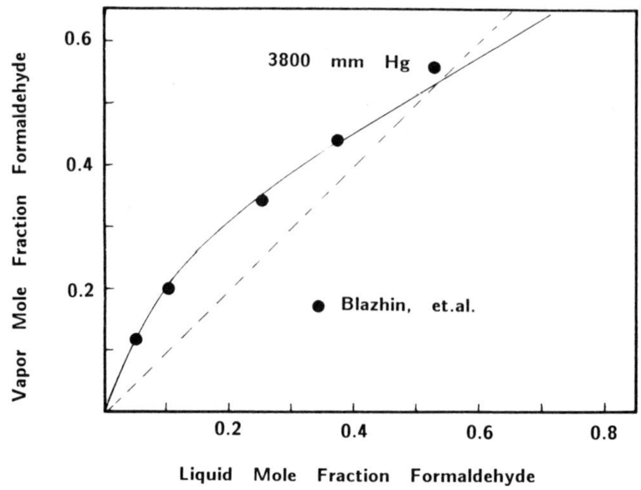

FIGURE 11. Prediction of the VLE of formaldehyde + water at 3800 mmHg pressure.[79-83]

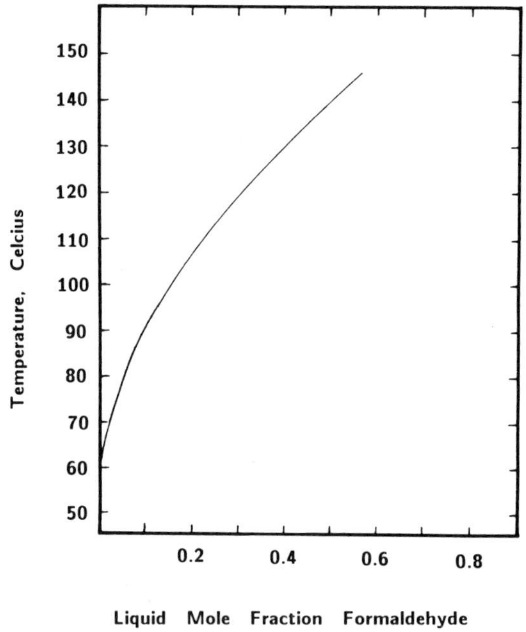

FIGURE 12. Prediction of the formaldehyde + water azeotrope composition as a function of temperature.[79-83]

methods for heats of mixing. One of these is based on the UNIFAC equation.[85] That prediction method appears to give the correct shape of the excess enthalpy curves as a function of composition, even for the asymmetric alcohol + hydrocarbon systems. Potentially, heat of mixing interaction parameters could be derived for functional groups not yet examined with accurate VLE data.

1. Empirical Activity Coefficient Equations

The excess enthalpy gives the effect of temperature on the excess Gibbs free energy. In

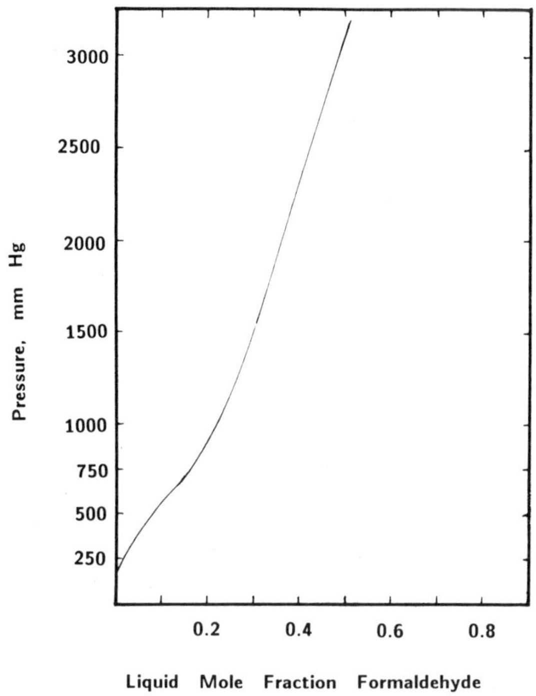

FIGURE 13. Prediction of the formaldehyde + water azeotrope composition as a function of pressure.[79-83]

other words, given a good set of VLE data at one temperature, a fairly safe extrapolation can be made to another temperature using the excess enthalpy. Attempts have been made, however, to by-pass the need for excess free energy at a base condition. This is done by using models of the Gibbs free energy and then differentiating, with the assumption that the parameters are not temperature-dependent. That assumption is not always valid. Hanks and Christensen[86] have shown that the Wilson and two-parameter NRTL equations are rather unreliable bases for such predictions. However, the LEMF version of the NRTL equation gave reasonable results.[55] In the LEMF equation, the value is set to -1. Interaction parameters are derived from heat of mixing data using the equation developed by Hanks and Christensen. Then the VLE calculations are made using the regular NRTL equation presented previously as Equation 42.

The UNIQUAC equation is an unreliable method for interconversion of heat of mixing and VLE, although it can be improved by changing the definition of the parameters so that the temperature in the denominator of the exponent is squared.[87]

Overall, however, this approach based on semiempirical equations is unreliable. Use of UNIFAC or analytical solutions of groups (ASOG) generally would be preferable. If an expression for excess free energy based on addition of excess enthalpy and entropy contributions were developed, it would be more reliable. However, unless the excess entropy term were predictive, even that approach would be unsuccessful.

2. Conformal Solution Theory

One approach that has some long-term potential, but has been insufficiently tested, involves conformal solution theory — the extension of the theorem of corresponding states to mixtures. Its derivations based on partition functions look formidable. Nevertheless, it is probably the most powerful method available for correlating excess properties over a wide temperature range. However, development of nonempirical mixing rules to handle polar and associating components has lagged. For that reason this method has not been accepted anywhere in industry for correlation of activity coefficients. It may have a role, however, in utilizing heat of mixing data.

Prof. B. D. Smith, together with several students at Washington University, speculated that it would be possible to predict excess free energy if the excess enthalpy plus the excess volume were measured at two temperatures. With complex computer programming they were in fact able to predict excess free energies accurately for several mixtures. An advantage of this method is that it does not even use a mixing rule, but computes excess free energy at each composition based on the enthalpy and volume data points at the same composition. Unfortunately, this promising approach has not been sufficiently tests. It would be possible to measure excess volume in the same experiment with excess enthalpy if its utility were demonstrated.

3. Recommendation on Prediction of VLE from Heat of Mixing

For the present, prediction of VLE from excess enthalpy is not recommended unless the group contribution parameters are not available and excess enthalpy data are available. Until more work based on fundamental approaches is undertaken, such predictions should be used only for scoping designs and viewed with great caution.

I. Henry's Constants

A gas can be defined as a chemical residing primarily in the vapor phase and being above its critical temperature at system conditions. With the system being above the critical temperature of a component, vapor pressures cannot be used unless the vapor pressure curve is extrapolated. That can be done successfully if the range is not too large. However, treatment of systems with extrapolated vapor pressures is severely limited when wide pressure ranges are encountered. Use of Henry's constants to establish the standard state is mathematically equivalent and has found wide acceptance.[89]

1. Unsymmetric Convention

Henry's constants, expressed in units of pressure, are commonly used to predict gas solubility. The equilibrium constant is essentially just the Henry's constant divided by total pressure. The lower the constant, the greater the solubility. The rigorous form of the equation is as follows:

$$K_1 = \frac{y_1}{x_1} = \frac{H_{1,2}}{P\,\hat{\phi}_{1,P}} \exp \int_{P_2^*}^{P} \frac{\overline{V}_1^\infty}{RT}\, dP \tag{67}$$

The activity coefficient in this expression is defined as being unity at infinite dilution. It can vary with composition. However, for gases with low solubility it can be ignored. Because activity coefficients for both the solute and solvent are unity at the same time, rather than at opposite ends of the composition range, this approach is called the "unsymmetric convention". The approach can be straightforward for single-component solvents. For mixed solvents the procedure can be handled mathematically but not necessarily correctly. Furthermore, it is seldom that experimental multicomponent gas solubility data are available.

2. Conversion Factors

Sometimes gas solubility experimental data are expressed as Ostwald coefficients. Henry's coefficients, in atmospheres per mole fraction, are obtained by dividing 22,414 by the Ostwald coefficient and by the molar volume of the liquid. Apparent contradictions in reported gas solubilities sometimes can be reconciled through careful analysis. Comprehensive examples, including the use of Bunsen coefficients, are given in References 90 and 91.

3. Correlation of Henry's Constant Temperature Dependence

The typical correlation of Henry's constants has historically been the logarithm of the Henry's constant vs. reciprocal temperature. For some, but not all mixtures, this procedure results in a linear correlation. It is obviously inadequate for the systems which display a maximum in the Henry's constant. An improved procedure was developed by Jonah.[92] He proposes plotting the logarithm of the Henry's constant, divided by the liquid phase fugacity of the solvent, vs. reciprocal absolute temperature. At low pressures the liquid fugacity coefficient can be approximated by the vapor pressure. It is modified at higher pressures by the fugacity coefficient computations using EOS. This procedure needs additional verification on a wide variety of systems, but appears to have some merit. The author found that it worked very well for correlation of data on solubility of oxygen in water.

4. Henry's Constants in Nonpolar Solvents

The solubility of gases in nonpolar solvents was the first problem to be theoretically treated. Two significant papers were by Preston and Prausnitz[93] and Gunn et al.[94] A Japanese paper[95] appeared to make another step forward, giving excellent reproduction of experimental data. The latter paper requires determination of two "characteristic critical properties". These serve as effective binary interaction parameters and are even temperature-dependent. However, the correlation has enough flexibility to handle variations in the Henry's constant including maxima and S-shaped curves with respect to temperature. This power may compensate for the need to determine characteristic parameters. Extension of this correlation to mixed solvents and polar solvents would be most helpful.

The correlation recommended at this time is the perturbed-hard-chain (PHC) EOS.[96,97] As originally developed, it could handle a wide range of components from polymers down to gases. A computer program is also given by Donohue.[98] Perhaps an indication of the power of this EOS is the fact that only one binary interaction parameter is needed, and that parameter is not even dependent on temperature! The interaction parameter is a linear function of carbon number of the solvent up to about 10 carbons. Then it flattens out and is constant beyond about 15 carbon atoms. This approach is fairly general, but more work is needed on application to mixed and polar solvents. It is being carried on by Donohue.

5. Tassios Method

Tassios (see Kung et al.[99]) has used the simple approach of dividing the Henry's constant by a hypothetical liquid phase fugacity (essentially an extrapolated vapor pressure) to derive an activity coefficient that can be used in the symmetric convention. That value then can be used to calculate the parameters for a one-parameter Wilson equation. For 40 binary solvent systems, this method is reasonably good, with an average error of 5.9%. It was claimed to be somewhat better than methods by Prausnitz and O'Connell. The method is unreliable, however, for aqueous systems. For pure solvents Antunes and Tassios[100] also have published a somewhat limited but still accurate correlation based on Tassios' modification of UNIFAC. It involves oxygen, nitrogen, and methane in water or in alkanes. Some amplification of that work will be necessary to develop a general tool, and it must be integrated somehow into the work on binary mixtures.

6. Gas Solubilities in Water

Aqueous systems are frequently encountered, but the molecular nature of water makes prediction of gas solubilities difficult. Fortunately, "scaled particle theory" has come to the rescue.[101] The Henry's constants in water are highly nonlinear, and in fact go through maxima with respect to temperature. The semitheoretical approach presents a reasonable means of carrying the predictions to high temperature (300°C). Only nonreactive gases can be treated in this way. For example, the predicted solubility of carbon monoxide is lower

Table 8
HENRY'S CONSTANTS, IN BARS, FOR COMMON GASES IN WATER

t (C)	CO_2	H_2	CH_4	N_2	O_2
0	1,190	58,000	23,900	59,700	32,100
10	1,520	62,700	30,400	70,900	38,300
20	1,880	66,600	37,100	81,400	44,200
25	2,230	68,200	40,400	86,200	47,000
50	3,080	73,600	56,000	106,000	59,000
75	4,090	74,800	67,700	118,000	66,700
100	4,990	72,700	73,900	120,000	69,500
125	5,700	68,100	74,900	115,000	68,400
150	6,170	61,900	71,100	105,000	63,900
175	6,400	54,700	64,300	91,600	57,300
200	6,380	47,300	55,600	77,200	49,600
225	6,150	39,800	46,200	62,800	41,500
250	5,740	32,700	36,900	49,300	33,600
275	5,170	26,000	28,100	37,200	26,200
300	4,480	19,800	20,400	26,600	19,400

Note: See Schulze, G. and Prausnitz, J. M., *Ind. Eng. Chem. Fundam.*, 20, 175, 1981 for additional constants relating to four noble gases.

than observed solubility in the higher temperature range; this is due to chemical reaction between carbon monoxide and water. Table 8 gives useful values of Henry's constants, in units of bars, for the more common gases. The original reference[101] also includes values for argon, helium, neon, and xenon.

7. Group Contribution Methods

A UNIFAC EOS for gas solubility prediction has been developed by Danish researchers.[102] They conclude that as a general rule, predicted gas solubilities will not be as accurate for mixed as for pure solvents.

A group contribution EOS useful for predicting gas solubilities was developed by Skjold-Jørgensen,[103] with local composition mixing rules. It is relatively complicated, but is presented with extensive tables of parameters for 23 different groups. These are limited to hydrocarbons, alcohols, ketones, and common gases. For simple hydrocarbon systems, the ordinary Soave equation does almost equally well, even when the interaction parameters are all set to zero. However, the new method does rather well when applied to solvent mixtures and pure components. Not surprisingly, when the mixtures contain a large amount of water, the predicted gas solubilities are poor. If this approach were extended to other groups and given capabilities for handling water, it could have considerable utility in handling polar compounds.

8. Mixed Solvents

Prediction of gas solubility in mixtures of polar solvents is particularly difficult. Methods are still evolving which require comparison and further testing. These include methods developed by Brandani and Prausnitz[104] as well as by O'Connell.[105]

9. Data Sources for Gas Solubilities

Most prediction methods encounter serious problems with water as the solvent. Fortunately, however, there is a significant amount of experimental data on solubilities in water. In addition to previously mentioned bibliographies by Hala et al., Reference 106 gives a considerable amount of information; 61 gases are examined up to 100°C. Only data at relatively low pressure were considered; however, the review does give a critical selection of available data, and references to the original sources are included. Gases which react strongly with water, such as hydrogen chloride and sulfur trioxide, were not included. An earlier review giving data in nonaqueous solvents was presented by the same authors.[107] Data reviews by Himmelblau are given in References 108 and 109. Three other papers give the solubility of gases in heavy hydrocarbons.[110-112]

10. Problems near the Critical Point

As the solvent critical point is approached, most of the preceding methods experience deteriorating accuracy. In fact, the data base is sparse at high temperature. Fortunately, it has been observed that all gas solubilities in a solvent tend toward a common value as the solvent critical temperature is approached.[113] The numerical value of the Henry's constant at the critical point is just the critical pressure of the solvent times the fugacity coefficient of the solute.[114] The fugacity coefficient can be computed with reasonable accuracy using an EOS. This simple procedure can be of considerable assistance, since the slope of the Henry's constant changes rapidly with temperature in the critical region. Thus, it is very helpful to know the end point.

11. Virial Coefficients from Henry's Constants

The small solubility of light gas in a heavy-hydrocarbon liquid is conceptually and theoretically similar to the solubility of small amounts of the hydrocarbon in the gas phase. An interesting theoretical development, PHC theory, has in fact linked Henry's constants with second virial coefficients.[115] Some experimental data are needed *a priori* to establish the single unknown equation parameter. The equation appears to be reasonably powerful in extending a single point with respect to temperature. The interaction parameter derived gives a sensitive prediction of cross second virial coefficients, but Henry's constants cannot be predicted accurately based on virial coefficients. Six characteristic parameters are needed for each pure component; they are given in Reference 115 for 44 chemicals, including some gases and hydrocarbon solvents.

V. EQUATIONS OF STATE

EOS can perform a myriad of useful tasks. They can give the compressibility factors for gases. Through various thermodynamic relations they can be used to calculate vapor pressure, heat of vaporization, fugacity, enthalpy, entropy, and free energy. All of the departures from ideality in the thermodynamic properties can be computed. Furthermore, mixture properties can be obtained — even phase equilibria. EOS can be used to predict the critical locus and critical points. Simultaneous chemical and phase equilibria problems also can be solved. Many problems in thermodynamics would be trivial if EOS accurately reflected nature for all molecules and all temperature and pressure conditions. Unfortunately, that "holy grail" of thermodynamics has not yet been found. Nevertheless, there is a consensus among thermodynamics experts that EOS will overtake activity coefficients as the method of choice for prediction of phase equilibria. Therefore, some centers of research are developing new EOS at a prodigious pace.

The EOS approach is particularly helpful as the pressure of operations increases. There is a trend toward operation of some distillation towers at higher pressures. This permits use of the heat of condensation in a high-pressure tower to reboil another tower which is operated at lower pressure. Further, it is necessary to operate some reactors and recovery equipment at conditions near the critical temperature of one or more components. Prediction of the phase equilibrium and enthalpy of liquid and vapor phases becomes vital to design of the reactor and auxiliary equipment.

The emphasis in this section is on the EOS being used in industry, with indications of their applicability and probable error. The equations discussed are the BWR and BWRS, Lee-Kesler-Plocker, Redlich-Kwong (and Zudkevitch-Joffe modification), Soave, Soave-Mathias, Peng-Robinson, NBS-Boulder, and equations using Gibbs free energy mixing rules. Applications are described for prediction of multiple liquid phases and solid phases, including gas hydrates. Their use for thermodynamic description of synfuels and petroleum properties is described in a later section of this chapter.

A. Benedict-Webb-Rubin-Starling (BWRS)

The virial EOS cannot be used for calculation of phase equilibria because it does not converge properly for liquids. However, an empirical extension of the virial EOS by Benedict, Webb, and Rubin did permit successful computation of phase equilibria.[124,125] It has eight parameters, empirically derived from liquid fugacity and volumetric data. The most accepted version of the equation is an extension by Starling[126] and Starling and Powers.[127] The new equation, presented here, has 11 parameters.

$$P = RT\rho + \left(B_0RT - A_0 - \frac{C_0}{T^2} + \frac{D_0}{T^3} - \frac{E_0}{T^4}\right)\rho^2 + \left(bRT - a - \frac{d}{T}\right)\rho^3 + \alpha\left(a + \frac{d}{T}\right)\rho^6 + \frac{c}{T^2}\rho^3(1 + \gamma\rho^2)\exp(-\gamma\rho^2) \qquad (68)$$

Values for the parameters are presented for 11 substances in Starling's book. The parameters were determined from a data base of 16,000 density, vapor pressure, and enthalpy data points, using multiproperty regression. The data ranged in temperature from −340 to +720°F, and pressure ranged up to 700 atm. Starling also includes mixing rules, binary interaction parameters, and generalized mixing rules. The average deviations between the BWRS values and the experimental data used for parameter development were

Densities	1.07%
Enthalpies	1.49 Btu/lb
Vapor pressures	1.35%
K values	6.1%

Application of the BWRS equation to the natural gas processing industry, for which it was originally intended, has been very successful. However, it receives little attention in the current scientific literature; in fact, it was not even mentioned in a recent book reviewing EOS.[128] This may be more a reflection of the equation's empirical nature, however, than of the extent of its use. Cryogenic systems are particularly well handled by the BWRS equation.

It is not surprising that significant efforts have been made to extend the applicability of this equation. Parameters for other hydrocarbons, including butadiene and a number of acetylenes, have been compiled, and are conveniently presented in Table E5 of Reference 43. In Japan, the BWR equation has been extended to C_{20} paraffins and their mixtures.[129] Other work has been done by oil and gas production companies in the U.S. to expand the applicability of the BWR and BWRS equations. Their interest has been in the variety of heavy hydrocarbons typically encountered during carbon dioxide flooding of old oil fields.

Polar substances, including water, have been represented by the BWR equation with addition of three polar parameters. Data are given for 92 substances.[130] Independent confirmation of the validity of the equation in handling phase equilibria involving polar substances will be required before this extension can be given a blanket recommendation.

B. Lee-Kesler-Plocker

An equation that is analogous to the BWRS was developed by Lee and Kesler[131] and modified by Plocker et al.[132] The equation is a corresponding-states type, using a reference volume. It has made obsolete the tables developed earlier by Pitzer. In comparison with the BWRS equation, the Plocker modification worked much better for both VLE correlation and enthalpy predictions. This would not be surprising, given its stronger theoretical basis.

This equation is applicable for hydrocarbon mixtures, with a few minor extensions. In other words, its applicability is just as constrained as that of the BWRS equation. However, it has developed an informal reputation in some circles for giving superior enthalpy predictions; therefore, it finds use in design of both gas processing and ethylene plants. The equations and coefficients can be found in Reference 132, and it is reported that a computer program may be available from the authors. To reduce the equation dependence on critical properties, which are often unmeasured, a new parameter has been proposed by Kesler et al.[133] It is a new type of acentric factor, evaluated at 85% of the absolute bp temperature. Although preliminary tests suggest it would work well, its superiority over other methods requires additional confirmation.

C. Redlich-Kwong

Repetitive flash calculations are required in many distillation column process simulators. This is particularly true when the phase equilibrium K values are recomputed for every state. Empirical equations with many parameters, such as the BWRS, become relatively inefficient in such circumstances, yet simpler equations such as those of van der Waals and Berthelot did not have sufficient accuracy. The Redlich-Kwong EOS,[134] published in 1949, became a significant milestone in thermodynamics research. It has enjoyed considerable success because of its ability to reproduce accurately second virial coefficient curves of simple fluids at low-pressure conditions. It also provides reasonable predictions of the high-pressure (supercritical) region for simple fluids.[135]

This equation is classified as a "cubic" equation because it is explicit in pressure and is third-order in volume. Such an equation must be at least cubic to be able to satisfy the requirement that the ideal gas condition be predicted as the volume goes to infinity. Also, it must represent vapor and liquid volumes accurately. Finally, it must satisfy the condition that the first and second derivatives of pressure with respect to volume both must equal zero at the critical point. Below the pure component or mixture critical point, there are three roots to the equation. The largest gives the vapor volume, the smallest gives the liquid volume, and the intermediate root has no physical significance. Above the critical point there is only one root.

The Redlich-Kwong equation is simply

$$P = \frac{RT}{V - b} - \frac{a}{\sqrt{T}\, V(V + b)} \tag{69}$$

where $a = 0.42748\ R^2T_c^{2.5}/P_c$ and where $b = 0.08664\ RT_c/P_c$. The compressibility is given by

$$Z = \frac{V}{V - b} - \frac{a}{RT^{1.5}\,(V + b)} \tag{70}$$

The fugacity coefficients, needed in order to calculate K values (ratio of the liquid to vapor fugacity coefficients) are computed from

$$\ln \hat{\phi}_i = \frac{b_i}{b}(Z - 1) - \ln\left[Z\left(1 - \frac{b}{V}\right)\right] + \frac{1}{bRT^{1.5}}\left[\frac{ab_i}{b} - 2\sqrt{a\,a_i}\right]\ln\left(1 + \frac{b}{V}\right) \quad (71)$$

The original mixing rules are as follows:

$$a_{mixture} = \sum_i \sum_j y_i y_j a_{ij} \quad (72)$$

$$a_{ij} = (a_i a_j)^{1/2} \quad (73)$$

$$b_{mixture} = \sum_i y_i b_i \quad (74)$$

The simplicity of these mixing rules illustrates one of the reasons for the popularity of cubic EOS. They have an energy parameter and a size parameter to represent pure component properties. They are, as has been shown, temperature-dependent. The minimal number of pure component parameters results in a minimal number of mixing rules and thus simplifies the problem of computing phase equilibria and other mixture properties.

Roughly 200 papers have been published on the use of the Redlich-Kwong EOS. Many of those have proposed modifications. Often, the original equation continues to do the best job of predicting fugacity coefficients. Therefore, the original remains a respectable equation for computation of the gas phase correction in the activity coefficient approach to phase equilibria. However, significant improvements have been discovered for the EOS approach to phase equilibria. Many of these relate to the method in which the parameters are derived.

The Redlich-Kwong equation incurs greater error as molecules become larger. The reason is that it contains only two corresponding-states parameters, T_c and P_c. Recognizing this problem, Wilson[138,139] introduced the acentric factor into the expression for one of the parameters. Although Wilson's equation had the wrong form for the region above the critical point, his basic idea was good. It was exploited by Soave, as will be described shortly.

D. Zudkevitch-Joffe Modification of the Redlich-Kwong

One of the better methods, developed by Zudkevitch and Joffe,[211] fits the parameters simultaneously to the liquid phase fugacity (corrected vapor pressure) and liquid molar volume. This is particularly helpful when dealing with heavy hydrocarbons for which critical properties are unknown. The equation has been used extensively by the Exxon Corp. with some surprisingly good results. As reported by Gray,[136] excellent results were obtained for hydrogen + hydrocarbon systems when the Chueh-Prausnitz method[137] was employed for treatment of hydrogen. The carbon dioxide + methane system, which is quite nonideal, was fitted well; K value errors averaged less than 3%. Finally, it is possible to "fudge" the parameters to obtain good fits of the solubilities in both phases for hydrocarbon + water liquid-liquid systems.

One of the problems with use of EOS is accurate prediction of dew points. This is particularly troublesome in oil field and petroleum processing applications. For example, the decision to invest tens of millions of dollars in carbon dioxide flooding of an oil field can be influenced by thermodynamic predictions of the expected recovery. In an extensive report on work at the Amoco Production Co., Yarborough[140] disclosed a fairly extensive modification to the Redlich-Kwong equation. As a preliminary to his paper, he indicated that the following requirements had to be satisfied: "Therefore, a phase-equilibrium prediction method generally applicable to petroleum reservoir fluids must be capable of being used with supercritical components including the nonhydrocarbons, nitrogen, CO_2, and H_2S, in addition to predicting the effect of the different molecular types of hydrocarbons on phase equilibria. The prediction method also must provide reasonable results in the retrograde, near critical, and critical regions.''

It was discovered that interaction parameters in the Zudkevitch-Joffe mixing rule could be set to zero for all hydrocarbon/hydrocarbon mixtures. However, mixtures with nonhydrocarbons required use of nonzero interaction parameters. Yarborough's tests represent a fairly extensive evaluation against a body of experimental data. Some of the data were unpublished at the time. Figure 14 illustrates the results obtained for a multicomponent mixture. Overall, the results obtained can be summarized as follows:

Component	Number of points	Avg % deviation in K value
Methane	434	5.57
Ethane	434	4.14
Propane	434	4.89
n-Pentane	434	7.94
n-Heptane	430	11.94
n-Decane	427	26.51
Toluene	277	14.87
Nitrogen	185	14.94
Carbon dioxide	160	6.75
Hydrogen sulfide	257	7.23

The greatest errors are for the heaviest components, where the greatest experimental error seems to be evident. These results are also useful to give some idea of probable errors that can be expected when making computations on similar mixtures. Yarborough discovered that through proper use of chromatographic analysis and density estimates it is possible to represent accurately the properties of fluids at least up to a carbon number of 32! Figure 15 represents the comparison of experimental data vs. prediction for a mixture containing molecules up to C_{32}. This figure shows the volume percent liquid as a function of pressure. Such predictions are very sensitive to the presence of small amounts of heavy materials, so the prediction is really quite good.

E. Soave

The Soave equation,[141-143] one of the more significant modifications of the Redlich-Kwong EOS, is as follows:

$$P = \frac{RT}{V - b} - \frac{a\alpha}{V(V + b)} \tag{75}$$

where $a = 0.42747\ R^2T_c^2/P_c$, $b = 0.8664\ RT_c/P_c$, and $\alpha = (1 + (0.48508 + 1.55171\omega - 0.15613\ \omega^2)\ (1 - T_r^{0.5}))^2$.

The Soave EOS is recommended in the API *Technical Databook — Petroleum Refining*. In developing the computer version of this equation, Graboski and Daubert[144] developed procedures for determining the parameters. They also extended its usefulness to include hydrogen, with the following term:

$$\alpha_{H_2} = 1.202\ \exp(-0.30228\ T_r)$$

A computer program giving phase equilibrium predictions by a modified Soave equation is sold by the Gas Processors Association (GPA).[146] The Soave and other modifications of the Redlich-Kwong equation are computationally efficient. By contrast with the more sophisticated local composition PHC (LC-PHC) EOS,[147] the Soave equation uses much less

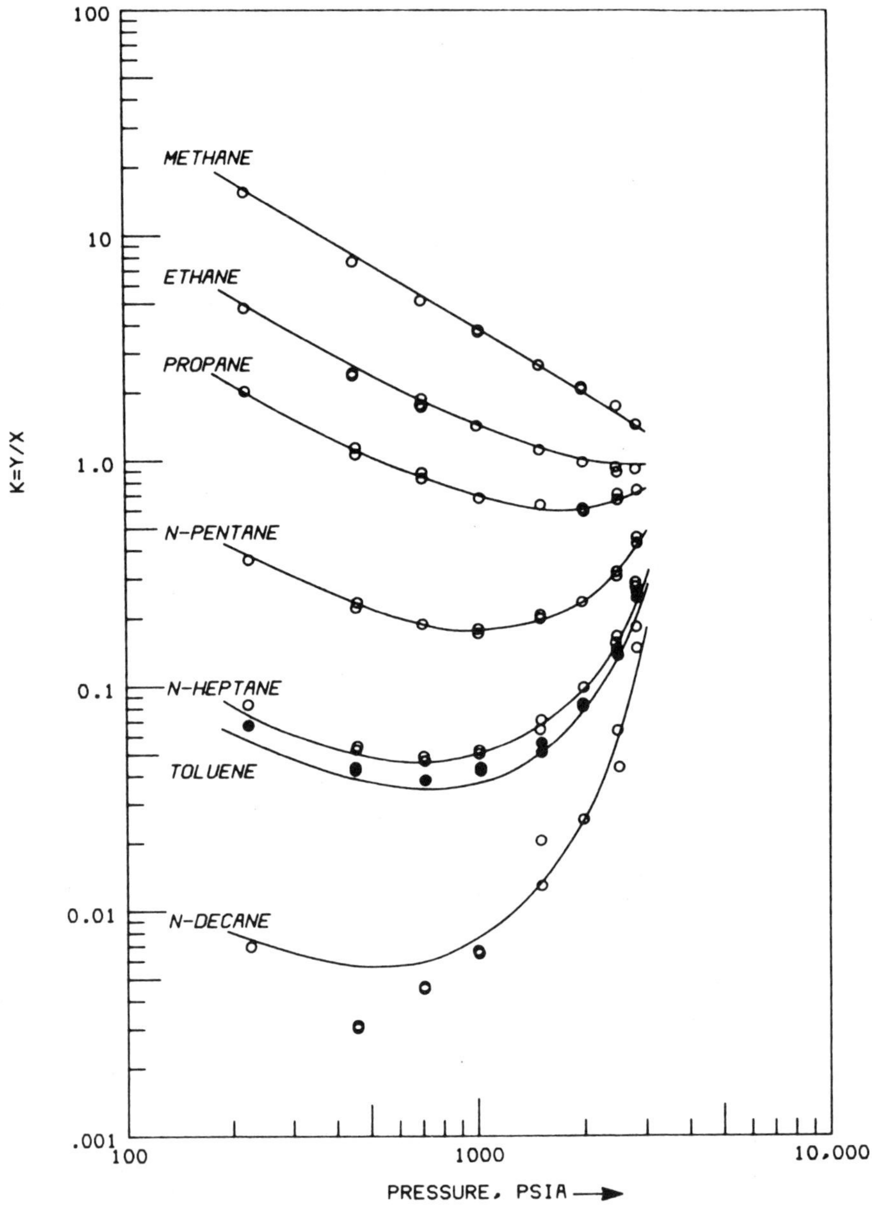

FIGURE 14. Prediction of K values for a multicomponent hydrocarbon system using the Zudkevitch-Joffe modification of the Redlich-Kwong EOS. Predictions by L. Yarborough of Amoco Production Co. (Reprinted from Yarborough, L., in *Equations of State in Engineering and Research,* Advances in Chemistry Series, NO. 182, Chao, K. C. and Robinson, R. L., Jr., Eds., American Chemical Society, Washington, D.C., 1979, 401. With permission.)

computer time.[148] The following relative amounts of cpu time were consumed by the LC-PHC equation in making predictions, relative to a value of 1.0 for the Soave equation:

	2 components	**8 components**
Fugacity coefficients	4.8	16.0
Partial derivatives of fugacity coefficients wrt composition	4.0	11.5

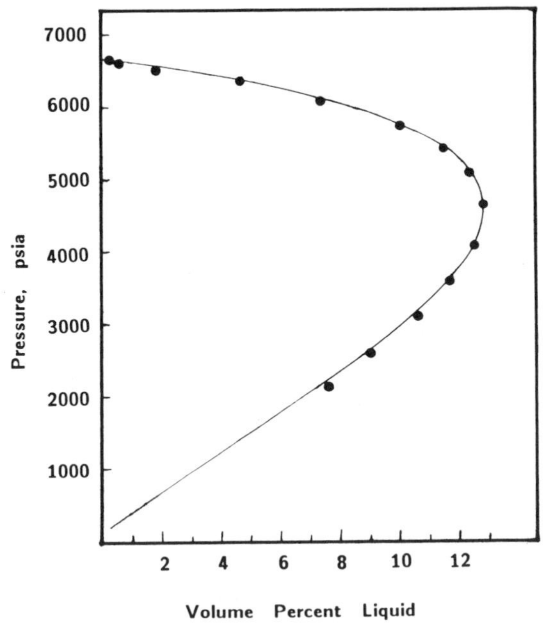

FIGURE 15. Comparison of measured and predicted phase distribution for a multicomponent hydrocarbon mixture containing 7 to 31 carbon atoms per molecule. Temperature = 335°F. Data and correlation work using Zudkevitch-Joffe equation by L. Yarborough of Amoco Production Co. (Reprinted from Yarborough, L., in *Equations of State in Engineering and Research,* Chao, K. C. and Robinson, R. L., Jr., Eds., Advances in Chemistry Series, No. 182, American Chemical Society, Washington, D.C., 1979, 401. With permission.)

The conclusion drawn from this information is not particularly new; cubic EOS are computationally efficient. Since they are also quite powerful, they are used very extensively in industry.

It was shown[144] that an empirical binary interaction parameter is required for all mixtures containing other than hydrocarbons. Substantial improvements were also made for hydrogen, an important chemical in refinery operations. Nevertheless, for seven ternary hydrocarbon systems, the average errors for the three K values were 13.7, 13.7, and 28.0%. These errors are not usually acceptable in many industrial applications. The phase equilibrium errors can result in even larger material balance errors.

Huron et al.[145] have demonstrated that with use of appropriate calculation methods it is possible to calculate the critical locus, avoiding trivial solutions. They agree that no binary interaction parameter is needed for hydrocarbon + hydrocarbon mixtures. However, with polar compounds such a parameter (K_{12}) is required. This can be obtained from VLE data. Unfortunately, no correlations have been found which relate K_{12} to molecular parameters; it remains a totally empirical parameter.

Unpublished work within Amoco Chemicals Corp. has shown even more distressing results. For systems with polar components, the errors are significant. The best analogies are the very poor predictions of VLE for polar systems that are made with solubility parameters. As previously noted, the Redlich-Kwong and Soave equations were developed for hydrocarbon systems. They give a very poor fit of the vapor pressure of polar materials, so there is little chance of predicting accurately the VLE of mixtures. This has motivated development of a Soave equation specially modified to handle polar compounds.

F. Soave-Mathias

Recognizing the poor applicability of the Soave equation to polar compounds, Mathias made a minor modification to the α-parameter in that equation. He redefined it as follows:

$$\alpha_i^{0.5} = 1 + m_i(1 - T_{r_i}^{0.5}) - p_i(1 - T_{r_i})(0.7 - T_{r_i}) \tag{76}$$

where $m_i = 0.48508 + 1.55191\ \omega_i - 0.15613\ \omega_i^2$. The first two terms in this equation are the same as in Soave's equation. The next term introduces a totally empirical polar parameter p. Although not fully demonstrated in the paper by Mathias,[217] introduction of that additional parameter for polar compounds makes a tremendous difference. It becomes possible to represent the vapor pressures of virtually all compounds from low pressure to the critical point within a small percentage error. By contrast, the Soave equation by itself does a terrible job of predicting the bp of water.

Enthusiasm for the Mathias modification would not be so great, were it not for the fact that it seems to work remarkably well. Even the mixture predictions are tremendously enhanced, relative to the Redlich-Kwong, Soave, or the Peng-Robinson equation which will be discussed subsequently. This is all the more remarkable since the mixing rules used are essentially the same as those used by Soave and others. The extra polar parameter is obtained by fitting the vapor pressure data on the pure compounds. For nonpolar compounds the polar parameter is zero and the equation is identical to that proposed by Soave.

G. Peng-Robinson

The Peng-Robinson equation[149-154] was developed to improve on the liquid density predictions of the Soave equation. In addition to that accomplishment, it gave better results in the critical region. The equation is as follows:

$$P = \frac{RT}{V - b} - \frac{a\alpha}{V^2 + 2bV - b^2}$$

Parameters:

$$a = 0.45724R^2T_c^2/P_c$$

$$b = 0.07780RT_c/P_c$$

$$\alpha = [1 + (0.37464 + 1.54226\omega - 0.26992\omega^2)(1 - T_r^{0.5})]^2 \tag{77}$$

A computer program developed by the authors of the equation is available through the GPA.[155]

The Peng-Robinson equation has been tested rather thoroughly on a wide variety of systems. It is used very extensively in the natural gas industry. Provided that the binary interaction parameter is fitted to the data, it can handle polar components such as hydrogen sulfide and water. In a survey by E. I. du Pont de Nemours & Co. of compressibility factors on six systems, the Peng-Robinson equation was slightly better than the Soave. The Lee Kesler equation gave somewhat erratic results. ''The modified Lee-Kesler equation did not always converge when applied near the critical region...The Lee-Kesler equations are extremely complex. This is important in process calculations where many iterations are usually required.'' The Peng-Robinson equation ''...seems preferable. It is simple, gives acceptable results, and does not give convergence problems when properly used in computer programs.''[156] The average deviation in compressibility factor ranged from 1.4 to 3.7%. An exception was pure hydrogen chloride, where the average deviation was 10%.

One of the significant questions faced when using these cubic EOS is whether the binary interaction parameter can be set equal to zero. One study has given some useful conclusions based on methane + higher hydrocarbon VLE data — namely, that with the interaction parameter set to zero, the Soave equation gave slightly better results than the Peng-Robinson. The equations had similar accuracy when the interaction parameters were adjusted. The temperature dependence of the interaction parameters was similar for both equations.[157] It would seem unwise to generalize the conclusions of this study to other types of systems without doing the computations.

Improvements in liquid density were in fact achieved by Peng and Robinson relative to the Soave EOS. Nevertheless, some liquid density predictions are still not very accurate. For example, it has been shown that densities of some heavy hydrocarbons are in error by 15 to 19%. The density of water predicted by Peng-Robinson is in error by an average absolute amount of 26%. For methanol the predicted density error is 20%, and for ammonia it is 16%. A very substantial improvement in capability for prediction of liquid densities is provided by the Patel-Teja EOS,[158] which has approximately the same capabilities for phase equilibrium prediction.

H. NBS-Boulder

One of the problems of "analytic" EOS is the nonanalytic nature of the critical region. The problem is that isotherms on the PVT surface become nearly flat as the critical point is approached. Various methods have been developed to account for this, one of which has been described in an unpublished NBS paper.[159] There are other unpublished papers and informal communications which describe the EOS developed at the NBS in affiliation with other researchers. The reader might conclude that this effort was unduly theoretical. However, the equation actually was developed as part of an effort supported cooperatively by a number of process companies. Preliminary tests of the EOS suggest that it has very attractive properties. There is one interaction parameter which must be fitted to data whenever polar compounds are involved. Unfortunately, it is also temperature-dependent. However, the equation itself appears quite powerful. With time, more will be available on this development. The computer program, developed specifically for supercritical extraction problems, is available from Dr. J. F. Ely, [(303) 497-5467].

I. Multiple-Phase Computations

One of the advantages of EOS is their theoretical capability of predicting a variety of different phase diagrams. In practice, the critical region and phase equilibria other than VLE can be difficult to predict; convergence problems and multiple roots of the cubic equations interfere. This problem has been addressed in detail by Michelsen.[160] Topliss[148] pursued Michelsen's ideas further in his dissertation. That work formalized the methods for developing and testing needed derivatives in EOS. A computer program is available from Dr. R. Topliss, c/o Amoco Chemicals Corp., P.O. Box 400, Naperville, IL. 60566.

Reproduced here, with permission, are two of the predictions made using the formalism developed by Topliss. In Figure 16 the methane + water equilibrium is illustrated as a function of pressure. Not only is the vapor-liquid envelope successfully represented, but a gas-gas equilibrium at high pressures is predicted. There are no data to verify the shape of that envelope. In Figure 17 there is a successful prediction of the benzene + water LLVE as a function of temperature.

Under circumstances of high pressure, particularly when the system is near the critical point of one or more components, it is necessary to use an EOS. This is particularly true near the three-phase critical end point for hydrocarbon water mixtures. The Peng-Robinson equation has been used with some success in that application.[161] However, experimental data are necessary to obtain the interaction parameter. Another paper[162] gives the techniques for successful application of the Peng-Robinson equation to LLVE. The examples are taken from typical reservoir-oil-carbon dioxide mixtures.

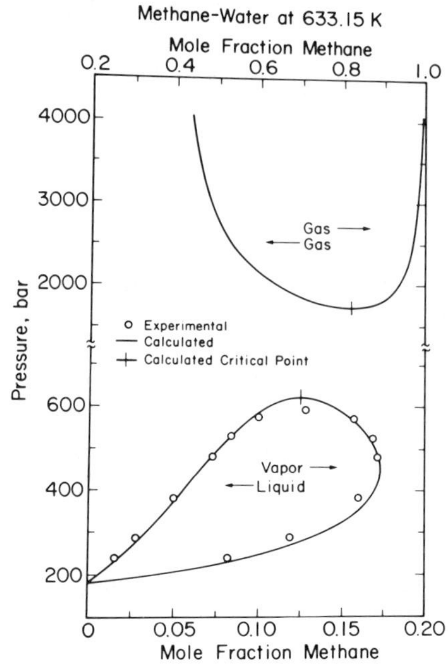

FIGURE 16. Prediction of methane + water phase equilibria at 360°C. Note the excellent reproduction of VLE. Of added interest is the range of predicted gas-gas equilibria above 2000 bar. (Reproduced by permission of Dr. R. Topliss.)

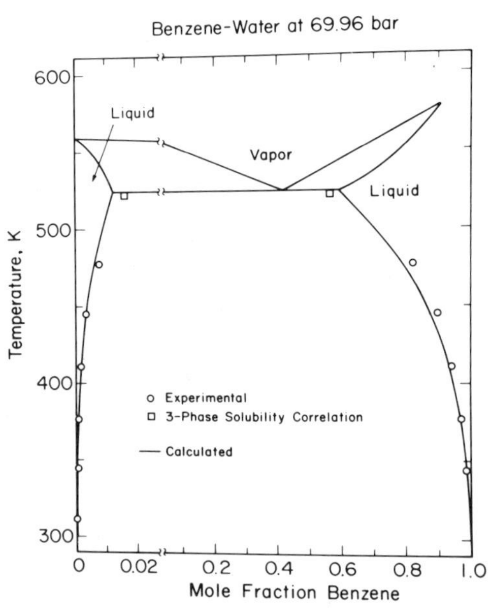

FIGURE 17. Prediction of benzene + water phase equilibria at 70 bar pressure. (Reproduced by permission of Dr. R. Topliss.)

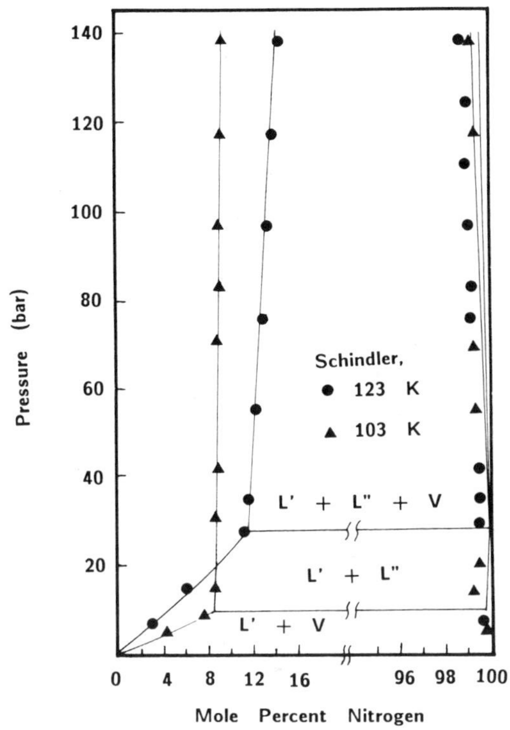

FIGURE 18. Prediction of LLVE for nitrogen + propane with the Peng-Robinson equation and fitted binary interaction parameter. (Adapted from Kremer, H. and Knapp, H., *Hydrocarbon Process.*, April, 79, 1983.)

Two liquid phases can form between light hydrocarbons and either carbon monoxide or nitrogen. This can occur at conditions where the gases are normally separated by cryogenic distillation. Therefore, prediction of the conditions at which this occurs can be of considerable practical importance. Reference 163 illustrates the use of various EOS for generation of such phase envelopes. Figure 18 illustrates the prediction of LLVE for nitrogen + propane as a function of pressure and composition. Two different isotherms are shown, together with the predictions by the Peng-Robinson equation. A single interaction parameter with no temperature dependence was employed.

J. EOS for Polar Fluids

In the previous discussion of EOS it has been inferred that the BWR and BWRS equations are valid for mixtures exhibiting no liquid phase nonideality. Introduction of a binary interaction parameter for the Redlich-Kwong and similar cubic equations allows predictions which even account for formation of two or three liquid phases. However, those equations are not really suited to handling polar components. They do not have the functional form necessary to represent accurately the sometimes complex molecular interactions that occur between polar and associating molecules. Therefore, specialized equations have been developed for polar fluids.

1. The Won Equation

A modified Soave equation of state was developed by Won (see Won and Walker[164]) to handle polar compounds. The reader should refer to the original paper for the complex equations and discussion of data reduction procedures. The result of that work was only modestly encouraging. The equation was able to reproduce fugacity coefficients and enthalpies of water. Limited success was also obtained in fitting water solubilities in hydrocarbons. Substantial improvements over the Soave equation were obtained in predicting K values for the methanol + *n*-hexane system. Nevertheless, the values in the infinite dilution regime were still not handled well. Some improvements over the Soave equation were found in prediction of the enthalpy departure for pure, gaseous methanol. The authors summarize with this statement: "This equation of state, however, is not recommended for liquid mixtures rich in the polar component." It is suspected that new equation forms are required to handle such systems.

2. Cubic Chain-of-Rotators (CCOR) EOS

It has been shown that the repulsion term in most cubic EOS is in error. That term, RT/(V − b), is inconsistent with the molecular modeling results for hard spheres, believed to be correct.[165] An improved expression, developed by Chao (see Kim et al.[166]) accurately represents the results of that modeling. The derived term is

$$RT\,(1 + 0.77\,b/V)/(V - 0.42b)$$

It is obviously substantially different, but the overall equation gives outstanding results. Pure component vapor pressures can be represented with a high degree of accuracy. At low pressures, this equation rivals the Wilson activity coefficient equation. At higher pressures where activity coefficients cannot be used effectively, it represents experimental data very well. The original paper should be consulted for the full equation and information on how to find all the parameters. For each binary mixture, two interaction parameters are used. Constant values work well for nonpolar mixtures; for polar mixtures they are temperature-dependent.

At this writing, the CCOR equation had not been widely, if at all, implemented in industry. However, it has considerable promise in view of its ability to represent highly nonideal polar mixtures with two parameters. Further testing and distribution of computer code will help to promote its use.

3. Improved EOS Mixing Rules

In the region near the critical point it is difficult to make accurate predictions of mixture equilibria. This can be of serious concern in design of supercritical fluid flooding or extraction processes. After evaluation of the classical mixing rules used in the Peng-Robinson equation, Mansoori and Kwak[167] concluded that those rules were theoretically incorrect. New mixing rules were proposed, based on statistical mechanical considerations. More work will be needed in testing these new rules before general use can be recommended. It should be noted that the new mixing rules introduce a third unknown but nonzero parameter. This appears to give the advantage of improved phase equilibrium predictions over wide ranges of pressure. Certainly, use of theoretically derived mixing rules is preferable to the semiempirical approach used for so many years. The rules developed by Mansoori have the advantage of applicability to other EOS as well as the Peng-Robinson.

Another approach to the mixing rules was developed by Vidal[168] and Huron and Vidal,[169] who developed a local composition mixing rule. They have pointed out that the standard mixing rule fails for polar systems. As an example they note the acetone + water binary system which forms an azeotrope but has only one liquid phase. When experimental VLE data are fitted with the Soave or Peng-Robinson equation, the oversimplified classical mixing rules cause false prediction of two liquid phases. On the other hand, the Huron-Vidal rules permit an excellent representation of the experimental data.

In the Huron-Vidal mixing rules, the value of b from the EOS is given by a simple summation.

$$b = \sum_{i=1}^{n} b_{ii} x_i \tag{78}$$

This value is used in a modified NRTL equation to compute the other mixture parameter at each composition and temperature.

$$a = b \sum_{i=1}^{n} x_i \left(\frac{a_{ii}}{b_{ii}} - \frac{1}{\ln 2} \frac{\sum_{j=1}^{n} x_j G_{ji} C_{ji}}{\sum_{k=1}^{n} x_k G_{ki}} \right) \tag{79}$$

where G_{ji} is computed from

$$G_{ji} = b_j \exp\left(- \alpha_{ji} \frac{C_{ji}}{RT} \right) \tag{80}$$

The unknown parameters in these equations are α, C_{12}, and C_{21}. In many cases it is possible to use just two parameters, and one is sufficient when both materials are hydrocarbons.

These improved mixing rules developed by Huron and Vidal were derived from thermodynamic relationships between the excess Gibbs free energy and fugacity coefficients. This gives an interrelationship of the excess Gibbs free energy, which now can be predicted reasonably well, and the interaction parameters of the EOS, offering promise of predictions with EOS in the absence of mixture data. Ultimately, it may be possible to predict routinely the EOS parameters using UNIFAC or ASOG to compute the excess free energy part of the interaction parameter equation. Future progress in use of EOS may be related in part to the theoretical and practical progress already made in handling and predicting liquid phase excess free energies and activity coefficients. Arguments have been made that the Huron-Vidal approach is theoretically flawed at the low-pressure limit because the resulting equation does not represent accurately the second virial coefficient. However, the practical significance of that deficiency has not yet been quantified. Furthermore, it may be completely overcome using density dependence in the EOS.

Density-dependent mixing rules appear to have two chief characteristics. First, they give rather good results. Second, they use considerable amounts of computer time. The two centers of research on these mixing rules are with Prof. J. M. Prausnitz at the University of California-Berkeley and Prof. S. I. Sandler at the University of Delaware. It is clear that research on mixing rules will be followed carefully by practitioners, as there is much yet to be discovered.

4. Associating Systems

Union Carbide workers have had considerable success with an equation developed by Gmehling et al.[219] The equation is based on PHC theory, and uses chemical equilibrium constants to represent reversible reaction between the individual molecules which can dimerize, and formation of cross-dimers. It was shown by Buck[218] that the equation could be adapted successfully to handle highly challenging industrial problems. These included the water + acetic acid, methanol + ethanol + water, and ethanol + water + 1,4 dioxane systems. The predictions were very close to the experimental data.

Three parameters are needed for each pure component, with one additional required for associating substances. Mixtures of associating substances involve the creation of new compounds, so the number of mixture parameters can escalate rapidly. Nevertheless, the Gmehling equation may be of value in some circumstances.

The author has used an EOS developed by Wilson (see Wilson and Weiner[170]) to model the butane + acetic acid + water + oxygen + carbon oxides system. It was derived by working backward from an activity coefficient equation to an EOS through introduction of void spaces in the liquid. As might be expected, it was necessary to take into account the dimerization of acetic acid in the gas phase. Dimers were represented by using the properties of paraxylene which may have a shape similar to the acetic acid dimer. In spite of these complications, it was possible to predict three phase equilibria. The acetic acid + water binary system was well represented up to high pressure. However, the available data were insufficient to make a thorough test of the accuracy of the multicomponent predictions. The environment described by this EOS was unusually harsh. Nevertheless, there is hope that in the long range there will be more extensive use of EOS in situations where there is phase splitting.

K. Gas Hydrates by EOS

The natural gas processing industry has a problem with formation of solid hydrates, a form of clathrates, in their pipelines and processing equipment. These hydrates can form substantially above the fp of water, with increasing pressure increasing the temperature at which they are stable. The hydrates are not stoichiometric compounds; they are solid solutions of a gas in ice. The stabilizing gas, usually having a low molecular weight, contributes one molecule to the lattic of a crystalline ice structure.[171] Other properties of these hydrates have been studied in detail by Kobayashi at Rice University (see Schroeder et al.[172] and Song and Kobayashi[173]). EOS can be used to predict solid formation; the thermodynamic framework for their use is given in Reference 174. Computational methods are provided therein for multicomponent mixtures. The following components are hydrate formers: methane, ethane, propane, isobutane, hydrogen sulfide, carbon dioxide, nitrogen, ethylene, and propylene.

Quantitative predictions are best made with a special-purpose computer program available from the GPA [(918) 582-5112]. Details can be obtained from the author, Prof. Dendy Sloan at the Colorado School of Mines [(303) 273-3723]. The following description is given: "The program uses the method of van der Waals and Platteeuw for the hydrate phase, and the Soave-Redlich-Kwong equation of state for the fluid phases, with parameters obtained from experimental measurements." The computer program answers the following questions:

1. How much dissolved water is allowable in a hydrocarbon liquid or vapor phase before hydrates will form at a given temperature and pressure?
2. What is the allowable temperature or pressure before hydrates will form from a free water phase and a hydrocarbon vapor?
3. How much hydrate inhibitor, such as methanol, should be used to prevent hydrate formation from a free water phase and a vapor at a given temperature and pressure?
4. What are the conditions for hydrate formation when three other phases (free water-hydrocarbon liquid-vapor) are present?

The GPA offers another hydrate computer program, produced by D. B. Robinson and Associates, as part of a generalized phase equilibrium prediction package called Equi-Phase. It uses the Peng-Robinson equation instead of the Soave. The computer program has been expanded to include mixtures of different hydrate inhibitors, including methanol, sodium chloride, and ethylene, diethylene, and triethylene glycol. An advantage of this program is that it is continuously updated by the authors.[175-177] Their work includes liquid-liquid-vapor-solid systems. They would appear to be good consultants on problems of gas hydrates.

D. B. Robinson and Associates
9324 — 31 Avenue
Edmonton, AL, Canada T6N 1C4
(403) 463-8638

L. Summary

EOS are growing in popularity for prediction of phase equilibria. The BWR and BWRS equations are still used in the gas processing, oil refining, and oil production industries. The newer Lee-Kesler-Plocker equation is also used, but is likewise limited to hydrocarbon compounds. Much of the research interest is in simpler equations, typified by cubic EOS. The famous Redlich-Kwong equation gives good predictions of gas phase nonideality but is inadequate for phase equilibrium predictions. Better equations are the Zudkevitch-Joffe, Soave, and Peng-Robinson. Further improvements are unlikely to come through empirical improvements. The new repulsive term developed for the CCOR equation by Chao and co-workers is a good example of extension of highly theoretical work in a practical format.

Some very accurate predictions are now being made for various types of phase equilibria. These even include formation of gas hydrates and gas-gas equilibria. Predictions in the region of the critical point are taking on particular importance as the chemical industry explores the potential for supercritical extraction.

More work is needed to demonstrate the best method for handling polar compounds. Much of the current work is directed toward improved mixing rules. The Huron-Vidal approach which invokes the activity coefficient mixing rules appears to have promise and should be further evaluated. Notwithstanding some theoretical arguments against it, there is potential for use of UNIFAC and ASOG to make *a priori* phase equilibrium predictions in regions inaccessible by the activity coefficient methods. Other equations which merit investigation have been developed by Gmehling et al. and Wilson and Weiner.

VI. ELECTROLYTE SYSTEMS

A. Introduction

Aqueous mixture phase equilibrium problems usually can be handled routinely. However, introduction of soluble salts complicates the phase equilibrium, since salts form a variety of species with which engineers and chemists are sometimes unfamiliar. These different species interact with the solvent mixtures, breaking or creating azeotropes, causing phase splitting, or precipitating solids.

A typical electrolyte system simulation problem begins with definition of a number of species and specification of the temperature and pressure. From that information the number of moles of solids, if any, must be calculated. In the liquid phase the moles of water and molalities of all the ions are needed. Finally, presence or absence of a vapor phase must be determined, and if vapor is present its composition and amount also must be computed.

B. Applications

Applications of electrolyte equilibrium techniques[116] are

1. Crystallization — solubility, water activity, bp elevation.
2. Precipitation — solubility, complex formation, etc. Some problems may seem simple but in fact are not. Figure 19 shows data on chrome solubility. It is clear that use of a simple solubility product can lead to errors in excess of three orders of magnitude. This example, based on a problem discovered in industrial practice, had to be solved by the more sophisticated techniques such as those now included in the DIPPR publication, *Aqueous Electrolyte Thermodynamics*.
3. Neutralization — such as sulfuric acid with limestone.

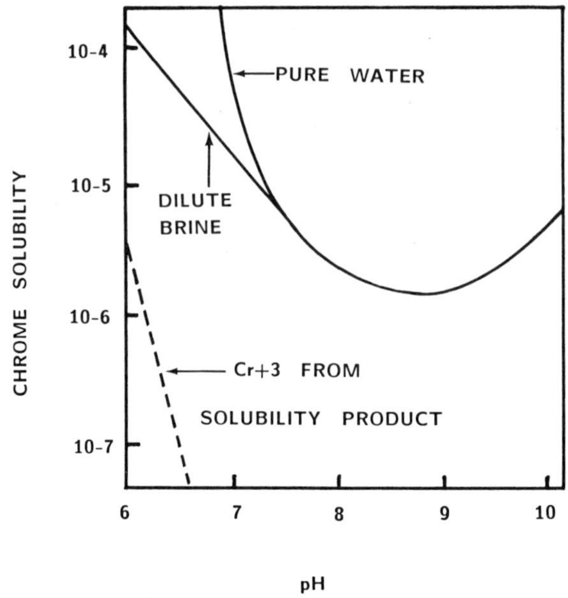

FIGURE 19. Correlation of actual chrome solubility, contrasted with the erroneous prediction that would be made using the solubility product. (Reprinted from Scrivner, N. C., E. I. du Pont de Nemours & Co. With permission.)

4. Hydrometallurgy — leaching, solution purification, metal recovery, hydrogen reduction of metals.
5. Flue gas scrubbing — SO_x/limestone.
6. Ion exchange — boiler feedwater softening, brine treatment, sulfuric acid regeneration.
7. Sour water cleanup — balancing of sour and basic gases to form nonvolatile ions.
8. Scale formation in wells — pressure letdown near well bore lowers solubility and water injection adds scale-forming ions.
9. Oil well completion fluids — salt solubility and density as a function of temperature and pressure.
10. Homogeneous catalyst solubility.
11. Breaking azeotropes by salt addition — e.g., ethanol + water.
12. Future applications — soil science, life sciences, mass transfer with chemical reaction, corrosion control with complexes, slurry stabilization.

C. DIPPR Electrolyte Handbook

While it is true that a great deal has been published on electrolytes, little had been done to synthesize the data and correlations into a workable manual until DIPPR funded a 3-year project that resulted in *Aqueous Electrolyte Thermodynamics*.[225] In this project the various prediction methods were compared; examples are given in Figures 20 and 21. It will be noted that this book considers electrolytes up to very high concentrations. The effect of temperature also has been addressed. Worked examples include the following systems:

- Sodium chloride solubility
- Water + chlorine
- Water + ammonia + carbon dioxide
- Water + sulfur dioxide
- Chrome hydroxides

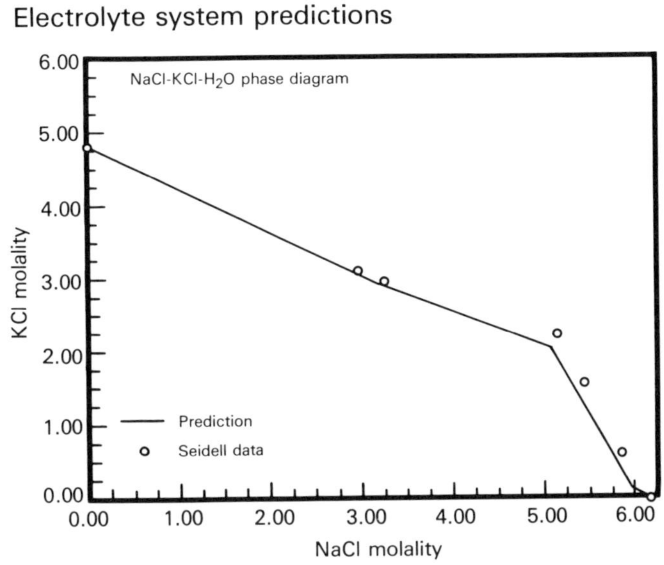

FIGURE 20. Example from DIPPR electrolyte manual. A prediction using manual techniques of the two-phase equilibrium between sodium chloride, potassium chloride, and water. (Adapted from Zemaitis, J. F., Rafel, M., Clarke, D., and Scrivner, N. C., *Aqueous Electrolyte Thermodynamics — Theory and Practice,* American Institute of Chemical Engineers, New York, 1986.)

- Gypsum solubility
- Phosphoric acid titrations

Annotated bibliographies of the data base for many important electrolyte systems were developed as part of this project at the NBS. Copies can be obtained from the NBS Office of Standard Reference Data (OSRD) [(301) 921-2581]. The manual also includes some computer programs and a listing of the computer programs available commercially. A very convenient listing also is given of thermodynamic properties for many common ions and aqueous undissociated species.

D. Consultants and Computer Programs

OLI Systems Inc.

Even with the methods published by AIChE/DIPPR, it is often helpful to be able to call on consultants. A number of people are competent in the area for specific types of problems. For nonroutine electrolyte problems the experts at OLI Systems Inc. are recommended.

OLI Systems Inc.
52 South Street, Morristown, NJ 07960
(201) 540-0291 and (201) 540-0353

Dr. Marshall Rafal, President of OLI, and Diane Clarke joined with Dr. Noel C. Scrivner in representing DIPPR to complete *Aqueous Electrolyte Thermodynamics*[225] after the untimely death of Dr. Joe Zemaitis who had begun the project. In its capacity as general consultant on electrolytes, OLI offers two very useful computer programs. ECES (Equilibrium Com-

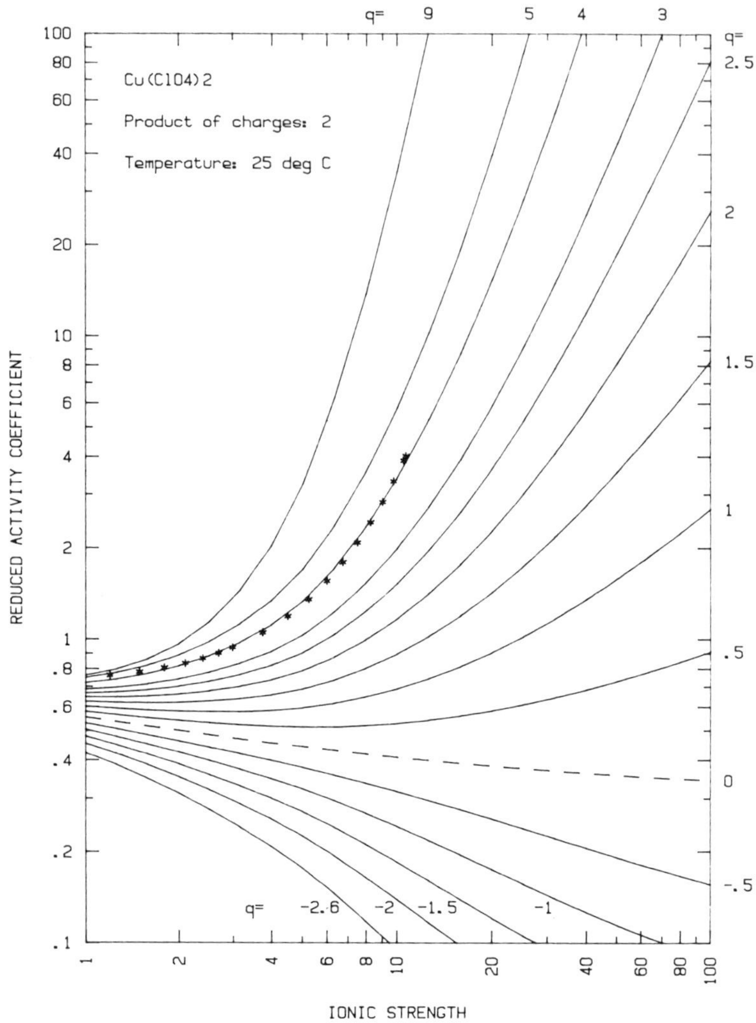

FIGURE 21. Example from DIPPR electrolyte manual. A comparison of predictions with experimental data for the reduced activity coefficient of cupric chlorate vs. ionic strength. (Adapted from Zemaitis, J. F., Rafel, M., Clarke, D., and Scrivner, N. C., *Aqueous Electrolyte Thermodynamics — Theory and Practice*, American Institute of Chemical Engineers, New York, 1986.)

position of Electrolyte Solutions) predicts liquid-vapor, liquid-solid, and vapor-liquid-solid equilibria for aqueous-based inorganic systems. The program can use various data bases and calculates all relevant K values, activities, stream enthalpies, ionic strengths, and pH. A leading industrial practitioner specializing in electrolytes reports that ECES works extremely well, provided that the species are chosen correctly. A key feature of this program is its wide range of applicability. The temperature range is 0 to 250°C. The pressure range goes from 0 to 200 atm. Finally, the program handles ionic strengths up to a molality of 30! An adjunct program is FRACHEM (Fractionation with Chemical and/or Ionic Reaction and Equilibria). In addition to electrolyte systems it can handle nonelectrolyte systems; thus it is able to handle reactive distillation. One example would be the production of ethyl acetate from acetic acid and ethanol. Another would be production of methyl *tert*-butyl ether (MTBE) from methanol and isobutylene in the CR&L process. FRACHEM also includes the ability to test the addition of an electrolyte to break an azeotrope. Detailed descriptions in the open literature on use of ECES and FRACHEM are given in Reference 117.

AspenTech

AspenTech [(617) 497-9010] has incorporated electrolyte system routines into its process simulator, ASPEN PLUS. In an example of design of a sour water stripper, AspenTech indicates that its data file includes "...information on the possible ionization reactions taking place, Henry's constant representations of H_2S and NH_3 in H_2O, and binary interaction parameters for the molecular and ionic species present." Data measured by Wiltec for the API were used as the basis for these parameters. AspenTech has been making its own scientific headway, developing improved techniques based on the NRTL equation for predicting the salting out of aqueous-organic mixtures.[118]

Simulation Sciences Inc.

Simulation Sciences Inc. [(714) 879-9180] also offers an electrolyte package in its process simulator. It includes the components and ionic species encountered in a variety of amine systems. The organization also has capabilities for various sour water systems, acid systems, and systems with solutions of inorganic salts.

Chemshare Corp.

Chemshare Corp. [(713) 627-8910] has the APISOUR program, specially designed to handle sour water systems. It can include hydrocarbons, hydrogen, water, caustics, organic acids, ammonia, hydrogen sulfide, and carbon dioxide. For less rigorous systems the APISOAVEK program uses the Soave EOS with special interaction parameters for hydrogen sulfide, carbon oxides, and nitrogen. For those licensing the Norton Company Selexol process for acid gas scrubbing, a computer program is available that contains proprietary data based on absorption of the acid gases, carbonyl sulfide, and mercaptans in Norton's solvent.

Gas Processors Association (GPA)

Various computer simulations are available from the GPA which address some aspects of the electrolyte problem. One is the OSU/GPA Sour Gas Equilibria Program. It makes the calculations necessary to predict the equilibrium between acid gas constituents and ethanolamine-treating solutions. Equilibrium constants are used to determine the extent of ion formation by the amine, water, hydrogen sulfide, and carbon dioxide. The program, written by Dr. R. N. Maddox, is available from the GPA [(918) 582-5112].

Other Computer Programs

The NBS/OSRD offers a computerized data base written in FORTRAN 77. It contains data on 350 binary aqueous electrolyte solutions at 25°C, and can be purchased at nominal cost [(301) 921-2228].

The U.S. Geological Survey offers a program called WATEQ2. It is limited to a molal ionic strength of 3.0 and is valid only at 25°C. The package contains an extensive data base, however. The Environmental Protection Agency developed a program of similar limited scope. It does predict solid formation but is limited to ionic strengths less than 1.0 *M* and 25°C. A second-generation version called REDEQL is reported to be available from the Civil Engineering Department of the Massachusetts Institute of Technology.

The Department of Energy also has been developing electrolyte computer programs. One of its contractors familiar with those programs is

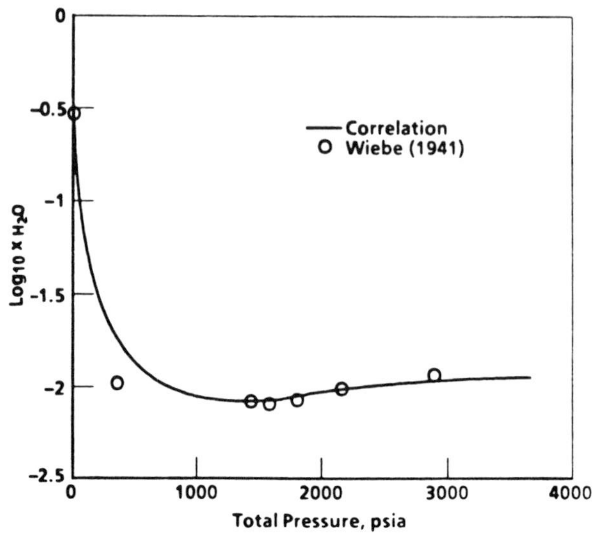

FIGURE 22. Correlation of the solubility of water in a carbon dioxide-rich phase at 75°C. (Reproduced from Cox, K. R. and Moore, M. A., *Fluid Phase Equilibria*, 24, 151, 1985. With permission.)

Dr. J. Dweck
6000 East Evans Avenue, Building I, Suite G-20
Denver, CO 80222
(303) 758-6862

E. Industrial Class Problems

1. Carbon Dioxide Flooding of Oil Fields

Real industrial problems can seriously challenge existing thermodynamic modeling capabilities. Take for example enhanced oil recovery by carbon dioxide flooding. In the ground the carbon dioxide is in equilibrium not only with an oil phase but also with naturally occurring brine. When the aqueous phase is not large compared to the carbon dioxide phase, the amount of water vaporizing into that phase can be important. An immediate effect can be precipitation of calcium carbonate as a solid. These events must be compensated for, since they can cause plugging of the strata or corrosion of pipes. Under downhole conditions the temperatures and pressures can be far beyond the usual range of data. Cox and Moore,[119] thermodynamicists at Shell Development Co., appear to have solved this problem successfully. Figures reproduced here with permission from the authors and Shell Development illustrate the success of their correlation effort. Figure 22 shows the solubility of water in the carbon dioxide phase as a function of pressure at 75°C, a very good prediction compared to the data. In Figure 23 there is an excellent correlation of the data for carbon dioxide solubility in the water phase. Finally, with no confirming data the authors have predicted the solubility of calcium carbonate in the water phase as a function of both temperature and pressure.

2. Modification of VLE

Addition of salts to aqueous-organic mixtures usually modifies the VLE. The effect was described by Johnson and Furter[120] and Jaques and Furter.[121] Research on methods to correlate such data is still underway. One method involves use of the UNIQUAC equation.[122] This

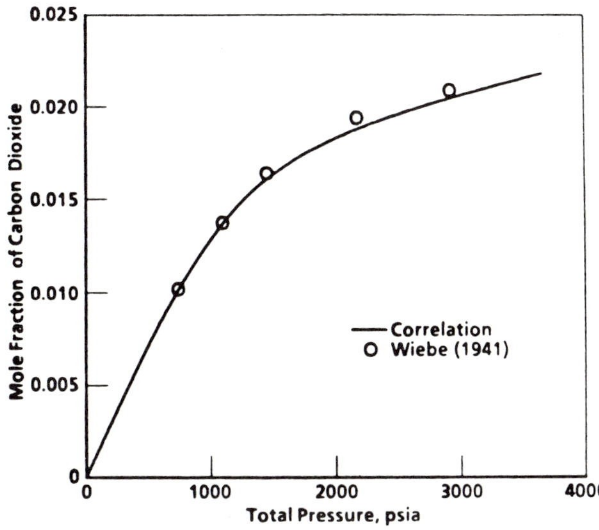

FIGURE 23. Correlation of the solubility of carbon dioxide in water at 75°C. (Reproduced from Cox, K. R. and Moore, R. A., *Fluid Phase Equilibria*, 24, 151, 1985. With permission.)

method requires in advance the UNIQUAC parameters which model the water + organic system. The system is treated as a binary mixture, with the mole fraction of the second component equal to the concentration of water plus the concentration of salt. An extra parameter is needed to predict the effect of the salt. With these additions it is possible to do a very good job of predicting the VLE as a function of composition and as a function of salt concentration; even partial miscibility can be predicted. However, extension of the method to multicomponent systems was not as successful; the predicted compositions differed significantly from experimental ones. It is possible that liberties taken with thermodynamic consistency for the binary cannot be masked when multicomponent systems are modeled.

The author had experience with development of data for a distillation column containing ammonia, water, methanol, and sodium carbonate. The approach was different, in that the system was treated on a salt-free basis. The activity coefficient for methanol in water at infinite dilution was measured at the concentration of salt expected in the process. Ebulliometry techniques described in Chapter 7 were used. The activity coefficient of water was then needed for a hypothetical solution of salt in methanol. It is reduced, relative to a salt-free system, because of the vapor pressure-lowering effect of the salt.

$$\gamma_w^{\infty *} = \frac{(P_w^* - \Delta P_w^*)}{P_w^*} \gamma_w^\infty \tag{81}$$

Thus, the vapor pressure-lowering effect of the salt was subtracted from and divided by the vapor pressure of water. This gave a correction factor that was used as a multiplier on the activity coefficient at infinite dilution. Under this circumstance the correction was smaller than the uncertainty in the literature value of the infinite dilution activity coefficient. The enhancement factor for ammonia, as a result of the salt, was measured. With these pieces of information the column was designed, constructed, and successfully operated.

An example of addition of a salt to enhance separation in the water + ethanol system was published by Lynd and Grethlein.[123] They propose extractive distillation using potassium acetate salt to modify the VLE. Figure 24 indicates the improvement in VLE when that salt

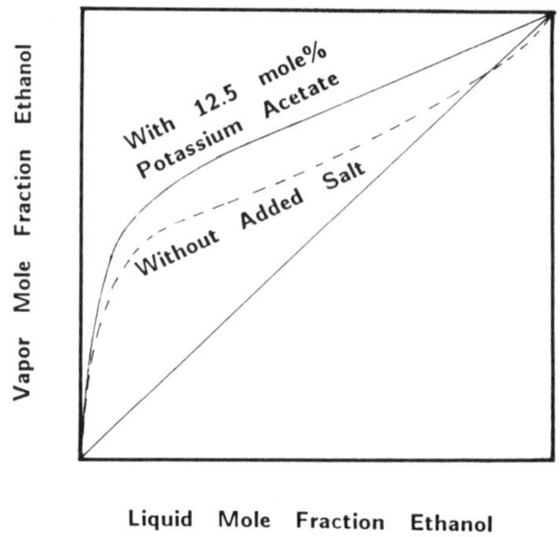

FIGURE 24. Effect of potassium acetate on the VLE of the ethanol + water mixture. Note the removal of the azeotrope. (Adapted from Lynd, L. R. and Grethlein, H. E., *Chem. Eng. Prog.*, November, 59, 1984.)

is added. The authors computed energy savings without using a correlation. That is a dangerous shortcut, because the graphical McCabe-Thiele analysis is not accurate when chemicals differ sharply in heat of vaporization. Mixtures of water with organic compounds usually differ significantly in molar heat of vaporization.

F. Summary of Recommendations on Electrolyte Systems

In this section the types of problems involved with electrolytes have been explained without details on how to solve such problems. For those systems not involving organic compounds the best source is the DIPPR manual, *Aqueous Electrolyte Thermodynamics*.[225] It shows how to handle even complex systems under extreme conditions. Several appropriate computer programs and consultants are available to help solve difficult problems. In fact, virtually every electrolyte problem is difficult enough to require a computer solution. For systems with organic compounds there are correlation techniques available; however, there are not yet any good correlations to predict in advance the effect of salts on aqueous-organic mixtures. The research has come a long distance but further work is needed.

VII. POLYMER-SOLVENT EQUILIBRIA

The manufacture of polymers is often done by a solution process, implying the presence of a solvent. Part of the manufacturing process involves stripping of that solvent and drying. This is just one of a number of situations where the activity, or volatility, of the solvent must be known for process design. Typically, the molecular weight and molar volume of the polymer are much greater than those of the solvent. Therefore, it is usually better to compute activity coefficients based on weight fractions than by means of the usual convention of mole fractions.

Much has been written about polymer-solvent thermodynamics. However, the equations needed to correlate these weight-fraction activity coefficients are not necessarily complicated. One approach, published in 1985,[179] gives a simple equation called by its authors the ASOG-VSP model. Perhaps the term ASOG should not have been applied, because the final equation does not resemble ASOG at all. There are no group contribution interaction parameters. There is one adustable parameter which is the weight-fraction-based activity coefficient for the solvent at infinite dilution in the polymer. This is fortuitous because that is a quantity easily measured. The experimental technique used to do that by gas chromatography is explained in Chapter 7. The weight-fraction activity coefficient is given by the following simple equation, where w_1 is the weight fraction of solvent:

$$\Gamma_1 = \frac{\exp(A/(w_1 + A))}{w_1 + A} \tag{82}$$

where A is defined as

$$A = 2.718(1 - w_1)/\Gamma_1 \tag{83}$$

This amazingly simple equation assumes that enthalpic effects are completely outweighed by entropic effects in mixing of solvents with polymers. It does take into account the effect of different free volumes between the solvent and the polymer.

This equation is more accurate for more systems than the traditional Flory-Huggins equation;[180] furthermore, density data are not required. It is also more reliable than the UNIFAC-FV model.[181] The comparison made between the different methods indicated that the ASOG-VSP model gave predictions within 5% of the experimental data some 80% of the time. That compares to 54% of the time with Flory-Huggins and 22% with UNIFAC-FV.

The UNIFAC-FV model has the advantage of being almost completely predictive, although densities are required. It is used whenever the experimental measurement of a data point cannot be justified. Presumably, its parameters could be manipulated to give results comparable to the ASOG-VSP model; however, this would be at the expense of more complicated computations. It is anticipated that improvements will be made to both models, and that the traditional Flory-Huggins model also will continue to be used. There are a large number of interaction parameters published for that model.[182]

In the less common case of supercritical gases in polymers, the procedure described in Reference 183 should be used. Being based on PHC theory, it has a reasonable basis in fundamentals.

VIII. SOLID-LIQUID EQUILIBRIA

The condition for equilibrium between a solid and a liquid in any kind of mixture is that the fugacities be the same for each component in both phases. The thermodynamic basis for the equations that follow has been described by Prausnitz.[184] He makes the reasonable assumption that solvent solubility in the solid is negligible, that the triple point and mp are very close to each other, and that some heat capacity terms are negligible. With those assumptions, the following simple equation results:

$$\ell n\left[\frac{1}{x_2\gamma_2}\right] = \frac{\Delta H^f}{RT}[1 - T/T_m] \tag{84}$$

A. Ideal Mixtures

For ideal mixtures, knowledge of the heat of fusion and the mp are the only required data. The activity coefficients are unity. To calculate the equilibrium curve for a mixture forming a eutectic, compositions are varied from both ends of a binary diagram until the curves on a temperature-composition plot cross. It is clear from this equation that the solid with the higher melting temperature has the lower solubility, assuming equal heats of fusion. Similarly, if two liquids have the same mp, the one with the lower enthalpy of fusion has

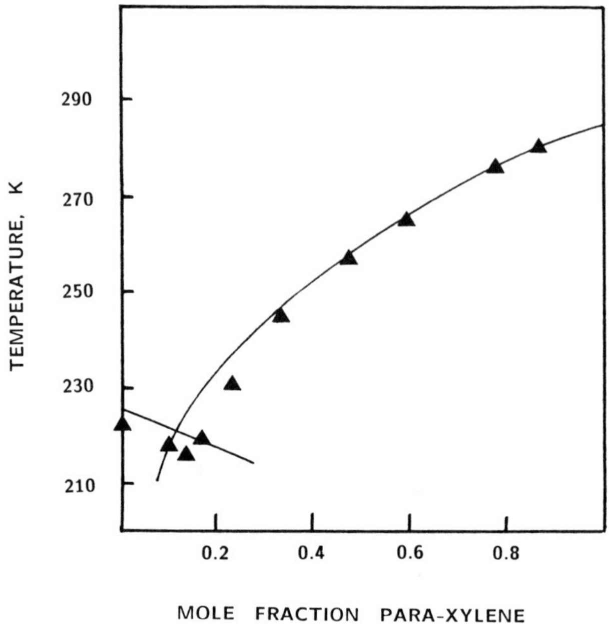

FIGURE 25. Prediction of the fp curve for para-xylene + meta-xylene ignoring activity coefficient effects.

the higher solubility. Solubility rises with increasing temperature at a rate almost proportional to the heat of fusion.

One of the most important industrial crystallization processes is the recovery of para-xylene from other xylene isomers and ethylbenzene. The para-xylene is removed, down almost to the eutectic point. Then it is oxidized to make terephthalic acid, the key raw material in polyester. The mixture should be ideal, within 1 or 2%, so the activity coefficient has been set equal to unity.

In the following example, the para-xylene + meta-xylene isomer system is examined. The heats of fusion are 4087 and 2763 cal/g mol for para- and metaxylene, respectively. The mp are 286.4 and 225.3 K. Using those data, a melt curve diagram was generated and is illustrated in Figure 25, where it is compared with experimental literature data. The prediction agrees rather well with experiment, and more recent proprietary data suggest that the agreement is even better than indicated in this figure.

B. Nonideal Mixtures

Most mixtures with a solid phase exhibit nonideal mixing behavior, therefore some computation of the activity coefficient is needed. If liquid phase activity coefficients have been correlated for the mixture of interest, they can be used directly.[186] Usually, however, the data are not available.

It has been found that activity coefficients can be predicted with sufficient accuracy using the ASOG or UNIFAC methods.[187] Neglecting several minor correction terms, the solubility is then computed easily. For example, the highly nonideal behavior in alcohol + hydrocarbon mixtures is very accurately predicted. Even systems containing acetic acid appear to be predicted with accuracy. This is illustrated in Figure 26 for the acetic acid + benzene system. The solid line was predicted by UNIFAC while the dashed line represents ideal behavior. For 17 highly nonideal mixtures, both the temperature and composition of eutectic mixtures were predicted well enough for preliminary design; actual data, of course, would be measured for final design. There were preliminary indications in that article that even multicomponent predictions could be made reasonably well.

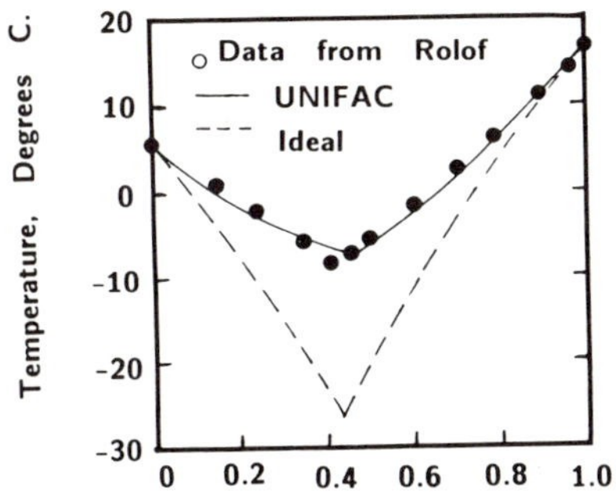

FIGURE 26. Prediction of SLE for the acetic acid + benzene system, using the UNIFAC method. The incorrect prediction assuming ideality is also illustrated. (Adapted from Gmehling, J. G., Anderson, T. F., and Prausnitz, J. M., *Ind. Eng. Chem. Fundam.*, 17, 269, 1978.)

As another example of the potential accuracy of this approach, computations were made for naphthalene dissolved in a variety of solvents at 40°C.[43] Activity coefficients were calculated by UNIFAC. Other methods tried, such as the solubility parameter approach, gave poor results.

	Solubility (mol %)	
Substance	**Experimental**	**Predicted**
Methanol	4.4	4.8
Ethanol	7.3	5.4
1-Propanol	9.4	9.3
2-Propanol	7.6	9.3
1-Butanol	11.6	11.1
n-Hexane	22.2	25.9
Cyclohexanol	22.5	20.5
Acetic acid	11.7	12.5
Acetone	37.8	35.8
Chloroform	47.3	47.0

The accuracy of these computations is remarkably good, when one considers that no solid-liquid data whatsoever were used in developing the parameters; they came entirely from group contribution methods based on VLE.

The ASOG method also can be used for such predictions.[88] Two articles from Japan, where the ASOG method is popular, confirm the accuracy of the estimates.[188,189]

When the solid activity coefficients cannot be predicted, the activity coefficients can be derived at eutectic points simply by rearranging Equation 84. Parameters in the Wilson, NRTL, or UNIQUAC equation can be calculated based on that data point. Then the entire freezing curve can be computed, taking into account both temperature and composition effects. The same procedures can be extended to multicomponent systems, provided that local composition activity coefficient equations are used. If the system simultaneously exhibits a second liquid phase, the Wilson equation should not be used. Reference 43 discusses very complex systems involving solids.

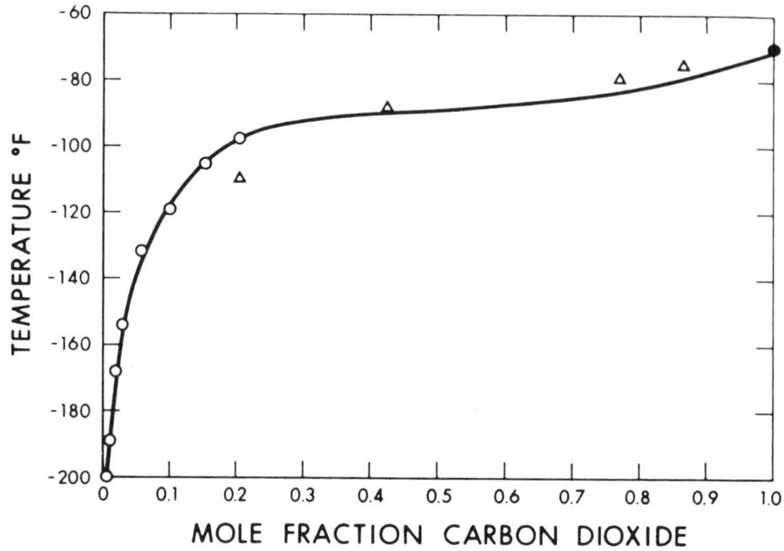

FIGURE 27. Carbon dioxide content of the liquid phase at the solid-liquid-vapor condition for the methane + carbon dioxide system. Prediction made with the Peng-Robinson equation. (Reprinted from Peng, D. -Y. and Robinson, D. B., in *Equations of State in Engineering and Research,* Chao, K. C. and Robinson, R. L., Jr., Eds., Advances in Chemistry Series, No. 182, American Chemical Society, Washington, D.C., 1979 With permission.)

C. EOS Methods for Predicting SLE

The EOS work by Peng and Robinson[178] and associates has encompassed the problem of solid formation. Their calculations are based on the fundamental principle that the fugacities must be equal in all phases. The fugacity of the solid at the system pressure is equal to the gas phase fugacity, calculated at the vapor pressure of the solid at the solid formation temperature and multiplied by the Poynting correction factor. Special computational procedures are required to converge the equations and determine the presence, if any, of a solid phase. These are described in detail in the reference. As an example of the applicability of this approach, Figure 27 shows the predicted content of carbon dioxide in the liquid phase for a solid-liquid-vapor system. The figure shows the effect of temperature as a function of mole fraction of carbon dioxide with methane as the second component.

IX. PROPERTIES OF PETROLEUM FRACTIONS AND SYNCRUDE

Petroleum fractions are at once easy and difficult to handle. They are easy because the interactions are relatively straightforward and can be modeled well. They are difficult because of problems in stream characterization. Synthetic petroleum also must be considered today, in spite of the temporary demise of the synthetic fuels industry. Thus, one also must look at oil fractions from coal liquefaction, shale oil retorting, and tar sand recovery. Those materials are different from each other because of differing hydrogen-to-carbon ratios. Components in syncrude often contain large amounts of oxygen and nitrogen, thus complicating the characterization and prediction methods developed for petroleum processing.

A. Crude Oil Characterization

1. Pseudocomponent Method

Without characterization it is impossible to represent a crude fraction. Each different crude source has its own unique set of properties. After elimination, through hydroprocessing, of the contained sulfur and nitrogen, the fraction contains a complex mixture of aliphatic, aromatic, and partially unsaturated hydrocarbons. They are characterized primarily by distillation. That gives the bp and liquid density of a number of fractions up to an end point of 1050°F. Each separate distillate fraction is treated as a separate pseudocomponent. The residue is essentially unexamined. The methods used to treat the pseudocomponents are described in the API *Technical Databook — Petroleum Refining*.[190]

An analytical advance reported in 1985 improves on the more traditional methods of characterization. The methods and equipment are all available, but applied together give an improved assessment of the crude or synfuel fraction.[191] High-boiling but distillable fractions are separated into about eight narrow-boiling fractions by distillation. The fractions are then used to find vapor pressure, elemental composition (from combustion and gravimetric analysis), hydrogen distribution (from H NMR), hydroxyl and amine concentrations (from IR analysis), and number-average molecular weight (from cryoscopy).

Both industrial and academic researchers agree that the current analytical methods must be upgraded. It is becoming increasingly important to characterize the residue, for example. Starling[192] argues that an actual phase equilibrium data point is needed in order to derive interaction parameters. He also argues for multiproperty analysis, demonstrating that an error of just 1% in the derived critical temperature of a fraction can lead to an error of over 15% in the predicted vapor pressure. Starling has made the fairly significant discovery that measured liquid viscosities are very sensitive to the characterization parameters. In fact, viscosities can be used to replace bp or vapor pressures in characterization of mixtures. This is very useful for fractions which have extremely low vapor pressures. Under normal circumstances, addition of viscosity data to the vapor pressure and liquid density data improves the characterization parameters.[193]

2. *Computer Programs*

Most of the general process simulators have the capability of using petroleum fractions as discrete, pure component data input. The Simulation Sciences computer program can accept either ASTM or TBP boiling curve data. The data are analyzed with the Cavett method to obtain critical properties of each cut. Vapor pressures are determined by the Yen-Alexander method. The Edmister method is used for the acentric factor, and the API technical data book method[190] is used for molecular weight. Alternatively, the Lee-Kesler method can be specified for these properties. When the information available goes beyond the required density and bp, the percentage of paraffins, olefins, naphthenes, and aromatics can be designated for each component.

3. *Continuous Thermodynamics*

So-called continuous thermodynamics methods are being developed, but at the time of this writing had not yet been widely implemented by the industry. In this approach the fractions are not modeled in increments but by distribution functions. The technique seems to offer the promise of improved predictions with less computation time.[194-196] This approach was too new as of this writing to assess its long-term impact. It probably will be implemented slowly, but the research in this area will be watched carefully.

B. Estimation of K Values

The old graphical methods and "convergence pressure" approaches are now almost obsolete. Work by Sim and Daubert[197] shows that the Soave and Peng-Robinson equations are nearly equivalent and give the best results. These EOS are more accurate than the Chao-Seader equation or the Maxwell correlation. Improvements to the Soave equation[198] are included in the procedures recommended in the API *Technical Databook — Petroleum Refining*.[190] Those procedures include the interaction parameters needed when dealing with gases commonly encountered in processing of such mixtures. Those include hydrogen sulfide, carbon oxides, and nitrogen. The correlation is completely detailed, together with instructions in Section 8 of the data book. It is so widely distributed that the results are not reprinted here.

It was found that an average error of 9% can be expected in both binary and multicomponent K values. Predicted dew and bubble point pressures for the systems studied by Daubert et al. were in error by 4.3%. Systems with large amounts of hydrogen sulfide and nitrogen exhibit larger errors. The best accuracy is obtained from pressures ranging from atmospheric to 2000 psia.

C. Synfuels

1. Data Base

The physical property data base for synfuels is distressingly inadequate. The most comprehensive survey of data sources was completed by the Bechtel Corporation for the API.[199] Fortunately, the 450 references are indexed by properties and type of data. Reference 200 gives data on phase equilibrium for coal liquefaction processes. It is hoped that, before large-scale synthetic fuel plants are constructed, the data base will be upgraded.

2. Composition Differences between Synfuels and Crude Oil

It must be recognized that synfuels are not just another form of crude oil; they differ in composition rather greatly, as illustrated by the following data:[201]

	(wt %)			
	Saturates	**Olefins**	**Aromatics**	**H/C ratio**
Petroleum crude	80	0	20	1.9
Coal liquids	35	5	60	1.5
Shale oil	30	40	30	1.6
Tar sands liquids	40	0	60	1.5

It is also very instructive to consider the large difference in types of impurities in the oils.

	(wt %)			
	Sulfur	**Nitrogen**	**Oxygen**	**% Compounds (S, N, O)**
Petroleum crude	1.8	0.1	0	10
Coal liquids	0.2	0.3	2.0	25
Shale oil	0.8	1.8	1.6	40
Tar sands liquids	4.2	0.4	0.5	25

It is clear that there are some added components that can affect the equilibria significantly. In some cases all the incremental oxygen will be in the form of phenols. The fact that they are very polar molecules which cannot be treated by conventional means has not escaped the attention of industry.[202]

3. Problems in Modeling Synfuel Systems

An example of the problems encountered with noncrude-derived organic fuels is reported in Reference 203. Analysis of a creosote oil showed that phenanthrene was the most representative constituent. However, the solubility of hydrogen in the creosote varied from two to four times the solubility in pure phenanthrene.

4. Pseudocomponent Property Simulation

A serious effort was made to measure and correlate both vapor pressures and heats of vaporization for narrow-boiling coal liquid fractions.[204] The data were taken on Exxon Donor Solvent liquid fractions. The correlations found most acceptable were the Brulé modification of the BWR equation, an equation by Wilson et al.,[205] and the Grayson-Streed equation.[206] Wilson's equation has fewer parameters than the modified BWR equation, and can predict critical temperatures successfully. An optimized Watson equation gave a better representation of the calorimetric heats of vaporization than could be obtained from differentiation of the vapor pressure curve. At the normal bp, the heat of vaporization (kilojoules per gram mole) is given by

$$\Delta H^v(NBP) = 0.08894\ T_b(K) - 1.5 \tag{85}$$

The temperature dependence of the heat of vaporization is found by

$$\Delta H^v(T) = \Delta H^v(NBP)\ [\{T_c - T\}/\{T_c - T_b\}]^{0.419} \tag{86}$$

The critical properties are often used as the basis for prediction of transport properties. For example, work by Teja and co-workers[207] gives equations for viscosity and thermal conductivity of synfuels as a function of critical temperature and pressure, plus acentric factor. Two reference fluids are also used. A similar approach, developed cooperatively at three institutions, likewise uses corresponding states to compute thermal conductivities.[208]

A fairly general correlation is thus needed to predict accurately the critical properties from the bp/specific gravity/molecular weight characterizations normally made. One very useful correlation[209] was developed which extended other correlations by having applicability to polar and associating components. The Starling correlations are as follows:

Critical temperature

$$\begin{aligned}\ln (T_c) = & -0.00093906T_b + 0.030950 \ln (MW) + 1.11067 \ln (T_b) \\ & + MW(0.078154SG^{1/2} - 0.061061SG^{1/3} - 0.016943SG)\end{aligned} \tag{87}$$

Critical volume

$$\begin{aligned}\ln (V_c) = & \ 80.4479 - 129.8038SG + 63.1750SG^2 - \\ & 13.1750SG^3 + 1.10108 \ln (MW) + 42.1958 \ln (SG)\end{aligned} \tag{88}$$

Critical pressure

$$\ln (P_c) = 3.95431 + 0.70682(T_c/V_c)^{0.8} - 4.8400MW/T_c - 0.15919T_b/MW \tag{89}$$

The critical temperature computed in Equation 87 and the critical volume calculated in Equation 88 must be used in computation of the critical pressure. Acentric factor

$$\begin{aligned}\omega = & \{0.922170 \times 10^{-3}T_b + 0.507288T_b/MW \\ & + 382.904/MW + 0.2420 \times 10^{-5}(T_b/SG)^2 \\ & - 0.2165 \times 10^{-4}T_b \times MW + 0.1261 \times 10^{-2}SG \times MW\end{aligned}$$

$$+ 0.1265 \times 10^{-4}MW^2 + 0.2016 \times 10^{-4}SG \times MW^2$$
$$- 80.6495T_b^{1/3}/MW - 0.3780 \times 10^{-2}T_b^{2/3}/SG^2\}\{T_b/MW\} \quad (90)$$

Another correlation was developed at Simulation Sciences Inc. It is limited in applicability to petroleum and coal-tar liquids. However, it has the advantage of not requiring molecular weight as an input. The procedure is based on bp ranging up to to 1300°F and specific gravities as high as 1.436. Normal alkanes are used as reference fluids.[210]

5. Mixture Predictions

Once the pure component properties are successfully modeled, equilibrium properties must be considered. It has been shown[203] that modifications of the Redlich-Kwong equation predict well the solubility of hydrogen in model compounds. By contrast, the Chao-Seader and Grayson-Streed equations perform poorly. A comparison was made[205] between the Zudkevitch-Joffe modification of the Redlich-Kwong equation[211] and the Grayson-Streed version of the Chao-Seader correlation. The conclusion was that for coal liquids the Zudkevitch-Joffe equation is far superior. An advantage of the equation is that its parameters are obtained from vapor pressure and liquid density data which are known. For heavy hydrocarbons, such as those derived from alternate energy sources, the critical properties used by other correlations are usually unknown. Another advantage of this equation, as pointed out by Gray,[212] is that it fits the critical region phenomena very well. In fact, Gray notes that its performance is substantially better than he had expected.

The Peng-Robinson equation also appears to work well in synfuel systems. Using model compounds based on coal-derived hydrocarbons, the authors showed that very unusual equilibrium curves could be correlated.[213] This is particularly relevant since the systems all included methanol, which is polar and associating. Various naphthalenes, tetralin, and 1-naphthol also were tested in mixtures with methanol.

A simulation of the HCOAL process comparing Chao-Seader and a modified Redlich-Kwong EOS revealed very large differences in converged flow rates. For example, the vapor flow from one flash drum differed by over a factor of two. A stripped oil-solids slurry stream differed by 62%. Finally, the flow of a sour gas stream varied by 48%.[214] When the very high capital cost of synfuel processes is considered. It is obvious that physical property predictions will have a tremendous economic impact on that industry.

6. Enthalpy Predictions

A sophisticated correlation for pure compounds appears to predict accurately the heat of vaporization from the triple to the critical point.[214] It is applicable for all hydrocarbons, from methane to three-ring coal model compounds. Data required include the critical temperature and acentric factor. For multiring compounds, those properties usually must be estimated. For the examples given, which were only for pure compounds, the average errors were quite small. The compounds studied, with percentage errors, included:

	(% error)
Diphenyl methane	0.80
Acridine	5.29
Thianaphthene	1.20
Bicyclohexyl	1.58
2-Methylnaphthalene	2.12
9-Fluorenone	6.93
Carbazole	2.42

A Kesler-Lee[215] modified Johnson-Grayson technique[216] gave good enthalpy predictions for mixtures including ordinary petroleum liquids. For example, in one study the average errors ranged from 0.5 Btu/lb for #2 fuel oil up to 4.0 Btu/lb for aromatic naphtha. However, the same prediction method gave poor results for some syncrudes. The average error was 12.1 Btu/lb for Utah syncrude distillate and 35.7 Btu/lb for Synthoil distillate. Low-molecular-weight fractions such as SRC I naphtha had a lower average error of 4.1 Btu/lb. These numbers suggest that a significant amount of characterization and correlation development should be carried out prior to commercialization of synfuel processes.

D. Recommendations on Petroleum and Synfuels

Characterization of petroleum or synfuel fractions appears to be the key to adequate modeling. At present this is done with discrete fractions, but in the future it is possible that continuous thermodynamics will be applied. It appears that a modification of the Soave equation, documented in the API *Technical Databook — Petroleum Refining,*[190] the Peng-Robinson equation, and the Zudkevitch-Joffe equation provide completely adequate models of these mixtures with very large numbers of components. Remarkably good predictions can be made, even with very high-boiling components. An excellent monograph by C. Tsonopoulos et al.[231] explores the ramifications of extending thermodynamic methods developed for petroleum fractions to various synfuel processes. It is a fairly complete and well-documented discussion of the subject. Many of the conclusions are similar to those noted here, and there are further very helpful recommendations for a number of physical properties of synfuel fractions.

REFERENCES

1. **Reid, R. C., Prausnitz, J. M., and Sherwood, T. K.,** *The Properties of Gases and Liquids,* 3rd ed., McGraw-Hill, New York, 1977.
2. **Danner, R. P. and Daubert, T. E.,** *Manual for Predicting Chemical Process Design Data,* American Institute of Chemical Engineers, New York, 1983 (tables extant).
3. **Holley, C. E.,** Funding of research in thermodynamics, *TRC Current Data News,* Thermodynamics Research Center, Texas A & M University, College Station, 7(2).
4. **Cha, T.-H. and Prausnitz, J. M.,** Thermodynamic method for simultaneous representation of ternary vapor-liquid and liquid-liquid equilibria, *Ind. Eng. Chem. Process Des. Dev.,* 24, 551, 1985.
5. **O'Reilly, M. G. and Edmonds, B.,** Physical property data — can you afford it?, *Chem. Eng. (London),* January, 61, 1978.
6. **Abrams, D. S., Massaldi, H. A., and Prausnitz, J. M.,** Vapor pressures of liquids as a function of temperature. Two-parameter equation based on kinetic theory of fluids, *Ind. Eng. Chem. Fundam.,* 13, 259, 1974.
7. **Lydersen, A. L.,** Estimation of Critical Properties of Organic Compounds, Coll. Eng. Exp. Stn. Rep. 3, University of Wisconsin, Madison, 1955.
8. **Fishtine, S. H.,** The modified Lydersen method for predicting the critical constants of pure substances, *Z. Phys. Chem. Neue Folge,* 123, 39, 1980.
9. **Fedors, R. F.,** A method to estimate critical volumes, *AIChE J.,* 25, 202, 1979.
10. **Macknick, A. B., Winnick, J., and Prausnitz, J. M.,** Vapor pressures of liquids as a function of temperature. Two-parameter equation based on kinetic theory of fluids, *Ind. Eng. Chem. Process Des. Dev.,* 16, 392, 1977.
11. **Plank, R. and Riedel, L.,** A thermodynamic criterion at the critical point of matter, *Tex. J. Sci.,* 1, 86, 1949.
12. **Gupte, P. A. and Daubert, T. E.,** New corresponding-states model for vapor pressure of non-hydrocarbon fluids, *Ind. Eng. Chem. Process Des. Dev.,* 24, 674, 1985.
13. **McGarry, J.,** Correlation and prediction of the vapor pressures of pure liquids over large pressure ranges, *Ind. Eng. Chem. Process Des. Dev.,* 22, 313, 1983.
14. **Maxwell, J. B. and Bonnell, L. S.,** Vapor Pressure Charts for Petroleum Engineers, Esso Research and Engineering Company, Linden, N.J., 1955.

15. **Clapeyron, J.,** *L'école Polytech.*, 14, 153, 1834.
16. **Gupte, P. A. and Daubert, T. E.,** A method for the prediction of heats of vaporization of non-hydrocarbon compounds based on the Clausius-Clapeyron equation, *Ind. Eng. Chem. Process Des. Dev.*, 24, 761, 1985.
17. **Tamir, A., Dragoescu, C., Apelblat, A., and Wisniak, J.,** Heats of vaporization and vapor-liquid equilibria in associated solutions containing formic acid, acetic acid, propionic acid, and carbon tetrachloride, *Fluid Phase Equilibria*, 10, 9, 1983.
18. **Chueh, C. F. and Swanson, A. C.,** Estimating liquid heat capacity, *Chem. Eng. Prog.*, 69(7), 83, 1973.
19. **Chueh, C. F. and Swanson, A. C.,** Estimation of liquid heat capacity, *Can. J. Chem. Eng.*, 51, 596, 1973.
20. **Kopp, H.,** Uber die specifische warme starrer Körper, und Folgerungen bezuglich der Zusammengesetztheit, *Ann. Chem. Pharm.*, 76, 362, 1863.
21. **Bondi, A.,** *Physical Properties of Molecular Crystals, Liquids, and Glasses*, John Wiley & Sons, New York, 1968.
22. **Abrams, D. S. and Prausnitz, J. M.,** Statistical thermodynamics of liquid mixtures: a new expression for the excess Gibbs energy of partly or completely miscible systems, *AIChE J.*, 21, 116, 1975.
23. **Reichenberg, D.,** DCS Rep. 11, National Physical Laboratory, Teddington, England, August 1971.
24. **Reichenberg, D.,** New methods for the estimation of the vicosity coefficients of pure gases at moderate pressures (with particular reference to organic vapors), *AIChE J.*, 21, 181, 1975.
25. **Reid, R. C., Prausnitz, J. M., and Sherwood, T. K.,** *The Properties of Gases and Liquids*, 3rd ed., McGraw-Hill, New York, 1977.
26. **Van Velzen, D., Cardozo, R. L., and Langenkamp, H.,** A liquid viscosity-temperature-chemical constitution relation for organic compounds, *Ind. Eng. Chem. Fundam.*, 11, 20, 1972.
27. **Baroncini, C., DiFilippo, F., Latini, G., and Pacetti, M.,** Organic liquid thermal conductivity: a prediction method in the reduced temperature range 0.3 to 0.8, *Int. J. Thermophys.*, 2, 21, 1981.
28. **Missenard, A.,** *Conductivité Thermique des Solides, Liquides, Gaz et de Leurs Mélanges*, Editions Eyrolles, Paris, 1965.
29. **Missenard, A.,** Méthode additive pour la determination de la chaleur malaire des liquides, *C. R.*, 260, 5521, 1965.
30. **Tsonopoulos, C.,** An empirical correlation of second virial coefficient, *AIChE J.*, 20, 263, 1974.
31. **Sugden, S.,** The variation of surface tension with temperature and some related functions, *J. Chem. Soc. Trans. (London)*, 125, 32, 1924.
32. **Banares-Alcantara, R. and Westerberg, A. W.,** Development of an expert system for physical property predictions, *Comput. Chem. Eng.*, 9, 127, 1985.
33. **Chao, K. C. and Greenkorn, R. A.,** *Thermodynamics of Fluids: An Introduction to Equilibrium Theory*, Marcel Dekker, New York, 1975.
34. **Abbott, M. M.,** Cubic equations of state, an interpretive review, in *Equations of State in Engineering and Research*, Advances in Chemistry Series, No. 182, Chao, K. C. and Robinson, R. L., Jr., American Chemical Society, Washington, D.C., 1979.
35. **Starling, K. E.,** *Fluid Thermodynamic Properties for Light Petroleum Systems*, Gulf Publishing, Houston, 1973.
36. **Dymond, J. H. and Smith, E. B.,** *Second Virial Coefficients of Pure Gases and Mixtures*, Oxford University Press, New York, 1980.
37. **Chao, K. C. and Seader, J. D.,** A general correlation of vapor-liquid equilibria in hydrocarbon mixtures, *AIChE J.*, 7, 598, 1961.
38. **Grayson, H. G. and Streed, C. W.,** Vapor-Liquid Equilibrium for High Temperature, High Pressure Hydrogen-Hydrocarbon Mixtures, 6th World Petroleum Congr. Proc., Sec. III, Frankfurt/Main, 1963, 233.
39. **Redlich, O. and Kwong, J. N. S.,** On the thermodynamics of solutions. V. An Equation of state. Fugacities of gaseous solutions, *Chem. Rev.*, 44, 233, 1949.
40. **Soave, G.,** Equilibrium constants from a modified Redlich-Kwong equation of state, *Chem. Eng. Sci.*, 27, 1197, 1972.
41. **Peng, D. Y. and Robinson, D. B.,** A new two-constant equation of state, *Ind. Eng. Chem. Fundam.*, 15, 59, 1976.
42. **Prausnitz, J. M.,** *Molecular Thermodynamics of Fluid Phase Equilibria*, Prentice-Hall, Englewood Cliffs, N.J., 1969.
43. **Walas, S. M.,** *Phase Equilibria in Chemical Engineering*, Butterworths, Reading, Mass., 1985.
44. **Null, H. R.,** *Phase Equilibrium in Process Design*, Robert E. Krieger, Huntington, N. Y., 1980.
45. **Chien, H. H. Y. and Null, H. R.,** Generalized multicomponent equation for activity coefficient calculation, *AIChE J.*, 18, 1177, 1972.
46. **Wilson, G. M.,** Vapor-liquid equilibrium. XI. A new expression for the excess free energy of mixing, *J. Am. Chem. Soc.*, 86, 127, 1964.

47. **Murthy, A. K. S. and Zudkevitch, D.,** The Combined Effect of Heat of Mixing and Vapor-Liquid Equilibrium on Design, Performance, and Economics of Distillation, Proc. Third Int. Symp. Distillation, Publ. 56, Institution of Chemical Engineers (England), 1979, 1.1/51.
48. **Zudkevitch, D.,** Design data — importance of accuracy, in *Encyclopedia of Chemical Processing and Design,* McKetta, J. J. and Cunningham, W. A., Eds., Marcel Dekker, New York, 1982.
49. **Hankinson, R. W., Langfitt, B. D., and Tassios, D. P.,** A single parameter equation for the prediction of multicomponent vapor-liquid data from binary isobaric data, *Can. J. Chem. Eng.,* 50, 511, 1972.
50. **Tassios, D.,** A single parameter equation for isothermal vapor-liquid equilibrium correlations, *AIChE J.,* 17, 1367, 1971.
51. **Krumins, A. E., Rastogi, A. N., Rusak, M. E., and Tassios, D.,** Prediction of binary vapor-liquid equilibrium from one-parameter equations, *Can. J. Chem. Eng.,* 58, 663, 1980.
52. **Benson, M. S. and Zudkevitch, D.,** Experimental results from the design institute for physical property data I: phase equilibria, *AIChE Symp. Ser.,* 81(244), 1985.
53. **Renon, H. and Prausnitz, J. M.,** Estimation of parameters for the NRTL equation for excess Gibbs energies of strongly non-ideal liquid mixtures, *Ind. Eng. Chem. Process Des. Dev.,* 8, 413, 1969.
54. **Renon, H. and Prausnitz, J. M.,** Local compositions in thermodynamic functions for liquid mixtures, *AIChE J.,* 14, 135, 1968.
55. **Marina, J. M. and Tassios, D. P.,** Effective local compositions in phase equilibrium correlations, *Ind. Eng. Chem. Process Des. Dev.,* 12, 67, 1973.
56. **Fredenslund, A., Michelsen, M. L., and Sorensen, J. M.,** Liquid-Liquid Equilibrium Calculations Using Activity Coefficient Models, Proc. 2nd Int. Conf., KEFCE, DECHEMA, Berlin, March, 21, 1980, 433.
57. **Palmer, D. A. and Smith, B. D.,** New two-parameter local-composition equation capable of correlating partial miscibility, *Ind. Eng. Chem. Process Des. Dev.,* 11, 114, 1972.
58. **Null, H. R.,** *Phase Equilibrium in Process Design,* Robert E. Krieger, Huntington, N.Y., 1980, 211.
59. **Gautam, R. and Seider, W. D.,** Computation of phase and chemical equilibrium. II. Phase splitting, *AIChE J.,* 25, 999, 1979.
60. **Sorensen, J. M., Magnussen, T., Rasmussen, P., and Fredenslund, A.,** Liquid-liquid equilibrium data: their retrieval, correlation and prediction. II. Correlation, *Fluid Phase Equilibria,* 3, 47, 1979.
61. **Chakravarty, T., Sivasubramanian, M. S., and Babu, S. V.,** Representation and calculation of three liquid-phase equilibria of two ternary systems, *Fluid Phase Equilibria,* 19, 221, 1985.
62. **Nagata, I.,** Prediction of ternary phase equilibria from binary data, *Thermochim. Acta,* 56, 43, 1982.
63. **Nagata, I.,** Liquid-liquid equilibria for four ternary systems containing methanol and cyclohexane, *Fluid Phase Equilibria,* 18, 83, 1984.
64. **Nagata, I.,** On the thermodynamics of alcohol solutions. Phase equilibria of binary and ternary mixtures containing any number of alcohols, *Fluid Phase Equilibria,* 19, 153, 1985.
65. **Dymond, J. H. and Smith, E. B.,** *The Virial Coefficients of Pure Gases and Mixtures: A Critical Compilation,* Clarendon Press, Oxford, 1980.
66. **Pitzer, K. S., Curl, R. F., et al.,** Volumetric and thermodynamic properties of fluids — enthalpy, free energy and entropy, *Ind. Eng. Chem.,* 50, 265, 1958.
67. **Pitzer, K. S.,** Origin of the acentric factor, in *Phase Equilibria and Fluid Properties in the Chemical Industry,* ACS Symp. Ser., No. 60, Storvick, T. S. and Sandler, S. I., Eds., American Chemical Society, Washington, D.C., 1977, 1.
68. **O'Connell, J. P. and Prausnitz, J. M.,** Empirical correlation of second virial coefficients for vapor-liquid equilibrium calculations, *Ind. Eng. Chem. Process Des. Dev.,* 6, 245, 1967.
69. **Nothnagel, K.-H., Abrams, D. S., and Prausnitz, J. M.,** Generalized correlation for fugacity coefficients in mixtures at moderate pressures, *Ind. Eng. Chem. Process Des. Dev.,* 12, 25, 1973.
70. **Tarakad, R. R. and Danner, R. P.,** An improved corresponding states method for polar fluids: correlation of second virial coefficients, *AIChE J.,* 23, 685, 1977.
71. **Hayden, J. G. and O'Connell, J. P.,** A generalized method for predicting second virial coefficients, *Ind. Eng. Chem. Process Des Dev.,* 14, 209, 1975.
72. **Stein, F. P. and Miller, E. J.,** Extension of the Hayden-O'Connell correlation to the second virial coefficients of some hydrogen bonding mixtures, *Ind. Eng. Chem. Process Des. Dev.,* 19, 123, 1980.
73. **Tsonopoulos, C.,** Second virial coefficients of polar haloalkanes, *AIChE J.,* 21, 827, 1975.
74. **Tsonopoulos, C.,** Second virial coefficients of water pollutants, *AIChE J.,* 24, 1112, 1978.
75. **Katayama, T. and Ohgaki, K.,** Measurement of interaction second virial coefficients with high accuracy, *J. Chem. Eng. Jpn.,* 13, 257, 1980.
76. **McCann, D. W. and Danner, R. P.,** Prediction of second virial coefficients of organic compounds by a group contribution method, *Ind. Eng. Chem. Process Des. Dev.,* 23, 529, 1984.
77. **Barton, J. R. and Hsu, C. C.,** P-V-T-x properties of associated vapors of formic and acetic acids, *J. Chem. Eng. Data,* 14, 184, 1969.

78. **Brandani, V., DiGiacomo, V., and Foscolo, P. V.,** Isothermal vapor-liquid equilibria for the water-formaldehyde system: a predictive thermodynamic model, *Ind. Eng. Chem. Process Des. Dev.*, 19, 179, 1980.
79. **Kogan, L. V., Blazhin, Y. M., Ogorodnikov, S. K., and Kafarov, V. V.,** *Zh. Prikl. Khim. (Leningrad)*, 50, 2682, 1977.
80. **Blazhin, Y. M., Valuev, K. I., and Kogan, L. V.,** *Zh. Prikl. Khim. (Leningrad)*, 50, 39, 1977.
81. **Credali, L., Mortillaro, L., Galiazzo, G., Russo, M., and De Checchi, C.,** Pressione di vapore sul sistema H_2O-CH_2O liquido e solido, *Chim. Ind. (Milan)*, 47, 732, 1965.
82. **Olsson, B. and Svensson, S. G.,** *Trans. Inst. Chem. Eng.*, 53, 97, 1975.
83. **Ogorodnikov, K. and Morozova, A. I.,** *Zh. Prikl. Khim. (Leningrad)*, 36, 1322, 1963.
84. **Christensen, J. J., Hanks, R. W., and Izatt, R. M.,** *Handbook of Heats of Mixing*, John Wiley & Sons, New York, 1982.
85. **Nagata, I. and Ohta, T.,** Prediction of the excess enthalpies of mixing of mixtures using the UNIFAC method, *Chem. Eng. Sci.*, 33, 177, 1978.
86. **Hanks, R. W. and Christensen, J. J.,** The Prediction of Multicomponent Vapor-Liquid Equilibria from Binary Heats of Mixing, 86th Natl. Meet., American Institute of Chemical Engineers, April 1—5, 1979.
87. **Schotte, W.,** Temperature dependence of the UNIQUAC parameters, *Chem. Eng. Sci.*, 39, 190, 1984.
88. **Ochi, K., Hiraba, S., and Kojima, K.,** Prediction of solid-liquid equilibria using ASOG, *J. Chem. Eng. Jpn.*, 15, 59, 1982.
89. **Van Ness, H. M. and Abbott, M. M.,** Vapor-liquid equilibrium. IV. Standard state fugacities for supercritical components, *AIChE J.*, 25, 645, 1979.
90. **Adler, S. B.,** VLE predictions revisited. I, *Hydrocarbon Process.*, May, 109, 1983.
91. **Adler, S. B.,** VLE predictions revisited. II, *Hydrocarbon Process.*, June, 93, 1983.
92. **Jonah, D. A.,** A new correlation for Henry's constant for the solubility of gases in liquids, *Fluid Phase Equilibria*, 15, 173, 1983.
93. **Preston, G. T. and Prausnitz, J. M.,** A generalized correlation for Henry's constants in nonpolar systems, *Ind. Eng. Chem. Fundam.*, 10, 389, 1971.
94. **Gunn, R. D., Yamada, T., and Whitman, D.,** Corresponding states. III. Henry's constants for nonpolar binary mixtures, *AIChE J.*, 20, 906, 1974.
95. **Yorizane, M. and Miyano, Y.,** A generalized correlation for Henry's constants in nonpolar binary systems, *AIChE J.*, 24, 181, 1978.
96. **Donohue, M. D., Kaul, B. K., and Prausnitz, J. M.,** Perturbed-Hard-Chain Theory for Correlation of Thermodynamic Properties, Proc. 56th Annu. GPA Conv., 1977, 36.
97. **Donohue, M. D. and Prausnitz, J. M.,** Perturbed-hard-chain theory for fluid mixtures: thermodynamic properties for mixtures in natural-gas and petroleum technology, *AIChE J.*, 24, 849, 1978.
98. **Donohue, M. D.,** Statistical Thermodynamics of Solutions in Natural-Gas and Petroleum Refining, Ph.D. dissertation, University of California, Berkeley, 1977.
99. **Kung, J. K., Nazario, F. N., Joffe, J., and Tassios, D.,** Prediction of Henry's constants in mixed solvents from binary data, *Ind. Eng. Chem. Process Des. Dev.*, 23, 170, 1984.
100. **Antunes, C. and Tassios, D.,** Modified UNIFAC model for the prediction of Henry's constants, *Ind. Eng. Chem. Process Des. Dev.*, 22, 457, 1983.
101. **Schulze, G. and Prausnitz, J. M.,** Solubilities of gases in water at high temperatures, *Ind. Eng. Chem. Fundam.*, 20, 175, 1981.
102. **Sander, B., Skjold-Jørgensen, S., and Rasmussen, P.,** Gas solubility calculations. I. UNIFAC, *Fluid Phase Equilibria*, 11, 105, 1983.
103. **Skjold-Jørgensen, S.,** Gas solubility calculations. II. Application of a new group-contribution equation of state, *Fluid Phase Equilibria*, 16, 317, 1984.
104. **Brandani, V. and Prausnitz, J. M.,** Thermodynamics of gas solubility in liquid solvents and solvent mixtures, *Fluid Phase Equilibria*, 7, 259, 1981.
105. **O'Connell, J. P.,** Thermodynamics of gas solubility, in *Proc. 2nd Int. Conf. on Phase Equilibria and Fluid Properties in the Chemical Industry, Berlin*, DECHEMA, Frankfurt, 1980, 443.
106. **Wilhelm, E., Battino, R., and Wilcock, R. J.,** Low pressure solubility of gases in liquid water, *Chem. Rev.*, 77, 219, 1977.
107. **Wilhelm, E. and Battino, R.,** Thermodynamic functions of the solubilities of gases in liquids at 25°C, *Chem. Rev.*, 73, 1, 1973.
108. **Himmelblau, D. M.,** Partial molal heats and entropies of solution for gases dissolved in water from the freezing point to near the critical point, *J. Phys. Chem.*, 63, 1803, 1959.
109. **Himmelblau, D. M.,** Solubilities of inert gases in water — 0°C to near the critical point of water, *J. Chem. Eng. Data*, 5, 10, 1960.
110. **Chappelow, C. C. and Prausnitz, J. M.,** Solubilities of gases in high-boiling hydrocarbon solvents, *AIChE J.*, 20, 1097, 1974.

111. **Tremper, K. K. and Prausnitz, J. M.,** Solubility of inorganic gases in high-boiling hydrocarbon solvents, *J. Chem. Eng. Data,* 21, 295, 1976.
112. **Cukor, P. M. and Prausnitz, J. M.,** Solubilities of gases in liquids at elevated temperatures. Henry's constants for hydrogen, methane, and ethane in hexadecane, bicyclohexyl, and diphenylmethane, *J. Phys. Chem.,* 76, 598, 1972.
113. **Hayduk, W. and Laudie, H.,** Solubilities of gases in water and other associated solvents, *AIChE J.,* 19, 1233, 1973.
114. **Schotte, W.,** Solubilities near the solvent critical point, *AIChE J.,* 31, 154, 1985.
115. **Kaul, B. K., Donohue, M. D., and Prausnitz, J. M.,** Henry's constants and second virial coefficients from perturbed-hard-chain theory, *Fluid Phase Equilibria,* 4, 171, 1980.
116. **Scrivner, N. C.,** Some Problems in Electrolyte Solutions, Annu. Meet., American Institute of Chemical Engineers, San Francisco, 1984.
117. **Sanders, S. I.,** Case studies in modeling aqueous electrolyte solutions, *Comput. Chem. Eng.,* 9, 223, 1985.
118. **Chen, C.-C., Britt, H. I., Boston, J. F., and Evans, L. B.,** Local composition model for excess Gibbs energy of electrolyte systems, *AIChE J.,* 28, 588, 1982.
119. **Cox, K. R. and Moore, M. A.,** Effects of temperature and pressure on multiphase equilibria involving aqueous electrolytes, *Fluid Phase Equilibria,* 24, 151, 1985.
120. **Johnson, A. I. and Furter, W. F.,** Salt effect in vapor-liquid equilibrium. II, *Can. J. Chem. Eng.,* 38, 78, 1960.
121. **Jaques, D. and Furter, W. F.,** Effect of dissolved salt on vapor-liquid equilibrium with liquid composition held constant, *Ind. Eng. Chem. Fundam.,* 13, 238, 1974.
122. **Rousseau, R. W. and Boone, J. E.,** Vapor-liquid equilibrium for salt containing systems: the Uniquac equation and ternary solvents, unpublished manuscript.
123. **Lynd, L. R. and Grethlein, H. E.,** IHOSR/extractive distillation for ethanol separation, *Chem. Eng. Prog.,* November, 59, 1984.
124. **Benedict, M., Webb, G. B., and Rubin, L. C.,** An empirical equation for thermodynamic properties of light hydrocarbons and their mixtures, *J. Chem. Phys.,* 8, 334, 1940; 10, 747, 1942.
125. **Benedict, M., Webb, G. B., and Rubin, L. C.,** An empirical equation for thermodynamic properties of light hydrocarbons and their mixtures: fugacities and liquid-vapor equilbria, *Chem. Eng. Prog.,* 47(8), 419, 1951; 47(9), 449, 1951.
126. **Starling, K. E.,** *Fluid Thermodynamic Properties for Light Petroleum Systems,* Gulf Publishing, Houston, 1973.
127. **Starling, K. E. and Powers, J. E.,** Enthalpies of mixtures by modified BWR equation, *Ind. Eng. Chem. Fundam.,* 9, 531, 1970.
128. **Chao, K. C. and Robinson, R. L., Jr., Eds.,** *Equations of State in Engineering and Research,* Advances in Chemistry Series, No. 182, American Chemical Society, Washington, D.C., 1979.
129. **Nishiumi, H.,** Thermodynamic property prediction of C_{10} to C_{20} paraffins and their mixtures by the generalized BWR equation of state, *J. Chem. Eng. Jpn.,* 13, 74, 1980.
130. **Nishiumi, H.,** An improved generalized BWR equation of state with three polar parameters applicable to polar substances, *J. Chem. Eng. Jpn.,* 13, 178, 1980.
131. **Lee, B. I. and Kesler, M. G.,** A generalized thermodynamic correlation based on three-parameter corresponding states, *AIChE J.,* 21, 510, 1975.
132. **Plocker, U., Knapp, H., and Prausnitz, J. M.,** Calculation of high-pressure vapor-liquid equilibria from a corresponding states correlation with emphasis on asymmetric mixtures, *Ind. Eng. Chem. Process Des. Dev.,* 17, 324, 1978.
133. **Kesler, M. G., Lee, B. I., and Sandler, S. I.,** A third parameter for use in generalized thermodynamic correlations, *Ind. Eng. Chem. Fundam.,* 18, 49, 1979.
134. **Redlich, O. and Kwong, J. N. S.,** On the thermodynamics of solutions. V. An equation of state. Fugacities of gaseous solutions, *Chem. Rev.,* 44, 233, 1949.
135. **Abbott, M. M.,** Cubic equations of state, an interpretive review, in *Equations of State in Engineering and Research,* Advances in Chemistry Series, No. 182, Chao, K. C. and Robinson, R. L., Jr., American Chemical Society, Washington, D.C., 1979.
136. **Gray, R. D., Jr.,** Industrial experience in applying the Redlich-Kwong equation to vapor-liquid equilibria, in *Equations of State in Engineering and Research,* Advances in Chemistry Series, No. 182, Chao, K. C. and Robinson, R. L., Jr., Eds., American Chemical Society, Washington, D.C., 1979.
137. **Chueh, P. L. and Prausnitz, J. M.,** Vapor-liquid equilibria at high pressures: vapor-phase fugacity coefficients in non-polar and quantum-gas mixtures, *Ind. Eng. Chem. Fundam.,* 6, 492, 1967.
138. **Wilson, G. M.,** Vapor-liquid equilibria, correlation by means of a modified Redlich-Kwong equation of state, *Adv. Cryog. Eng.,* 9, 168, 1964.
139. **Wilson, G. M.,** Calculation of enthalpy data from a modified Redlich-Kwong equation of state, *Adv. Cryog. Eng.,* 11, 392, 1966.

140. **Yarborough, L.,** Application of a generalized equation of state to petroleum reservoir fluids, in *Equations of State in Engineering and Research,* Chao, K. C. and Robinson, R. L., Jr., Eds., Advances in Chemistry Series, No. 182, American Chemical Society, Washington, D.C., 1979.
141. **Soave, G. S.,** Application of the Redlich-Kwong equation of state to solid-liquid equilibria calculations, *Chem. Eng. Sci.,* 33, 225, 1979.
142. **Soave, G. S.,** Application of a cubic equation of state to vapour-liquid equilibria of systems containing polar compounds, *Int. Chem. Eng. Symp. Ser.,* 56, 1.2/1, 1979.
143. **Soave, G. S.,** Rigorous and simplified procedures for determining the pure-component parameters in the Redlich-Kwong-Soave equation of state, *Chem. Eng. Sci.,* 35, 1725, 1980.
144. **Graboski, M. S. and Daubert, T. E.,** A modified Soave equation of state for phase equilibrium calculations. I. Hydrocarbon systems, *Ind. Eng. Chem. Process Des. Dev.,* 17, 443, 1978; II. Systems containing CO_2, H_2S, N_2, and CO, *Ind. Eng. Chem. Process Des. Dev.,* 17, 448, 1978; III, Systems containing hydrogen, 18, 300, 1979.
145. **Huron, M.-J., Dufour, G.-N., and Vidal, J.,** Vapour-liquid equilibrium and critical locus curve calculations with the Soave equation for hydrocarbon systems with carbon dioxide and hydrogen sulfide, *Fluid Phase Equilibria,* 1, 247, 1977/1978.
146. **Erbar, J. H., Jagota, A. K., Muthswamy, S., and Moshfeghian, M.,** Predicting Synthetic Gas and Natural Gas Thermodynamic Properties Using a Modified Soave Redlich Kwong Equation of State, Project 752, GPA RR-42, Oklahoma State University, Stillwater.
147. **Whiting, W. B.,** High Pressure Vapor-Liquid Equilibria for Asymmetric Mixtures, Ph.D. dissertation, University of California, Berkeley, 1982.
148. **Topliss, R. J.,** Techniques to Facilitate the Use of Equations of State for Complex Fluid-Phase Equilibria, PH.D. dissertation, University of California, Berkeley, 1985.
149. **Peng, D.-Y. and Robinson, D. B.,** A new two-constant equation of state, *Ind. Eng. Chem. Fundam.,* 15, 59, 1976.
150. **Peng, D.-Y. and Robinson, D. B.,** Two and three phase equilibrium calculations for systems containing water, *Can. J. Chem. Eng.,* 54, 595, 1976.
151. **Peng, D.-Y. and Robinson, D. B.,** A rigorous method for predicting the critical properties of multicomponent systems from an equation of state, *AIChE J.,* 23, 137, 1977.
152. **Peng, D.-Y. and Robinson, D. B.,** Calculation of three-phase solid-liquid-vapor-equilibrium, in *Equations of State in Engineering and Research,* Advances in Chemistry Series, No. 182, Chao, K. C. and Robinson, R. L., Jr., Eds., American Chemical Society, Washington, D.C., 1979.
153. **Peng, D.-Y. and Robinson, D. B.,** Two and three-phase equilibrium calculations for coal gasification and related processes, in *Thermodynamics of Aqueous Systems with Industrial Applications,* ACS Symp. Ser., No. 133, Newman, S. A., Ed., American Chemical Society, Washington, D.C., 1980. 393.
154. **Robinson, D. B., Peng, D.-Y., and Ng, H.-J.,** Capability of the Peng-Robinson programs. I. VLE and critical property calculations, *Hydrocarbon Process.,* 57(4), 95, 1978.
155. **Peng, D.-Y. and Robinson, D. B.,** User's Guide for GPA Peng-Robinson Program, Gas Processors Association, Tulsa, Okla., not dated.
156. **Schotte, W.,** Collection of phase equilibrium data for separation technology, *Ind. Eng. Chem. Process Des. Dev.,* 19, 432, 1980.
157. **Legret, D., Richon, D., and Renon, H.,** Critical evaluation of methane-hydrocarbon high-pressure experimental vapour-liquid equilibrium data using equations of state, *Fluid Phase Equilibria,* 17, 323, 1984.
158. **Patel, N. C. and Teja, A. S.,** A new cubic equation of state for fluids and fluid mixtures, *Chem. Eng. Sci.,* 37, 463, 1982.
159. **Fox, J. R.,** On the construction of nonclassical equations of state, National Engineering Laboratory, National Bureau of Standards, Boulder Colo., unpublished.
160. **Michelsen, M. L.,** Calculation of phase envelopes and critical points for multicomponent mixtures, *Fluid Phase Equilibria,* 4, 1, 1980.
161. **Peng, D.-Y. and Robinson, D. B.,** Two and three phase equilibrium calculations for systems containing water, *Can. J. Chem. Eng.,* 54, 595, 1976.
162. **Long, X. N. and Li, Y.-K.,** Computation of multiphase equilibrium phenomena with an equation of state, *Fluid Phase Equilibria,* 17, 77, 1984.
163. **Kremer, H. and Knapp, H.,** Three-phase conditions are predictable, *Hydrocarbon Process.,* April, 79, 1983.
164. **Won, K. W. and Walker, C. K.,** An equation of state for polar mixtures: calculation of high-pressure vapor-liquid equilibria of polar solutes in hydrocarbon mixtures, in *Equations of State in Engineering and Research,* Advances in Chemistry Series, No. 182, Chao, K. C. and Robinson, R. L., Jr., Eds., American Chemical Society, Washington, D.C., 1979.
165. **Barker, J. A. and Henderson, D.,** *Mol. Phys.,* 21, 187, 1971.
166. **Kim, H., Guo, T. M., Lin, H. M., and Chao, K. C.,** Cubic Chain-of-Rotators Equation of State for Polar Fluids, Annu. Meet., American Institute of Chemical Engineers, San Francisco, 1984.

167. **Mansoori, G. A. and Kwak, T. Y.,** Statistical Mechanical Description of Supercritical Fluid Extraction and Retrograde Condensation, presentation at 9th Symp. on Thermophysical Properties, University of Colorado, Boulder, June 24 to 27, 1985.
168. **Vidal, J.,** Phase equilibria and density calculations for mixtures in the critical range with simple equations of state, *Ber. Bunsenges. Phys. Chem.*, 88, 784, 1984.
169. **Huron, M.-J. and Vidal, J.,** New mixing rules in simple equations of state for representing vapour-liquid equilibria of strongly non-ideal mixtures, *Fluid Phase Equilibira,* 3, 255, 1979.
170. **Wilson, G. M. and Weiner, M. D.,** A computer model for calculating physical and thermodynamic properties of synthetic gas process streams, in *Phase Equilibria and Fluid Properties in the Chemical Industry,* ACS Symp. Ser., No. 60, Storvick, T. S. and Sandler, S. I., Eds., American Chemical Society, Washington, D.C., 1977.
171. **Davidson, D. W.,** Clathrate hydrates, in *Water, a Comprehensive Treatise,* Vol. 2, Franks, F., Ed., Plenum Press, New York, 1973, chap. 3.
172. **Schroeter, J. P., Hildebrand, H. A., and Kobayashi, R.,** Hydrate Decomposition Conditions in the System Hydrogen Sulfide — Methane and Propane, GPA Rep. TP-10, Gas Processors Association, Tulsa, Okla., not dated.
173. **Song, K. Y. and Kobayashi, R.,** Measurement and Interpretation of the Water Content of a Methane-Propane (5.31 mol %) Mixture in the Gaseous State in Equilibrium with Hydrate, GPA Project 775, RR-50, Gas Processors Association, Tulsa, Okla., not dated.
174. **Parrish, W. R. and Prausnitz, J. M.,** Dissociation pressures of gas hydrates formed by gas mixtures, *Ind. Eng. Chem. Process Des. Dev.*, 11, 26, 1972; errata 11, 462, 1972.
175. **Ng, H.-J. and Robinson, D. B.,** The measurement and prediction of hydrate formation in liquid hydrocarbon-water systems, *Ind. Eng. Chem. Fundam.*, 15, 293, 1976.
176. **Ng, H.-J. and Robinson, D. B.,** Equilibrium Phase Composition and Hydrating Conditions in Systems Containing Methanol, Light Hydrocarbons, Carbon Dioxide and Hydrogen Sulfide, GPA and CGPA Joint Rep., Proj. 825-82, RR-66, Gas Processors Association, Tulsa, Okla., not dated.
177. **Ng, H.-J. and Robinson, D. B.,** The prediction of hydrate formation in condensed systems, *AIChE J.*, 23, 477, 1977.
178. **Peng, D.-Y. and Robinson, D. B.,** Calculation of three-phase solid-liquid-vapor equilibrium using an equation of state, in *Equations of State in Engineering and Research,* Advances in Chemistry Series, No. 182, Chao, K. C. and Robinson, R. L., Jr., Ed., American Chemical Society, Washington, D.C., 1979.
179. **Misovich, M. J., Grulke, E. A., and Blanks, R. F.,** A generalized correlation for solvent activities in polymer solutions, *Ind. Eng. Chem. Process Des. Dev.*, 24, 1036, 1985.
180. **Flory, P. J.,** *Principles of Polymer Chemistry,* Cornell University Press, Ithaca, N.Y., 1953.
181. **Oishi, T. and Prausnitz, J. M.,** Estimation of solvent activities in polymer solutions using a group-contribution method, *Ind. Eng. Chem. Process Des. Dev.*, 17, 333, 1978.
182. **Sheehan, C. J. and Bisio, A. L.,** Polymer/solvent interaction parameters, in *Rubber Chemistry and Technology,* 1966.
183. **Liu, D. D. and Prausnitz, J. M.,** Thermodynamics of gas solubilites in molten polymers, *J. Appl. Polym. Sci.*, 24, 725, 1979.
184. **Prausnitz, J. M.,** *Molecular Thermodynamics of Fluid Phase Equilibria,* Prentice-Hall, Englewood Cliffs, N.J., 1969.
185. **Timmermans, J.,** *The Physico-Chemical Constants of Binary Systems in Concentrated Solutions,* Wiley-Interscience, New York, 1959.
186. **Morimi, J. and Nakanishi, K.,** Use of the Wilson Equation to calculate solid-liquid phase equilibria in binary and ternary systems, *Fluid Phase Equilibria,* 1, 153, 1977.
187. **Gmehling, J. G., Anderson, T. F., and Prausnitz, J. M.,** Solid-liquid equilibria using UNIFAC, *Ind. Eng. Chem. Fundam.*, 17, 269, 1978.
188. **Ochi, K., Hiraba, S., and Kojima, K.,** Prediction of solid-liquid equilibria using ASOG, *J. Chem. Eng. Jpn.*, 15, 59, 1982.
189. **Unno, Y., Hoshino, D., Nagahama, K., and Hirata, M.,** Prediction of solid-liquid equilibria from vapor-liquid equilibrium data using the solution of groups model, *J. Chem. Eng. Jpn.*, 12, 81, 1979.
190. **Daubert, T. E., Danner, R. P., et al.,** *Technical Databook — Petroleum Refining,* American Petroleum Institute.
191. **Alexander, G. L., Creagh, A. L., and Prausnitz, J. M.,** Phase equilibria for high-boiling fossil-fuel distillates. I. Characterization, *Ind. Eng. Chem. Fundam.*, 24, 301, 1985.
192. **Starling, K. E.,** Development of an adequate thermodynamic base for future synthetic fuels and heavy hydrocarbon processes, in Thermodynamic Needs for the Decade Ahead. Theory and Experiment, Proc. Natl. Sci. Found. Workshop, Sandler, S. I., Ed., 1983, 16.
193. **Brulé, M. R. and Starling, K. E.,** Thermophysical properties of complex systems: applications of multiproperty analysis, *Ind. Eng. Chem. Process Des. Dev.*, 23, 833, 1984.

194. **Cotterman, R. L., Bender, R., and Prausnitz, J. M.,** Phase equilibria for mixtures containing very many components. Development and application of continuous thermodynamics for chemical process design, *Ind. Eng. Chem. Process Des. Dev.,* 24, 194, 1985.
195. **Kehlen, H., Ratzsch, M. T., and Bergmann, J.,** Continuous thermodynamics of multicomponent systems, *AIChE J.,* 31, 1136, 1985.
196. **Cotterman, R. L. and Prausnitz, J. M.,** Flash calculations for continuous or semicontinuous mixtures using an equation of state, *Ind. Eng. Cem. Process Des. Dev.,* 24, 434, 1985.
197. **Sim, W. J. and Daubert, T. E.,** Prediction of vapor-liquid equilibria of undefined mixtures, *Ind. Eng. Chem. Process Des. Dev.,* 19, 386, 1980.
198. **Graboski, M. S. and Daubert, T. E.,** A modified Soave equation of state for phase equilibrium calculations. I. Hydrocarbon systems, *Ind. Eng. Chem. Process Des. Dev.,* 17, 443, 1978; II. Systems containing CO_2, H_2S, N_2, and CO, 17, 448, 1978; III. Systems containing hydrogen, *Ind. Eng. Chem. Process Des. Dev.,* 18, 300, 1979.
199. **Hipkin, H. G., Lin, T. C., and Lu, K. C.,** Survey of Design Data Needs for Refining Synthetic Crudes, Final Rep. December 1981.
200. **Chao, K. C. and Lin, H. M.,** Phase Equilibrium in Coal Liquefaction Processes, Final Rep. AF-466, Electric Power Research Institute, Palo Alto, Calif., 1976.
201. **Hochman, J. M.,** Synthetic fuels, are they different?, *Chemtech,* 12, 500, 1982.
202. **Tsonopoulos, C.,** Heavy hydrocarbons and synthetic liquids, an industry view, in Thermodynamic Needs for the Decade Ahead: Theory and Experiment, Proc. Natl. Sci. Found. Workshop, Sandler, S. I., Ed., 1983, 35.
203. **Zudkevitch, D.,** Design data — importance of accuracy, in *Encyclopedia of Chemical Processing and Design,* McKetta, J. J. and Cunningham, W. A., Eds., Marcel Dekker, New York, 1982, 431.
204. **Gray, J. A., Holder, G. D., Brady, C. J., Cunningham, J. R., Freeman, J. R., and Wilson, G. M.,** Thermophysical properties of coal liquids. III. Vapor pressure and heat of vaporization of narrow boiling coal liquid fractions, *Ind. Eng. Chem. Process Des. Dev.,* 24, 97, 1985.
205. **Wilson, G. M., Johnston, R. H., Hwang, S. C., and Tsonopoulos, C.,** Volatility of coal liquids at high temperatures and pressures, *Ind. Eng. Chem. Process Des. Dev.,* 20, 94, 1981.
206. **Grayson, H. G. and Streed, C. W.,** VLE for High Temperature, High Pressure Hydrogen-Hydrocarbon Systems, 6th World Petroleum Congr. Proc., Sec. III, Frankfurt/Main, 1963, 233.
207. **Teja, A. S., Thurner, P. A., and Pasumarti, B.,** Calculation of transport properties of mixtures for synfuels process design, *Ind. Eng. Chem. Process Des. Dev.,* 24, 344, 1985.
208. **Baltatu, M. E., Ely, J. F., Hanley, H. J. M., Graboski, M. S., Perkins, R. A., and Sloan, E. D.,** Thermal conductivity of coal derived liquids and petroleum fractions, *Ind. Eng. Chem. Process Des. Dev.,* 24, 325, 1985.
209. **Watanasirl, S., Owens, V. H., and Starling, K. E.,** Correlations for estimating critical constants, acentric factor, and dipole moment for undefined coal-fluid fractions, *Ind. Eng. Chem. Process Des. Dev.,* 24, 294, 1985.
210. **Twu, C. H.,** An internally consistent correlation for predicting the critical properties and molecular weights of petroleum and coal-tar liquids, *Fluid Phase Equilibria,* 16, 137, 1984.
211. **Zudkevitch, D. and Joffe, J.,** Correlation and prediction of vapor-liquid equilibria with the Redlich-Kwong equation of state, *AIChE J.,* 16, 112, 1970.
212. **Gray, R. D., Jr.,** Industrial experience in applying the Redlich-Kwong equation to vapor-liquid equilibria, in *Equations of State in Engineering and Research,* Advances in Chemistry Series, No. 182, Chao, K. C. and Robinson, R. L., Jr., Eds., American Chemical Society, Washington, D.C., 1979.
213. **Thies, M. C. and Paulaitis, M. E.,** Vapor-Liquid Equilibrium of Model Coal-Derived Compounds with Methanol at Elevated Temperatures and Pressures, Spring Natl. Meet., American Institute of Chemical Engineers, Houston, March 26, 1985.
214. **Sivaraman, A., Magee, J., and Kobayashi, R.,** Correlation for prediction of latent heat of pure components incorporating renormalization group formulations with corresponding-states principle, *Fluid Phase Equilibria,* 16, 1, 1984.
215. **Kesler, M. G. and Lee, B. I.,** Improved predictions of enthalpy of fractions, *Hydrocarbon Process.,* 55(3), 153, 1976.
216. **Johnson, R. L. and Grayson, H. G.,** Enthalpy of petroleum fractions, *Pet. Refiner,* 40(2), 123, 1961.
217. **Mathias, P. M.,** A versatile phase equilibrium equation of state, *Ind. Eng. Chem. Process Des. Dev.,* 22, 385, 1983.
218. **Buck, E.,** Applying phase equilibrium thermodynamics, *Chemtech,* 570, 1984.
219. **Gmehling, J., Liu, D. D., and Prausnitz, J. M.,** High-pressure vapor-liquid equilibria for mixtures containing one or more polar components, *Chem. Eng. Sci.,* 34, 951, 1979.
220. **Terrill, D. L., Sylvestre, L. F., and Doherty, M. F.,** Separation of closely boiling mixtures by reactive distillation. I. Theory, *Ind. Eng. Chem. Process Des. Dev.,* 24, 1062, 1985.

221. **Cleary, W. and Doherty, M. F.,** Separation of closely boiling mixtures by reactive distillation. II. Experiments, *Ind. Eng. Chem. Process Des. Dev.*, 24, 1071, 1985.
222. **Rackett, H. G.,** Equation of state for saturated liquids, *J. Chem. Eng. Data,* 15, 514, 1970.
223. **Spencer, C. F. and Danner, R. P.,** Improved equation for prediction of saturated liquid density, *J. Chem. Eng. Data,* 17, 236, 1972.
224. **Mallan, G. M., Michaellan, M. S., and Lockhart, F. J.,** Liquid thermal conductivities of organic compounds and petroleum fractions, *J. Chem. Eng. Data,* 17, 412, 1972.
225. **Zemaitis, J. F., Rafel, M., Clarke, D., and Scrivner, N. C.,** Aqueous Electrolyte Thermodynamics — Theory and Practice, American Institute of Chemical Engineers, New York, 1986.
226. **Buck, E. and Frank, E. M.,** Gaps in the pure component experimental physical property data base, *Chem. Eng. Prog.*, 82, March, 1984.
227. **Hand, D. B.,** Dineric distribution. I. The distribution of a consolute liquid between two immiscible liquids, *J. Phys. Chem.*, 34, 1961, 1930.
228. **Othmer, D. F. and Tobias, P. E.,** Tie line correlation, *Ind. Eng. Chem.*, 34, 693, 1942.
229. **Aly, F. A. and Lee, L. L.,** Self consistent equations for calculating the ideal gas heat capacity, *Fluid Phase Equilibria,* 6, 169, 1981.
230. **Fakeeha, A., Kache, A., Rehman, Z. U., Soup, Y., and Lee, L. L.,** Self consistent equations for calculating the ideal gas heat capacity, enthalpy, and entropy. II. Additional results, *Fluid Phase Equilibria,* 11, 225, 1983.
231. **Tsonopoulos, C., Heidman, J. L., and Hwang, S.-C.,** *Thermodynamic and Transport Properties of Coal Liquids,* John Wiley & Sons, New York, 1986.

Section III: Process Development

Process development is characterized by a significant investment of manpower. There is usually at least one process engineer, and often a larger team of people doing experimental work. The objective is development of data for the final design of an economical process. It should meet such targets as low energy consumption, low raw material usage, high reliability, and good throughput. This is therefore the appropriate time to strengthen predictions and literature data with experimental data. Although sometimes there will be sufficient data in the literature, a new process almost by definition will involve mixtures that have not been studied hitherto.

To determine where new data should be measured, it is helpful to have a flow sheet with a scoping estimate of the installed cost of each item of equipment written on it. This will help to identify the economic sensitivities. If a single exchanger costs over $1 million installed, then it would be folly to design it without accurate transport properties. Separations usually must be known to a high degree of accuracy. Thus, phase equilibrium data are usually of extreme importance.

It is not really too difficult to determine when data from the literature should be remeasured. Typically, the Design Institute for Physical Property Data (DIPPR) Data Compilation gives an estimate of accuracy. One also can look at the raw data, consider its age, and consider the reputation of the experimentalist. If there are confirming data, they cover the temperature range of interest, and a low error band is assigned by DIPPR, then the data can be considered to be safe. The converse of any of these conditions could lead one to conclude that new data are necessary if the data are critical to the design.

As considerations of cost enter into data measurements, it is well to consider the opportunity cost of not having the data. The uncertainty in the plant design will increase, and a dollar value can be put on having a plant that is 10 or 30% overdesigned in certain sections. It is poor practice to build a plant that is overdesigned because it often then will be energy-inefficient when running below its true capacity. Examples have been given previously of unit operations that did not work at all because of poor phase equilibrium information. With the wrong information, the wrong unit operation even may be chosen. This section is dedicated to companies that want to be sure before they spend their money or even "bet the company".

Chapter 6

CONTRACT DATA MEASUREMENTS

TABLE OF CONTENTS

I. RATIONALE FOR CONTRACT PHYSICAL PROPERTY MEASUREMENTS

There is sometimes resistance to the idea of contracting out data measurements. The dollars spent are "real", in that they actually flow out of the company. They thus occupy a conspicuous place in project budgets. However, there are compelling arguments for finding a good contractor to perform the physical property measurements for a company.

People who make their living taking data have a reputation to maintain, so they cannot afford to send out bad data; thus, they try to be expert in what they do. Typically, they will have access to good analytical equipment, or know how to build what they need. They are likely to be adept at figuring out the best way to measure the needed data. Sometimes a new piece of apparatus will be necessary and, based on extensive experience, they will know how to do it correctly.

In most cases, there is a higher probability of getting good data from such a contractor than from taking the data in-house. The large volume of thermodynamically inconsistent vapor-liquid equilibrium (VLE) data in the literature is a testament to the fact that a certain level of expertise is necessary to measure it correctly. In theory it is simple, but in practice it is extremely difficult. When a member of a team developing a new chemical process undertakes to measure VLE data, the project is courting disaster. Often, more money is spent on data that are worthless than would have been spent to get data of assured accuracy.

Problems that can be encountered in the measurement of VLE data in recirculating stills include use of impure chemicals, inaccurate temperature and pressure measurements, inadequate temperature or pressure control, uneven boiling, failure to recirculate both liquid and vapor, loss of light ends out the condenser, inadequate mixing of recirculated liquid and vapor, errors in sampling, inaccurate analysis, and carryover of liquid droplets in the vapor stream to the condenser. In static cells there can be inadequate mixing, inadequate degassing, gas leaks, effect of vapor withdrawal on liquid composition, adsorption losses during vapor sampling, contamination of liquid samples or phase separation during the sampling process, and some of the other problems of obtaining constant and accurately measured temperature and pressure. Experience shows that companies can take good VLE data, but only when they do it in one location under the direction of an expert.

There is considerable cost associated with testing a new apparatus. Not only must the system of interest be measured, but also a standard test system, to make sure that the apparatus works properly. Usually, some flaw will be found with the design, and the procedure will have to be repeated after remodeling of the equipment. This is particularly true for measurements at extreme conditions. Thus, before setting up equipment in-house, these questions should be asked:

1. Is there adequate expertise in-house to guarantee success?
2. Will this measurement contribute to an ongoing program of in-house physical property measurements or is it an isolated problem?
3. Are the security concerns so overriding as to compel in-house work, even though some contractors have excellent reputations for maintaining confidentiality?
4. Is it vital to have the data immediately, or can the project development continue with an *a priori* prediction until the data are measured by a contractor?

Rush projects are typically difficult to contract out. A 3-month delivery of results would be typical with private contractors, 1 or 2 years with university contractors. Projects can be expedited, but at a cost. If there has been insufficient forethought, there may be no contractor able to give 1-month service on a special project. That problem normally should not arise if the approach recommended in this book is followed. During the process assessment phase, at the latest, the need for such data will be clear. Given the time frame (1 to 3 years) of most process development work, simple planning can avoid crisis situations. If the equipment is not already set up, it is not necessarily faster to try to take such data in-house, even in the case of a crisis.

II. UNIVERSITY, GOVERNMENT, AND PRIVATE CONTRACTORS

There are fundamental differences among university, government, and private contractors because they operate in fundamentally different environments.

A. University Contracts

University professors are usually anxious to have research money, but not so anxious to have strings attached. The problem professors face is that they need to give their students an educational experience. Professors interested in physical property measurements often have their students construct new pieces of apparatus, measure several sets of data, and then store the apparatus in the basement. Measurement of data for its own sake is simply not on the priority list of most professors. The quantitative evidence of that statement is to be found in the latest bibliography of VLE data.[1] By contrast with the days when virtually all such data were measured in the U.S., only one third of the data referenced in this bibliography was even published in English! With the odd exception, most of the physical property data in the free world are now published in English. Thus, almost two thirds of such data now are being measured in Communist countries!

Fortunately, there are a few people in the universities who have recognized that there is very little that is routine about physical property data measurements. Those who have expertise in the area and are willing to make arrangements for contract research are listed later in this chapter.

It is typically difficult, although not impossible, to maintain strict confidentiality in university settings. The obstacle is the requirement that graduate students must publish work done for thesis requirements. That problem can be circumvented by hiring undergraduates or others as technicians. Even in that circumstance, control and emphasis on confidentiality are probably less than in a private organization. If a company has enough patience and is unconcerned about confidentiality, graduate student degree programs may be acceptable. The students are often highly motivated, and are forced to do a credible job in order to get their degrees. They have a professor looking over their shoulders to be sure the data are of publishable quality. Nevertheless, the experience of the contractor and the timing should be considered carefully. If the company works with the professor on a contract basis, it may be wise to have the amount of the fee depend on the timing of completion of an acceptable final report.

University contractors should be looked upon as specialists in fairly narrow areas. They typically will have one or two very good pieces of apparatus. If your project fits into the constraints of that apparatus you may have an excellent match. If your pressure, temperature, or corrosion problems are outside the range of that apparatus, it is recommended that you look elsewhere for help.

While cost estimates from a private contractor will fall in a fairly predictable range, those from universities will not. Some professors will be so anxious to establish links with industry that they will cut overheads and give relatively low prices. Others, on the other hand, will not be attuned to the market and will drastically overprice their services. Moral: it pays to shop around when dealing with university contractors.

Overhead rates in universities vary greatly. If the university has an unbending high overhead rate it is probably not feasible to do contract research there. Quite often, universities set high overhead rates as an exercise in entitlements, trying to get as much money as possible from government grant programs. A simple but stern letter to the appropriate administrator often can result in a major cut in the overhead rate. The Design Institute for Physical Property Data (DIPPR) has had a fair amount of experience in dealing with this problem while trying to put part of its data measurement projects in universities. By using the proper terminology it is sometimes possible to by-pass overhead charges entirely.

Foreign universities frequently can offer high-quality data. Three very good investigators in Europe who might be considered are A. Fredenslund in Denmark and J. Gmehling and H. Knapp in West Germany. However, communications can be a problem, so detailed descriptions of their facilities have not been included here.

B. Government Laboratories

Government laboratories such as those at the National Bureau of Standards (NBS) can be helpful. However, the comments about university research apply. There are problems with timing and special problems with confidentiality. Services will not be inexpensive. However, the quality provided is often the highest available.

C. Private Contractors

There are only four significant private contractors in North America. These are NIPER, Phillips Petroleum, D. B. Robinson and Associates, and Wiltec Inc.

National Institute for Petroleum and Energy Research (NIPER) is the former government laboratory in Bartlesville, Okla. directed now by Dr. Bruce Gammon. Because it was turned into a private laboratory by government decree, it is now able to offer complete confidentiality. As will be noted later, it offers very high-quality thermal data.

One operating company which offers physical property measurements with confidentiality is **Phillips Petroleum Co.** It has five different calorimeters, presenting a wide range of interesting industrial applications. It also offers activity coefficients at infinite dilution by ebulliometry and gas chromatography. There is no outside experience at this time, but an on-site inspection suggested to the writer that Phillips Petroleum is well equipped and staffed to make high-quality physical property measurements.

A private company with considerable experience is **D. B. Robinson and Associates** whose experience goes to over 40 binary VLE diagrams. It has a good reputation and has done some valuable work for the Gas Processors Association (GPA). As noted elsewhere, Dr. Robinson would be a good consultant on gas hydrates.

Wiltec Inc., owned and operated by Dr. Grant M. Wilson, has played an important role in the chemical industry. A description of Dr. Wilson's practices is thus in order. He has a staff of about 15 full-time employees. Long-lead-time projects from American Institute of Chemical Engineers (AIChE), American Petroleum Institute (API), GPA, and Fluid Properties Research Inc. (FPRI) are used to fill the slack periods in contract work from companies. The standard time for completion is 3 months. For 2-month service there is a 25% surcharge and for 1-month service it is 50%. Dr. Wilson is a specialist in virtually every type of physical property measurement. Often, his is the only shop in the world willing to take on hazardous or extreme conditions. If he does not have the apparatus he will build it, although the sponsoring company may have to participate financially.

The most unique aspect of Wiltec is the vast amount of experience it has accumulated. Consider over 50 critical property measurements, 200 vapor pressure curves, hundreds of heat capacity measurements, 1000 densities, 500 PVT points, about 100 PTx VLE curves, and over 500 PTxy VLE curves! It is little wonder that Wiltec has ideas on how to measure difficult systems. The company has experienced practically every type of problem, and in most cases figured out a solution.

Finally, it should be noted that Dr. Wilson has made very significant contributions to theoretical developments in thermodynamics. He developed the Wilson equation, introducing local composition theory which still dominates activity coefficient models. He also laid the theoretical basis for analytical solutions of groups (ASOG). Out of the concept that activity coefficients could be plotted for groups evolved the ASOG and UNIFAC equations which are so enormously useful. Dr. Wilson is also a specialist in equations of state. The theoretical perspective he brings to industrial problems often can help to better define the optimal set of measurements that should be run.

The work produced by Wilson and co-workers has been so reliable that there is almost a temptation to rely on it exclusively. There is no question that Wilson has an impeccable personal and professional reputation. He does consulting and has been known to suggest that contract data measurements were not needed because he knew where good existing data could be found. However, there have been many times when he has been physically unable to satisfy everyone's needs.

There is a need for support for universities and other organizations trying to maintain some expertise in data measurements. Sometimes the data requirements may go beyond standard industrial needs, to the level of accuracy produced by the NBS and several other laboratories. Finally, there are sometimes valid reasons to have in-house permanent capabilities. Therefore, the list of contractors in Section III may be of some help in sorting out the options when the decision is being made to contract out physical property data measurements.

III. LIST OF CURRENT CONTRACTORS

Contractor capabilities are listed in alphabetical order by investigators' names. The information given was provided by them in response to a questionnaire. Some investigators failed to return the questionnaires, and so were not included. To assist in scanning the list, the following index is provided, in alphabetical order by name of institution.

Institution	**Investigator**
Brigham Young University	Christensen, James J.
	Rowley, Richard L.
Brooklyn College of CUNY	Fried, Vojtech
Delaware, University of	Sandler, Stanley I.
Georgia Institute of Technology and Fluid Properties Research Inc.,	Teja, Amyn S.
Missouri-Rolla, University of	Manley, David B.
National Bureau of Standards	Olien, Neil A.
New Jersey Institute of Technology	Tassios, Dimitrios
National Institute for Petroleum and Energy Research	Steele, William V.
Notre Dame, University of	Kohn, James P.
Oklahoma, University of	Atkinson, Gordon
Pennsylvania State University	Daubert, Thomas E.
Phillips Petroleum Company	Parrish, William R.
Purdue University	
Thermophysical Properties Research Laboratory	Taylor, R. E.
School of Chemical Engineering	Chao, Kwang-Chu
Rice University	Kobayashi, Riki

Robinson, D.B. and Associates	Robinson, Donald. B.
Texas A & M University	Holste, James C.
Wiltec Inc.	Wilson, Grant M.

Principal Investigator:	**Atkinson, Gordon**	
Institution:	Department of Chemistry University of Oklahoma	
Address:	Norman, OK 73019	
Telephone:	(405) 321-0155	
Associate Investigator	B. L. Atkinson	
Pure Component Properties:		
Density, liquid		
Expansibility, liquid	0—100°C	1 bar
Future range	200°C	500 bar
A compressibility apparatus covering the same range has been designed		

Principal Investigator:	**Chao, Kwang-Chu**
Institution:	School of Chemical Engineering Purdue University
Address:	West Lafayette, IN 47907
Telephone:	(317) 494-4088
Associate Investigator:	Ho-Mu Lin

Pure Component Properties:
- Vapor pressures

Mixture Properties:
- Infinite dilution activity coefficients by gas chromatography
- PTx VLE data
- Liquid-liquid equilibrium (LLE) data
- PTxy VLE data

Pressure measurement apparatus includes Heise gauges, Texas Instruments quartz bourdon tube, and Ruska dead weight tester. Temperatures are measured with thermocouples and a platinum resistance thermometer. Compositions are measured by gas chromatography and Karl Fischer (for water). Cells are constructed of stainless steel with sapphire windows.

Data have been measured for DIPPR. The normal period for completion of measurements is 1 to 2 years. Measurements are typically done by students as part of thesis work. There is no period of confidentiality.

Principal Investigator:	**Christensen, James J.**
Institution:	Chemical Engineering Department Brigham Young University
Address:	Provo, UT 84602
Telephone:	(801) 378-2589

Associate Investigators:	Reed Izatt, John Oscarson	
Pure Component Properties:		
Heat capacity, liquid and vapor		
Mixture Properties:		
Heats of mixing, isothermal flow (no vapor space)	800 K	400 bar
PTxy VLE data (recirculating bubble-cap type, directly connected to composition analyzer)		

Confidentiality of up to permanent duration is possible with payment of a surcharge of 20%. The principal investigator was a pioneer in development of calorimetry apparatus and techniques. He has measured the heat of mixing for over 100 binary systems. His apparatus is very specialized and capable of measuring unusual heat effects around the critical point.

Principal Investigator:	**Daubert, Thomas E.**
Institution:	Chemical Engineering Department Pennsylvania State University
Address:	165 Fenske Laboratory University Park, PA 16802
Telephone:	(814) 865-2574

Pure Component Properties:	
Critical properties, T, P, V	
Refractive index	
Vapor pressure	220—670 K
Heat capacity, liquid and vapor	(Being developed)

Work is done by students, with a period of confidentiality of 1 year. A small ebulliometer is used for subatmospheric measurements. A glass cell heated by boiling vapor is used for measurements up to 70 bar. As part of the vapor pressure curve the critical point can be determined if the compound is stable. Vapor pressure data have been measured under contract to DIPPR.

Principal Investigator:	**Fried, Vojtech**
Institution:	Chemistry Department Brooklyn College of CUNY
Address:	Bedford Avenue and Avenue H Brooklyn, NY 11210
Telephone:	(717) 780-5758

Pure Component Properties:
- Refractive index
- Heat of fusion
- Vapor pressure
- Viscosity, liquid
- Thermal conductivity, liquid
- Density, liquid
- PVT

Mixture Properties
- Heats of mixing — Tronac® isothermal calorimeter
- Excess volume — pycnometer and dilatometer
- PTxy VLE data — modified Gillespie still with direct connection to gas chromatograph
- Other properties — vapor pressure, viscosity, density

The investigator is interested in the behavior of binary liquid systems with partial miscibility. He is particularly interested in critical solution behavior. He has studied many systems and has measured data under contract to DIPPR. Both period of confidentiality and normal completion time are 1 year.

Principal Investigator: **Holste, James C.**
Institution: Thermodynamics Research Center (TRC), Texas A & M University
Address: College Station, TX 77843
Telephone: (409) 845-2930

Associate Investigators: P. T. Eubank, K. N. Marsh, B. M. Gammon, K. R. Hall

Pure Component Properties:		
Heat of combustion		
Critical properties (planned)		
Vapor pressure	100—500 K	2000 bar
Heat capacity, liquid and vapor	200—400 K	400 bar
Viscosity, liquid and vapor	273—373 K	1 bar
Thermal conductivity, liquid and vapor	200—400 K	1 bar
Density, liquid and vapor	100—500 K	2000 bar
Second virial coefficients	100—500 K	
PVT data	100—500 K	2000 bar

Mixture Properties:
- PVT data — many systems studied from gases to alcohols
- Heat of mixing — flow calorimeter
- Excess volume — dilatometer
- Single-phase properties — density, viscosity, enthalpy, free energy, and entropy
- Infinite dilution activity coefficients by gas chromatography
- PTx VLE data
- PTxy VLE data

Pressures are measured by transducer and by a dead weight tester. Temperatures are measured by platinum resistance thermometer. Composition measurements are by "gravimetric construction". Cells are made from 316 stainless steel and Hastelloy C.

The investigator comments that the department has ten types of apparatus in operation. Four more are under construction and five are in design. Measurements can be completed in as little as 2 months, depending on type. The issue of confidentiality is negotiable, with a surcharge of 100% for full confidentiality.

Principal Investigator: **Kobayashi, Riki**
Institution: Chemical Engineering Department
Rice University
Address: P. O. Box 1892
Houston, TX 77251
Telephone: (713) 527-4901 or 527-8101, ext. 3519.

Pure Component Properties:
- Vapor pressure
- Heat capacity, solid, liquid, and vapor
- Viscosity, liquid and vapor
- Density, liquid and vapor
- Heat of vaporization, calorimetric
- Second virial coefficients
- PVT data

Mixture Properties:
- PVT data
- Heat of mixing, DTA
- Infinite dilution activity coefficients
- PTx VLE data — vapor recirculation/elution

Temperature is measured by platinum resistance thermometer. Composition measurements are by chromatography and density, with direct connection to VLE cell. Cells are constructed of 316 stainless steel.

The principal investigator is widely known for his accurate high-pressure measurements, which span broad temperature ranges. He suggests a site visit to inspect the equipment; he has millions of dollars worth of very impressive apparatus. Work is done by degree-seeking students so there is limited confidentiality. His projects are necessarily long-term. Collaboration is most likely when the data needed represent extensions of his present interests and when the funding can be given to augment grants from other organizations.

Principal Investigator: **Kohn, James P.**
Institution: Department of Chemical Engineering
University of Notre Dame
Address: Notre Dame, IN 46556
Telephone: (219) 239-6491

Pure Component Properties:
- Flash point
- Flammability limits
- Refractive index
- Critical properties
- Heat of fusion
- Vapor pressure
- Viscosity, liquid
- Density, liquid
- PVT data — liquid volumes and isothermal compressibility to high pressure

Mixture Properties
- PVT — approximately 45 systems have been measured
- PTx VLE data — primarily cryogenic systems

Pressure is measured by dead weight tester and bourdon tube. Temperature measurements are made by platinum resistance thermometer. Composition measurements are by gas chromatography, refractive index, and IR. Cells are constructed of glass and 304 stainless steel.

This researcher is known primarily for his work at subambient temperatures. The normal period for completion of measurements is 2 to 6 months. Data are taken by either students or the principal investigator. Grants carry an overhead rate of 54%, while contracts have a 100% rate.

Principal Investigator: **Manley, David B.**
Institution: Chemical Engineering Department, University of Missouri—Rolla
Address: Rolla, MO 65401
Telephone: (314) 341-4416

Associate Investigators: Bruce Poling, Gary Bertrand, Orrin Crosser

Pure Component Properties:		
Critical properties — potential		
Vapor pressure — 1000 points	193—523 K	70 bar
Heat capacity, solid, liquid, and vapor — potential		
Viscosity, liquid and vapor — potential		
Density, liquid and vapor	223—423 K	400 bar
Mixture Properties:		
PVT data — four systems	425 K	700 bar
Heat of mixing — titration calorimetry (vapor space)		
PTx VLE data — static cell	253—373 K	700 bar
PTxy VLE data — three binary systems in recirculating apparatus		
Visual observation cell	425 K	2700 bar

Pressure measurement is made by dead weight tester. Temperature is measured by platinum resistance thermometer. Composition measurement is by gas chromatography. Cells are constructed of stainless steel and glass.

The experience of the principal investigator is primarily in measurement of accurate PTx VLE data for hydrocarbon systems. Simultaneous measurement of gas phase density improves accuracy in the data reduction. Normal period for measurements is 6 months.

Principal Investigator: **Olien, Neil A.**
Institution: Chemical Engineering Science Division
National Bureau of Standards (NBS)
Address: Boulder, CO 80303
Telephone: (303) 497-3257

Associate Investigators: Howard J. M. Hanley

Pure Component Properties:

Property	Temperature	Pressure
Critical properties — potential		
Vapor pressure	4—330 K	800 bar
Heat capacity, solid, liquid, and vapor	4—330 K	800 bar
Viscosity, liquid and vapor	70—650 K	350 bar
Thermal conductivity, solid, liquid, and vapor	70—900 K	800 bar
Density, liquid and vapor	4—900 K	350—800 bar

Mixture Properties:
- PVT data
- Viscosity
- Thermal conductivity
- Speed of sound
- Infinite dilution activity coefficients by gas chromatography
- PTx VLE data — vapor recirculation >100 systems
- PTxy VLE data — Archimedes cell >100 systems

Pressure measurements are made by dead weight tester. Temperature is measured by platinum resistance thermometer. Composition measurements are by gas chromatography, with direct connection to VLE cell. Cells are constructed of 316 stainless steel.

The NBS at Boulder is renowned for its accurate cryogenic system measurements. Its experience is considerable, with literally thousands of pure component data points and hundreds of VLE diagrams measured.

The NBS will take data under contract, using hired personnel. Technically, the period of confidentiality is zero, but there may be a period of effective confidentiality unless another company learns that the data have been taken and asks for them. The normal period for completion of data, depending on the system, is roughly 1 year. The NBS is not known to be inexpensive, but it does of course have a reputation for very high quality.

Principal Investigator:	**Parrish, William R.**	
Institution:	Phillips Petroleum Company	
Address:	266PL PRC Bartlesville, OK 74004	
Telephone:	(918) 661-9651	
Associate Investigagor:	G. C. Allred	
Pure Component Properties:		
Heat of fusion		
Vapor pressure	473 K	140 bar
Heat capacity, solid and liquid	1000 K	150 bar
Thermal conductivity (future)		
Density, liquid and vapor	423 K	400 bar
Heat of vaporization (calorimetric)	1000 K	150 bar

Mixture Properties:
- Heat of mixing — both flow and batch calorimeters
- Excess volume by vibrating tube densitometer
- Mixture heat capacity
- Infinite dilution activity coefficients — ebulliometry and gas chromatography
- VLE, PTx and PTxy

Pressure measurements below 1000 mmHg are done with a quartz capacitance gauge. Up to 140-bar pressures are measured with a dead weight gauge. Temperatures are measured with platinum resistance thermometers. Equilibrium stills are connected directly to gas chromatographs.

The normal period for completion of measurements is 2 months, with faster response time negotiable. The extent of confidentiality, which can be absolute, is also negotiable.

Principal Investigator:	**Robinson, Donald B.**	
Institution:	D. B. Robinson and Associates	
Address:	9419—20 Avenue Edmonton, AL, Canada T6N 1E5	
Telephone:	(403) 463-8638	
Associate Investigator:	Harish Kalra	
Pure Component Properties:		
Viscosity, liquid and vapor	310—670 K	30—200 bar
Density, liquid and vapor	175—475 K	1—200 bar
Mixture Properties:		
PTxy VLE data	200—475 K	30—200 bar

The cell is windowed and has either a stainless steel or sapphire cylinder. Both variable-

and constant-volume cells are available. Pressure is measured with a dead weight calibrated Heise Digigauge®. A transducer is also used. Calibrated thermocouples are used for temperature measurement, with an uncertainty of 0.1°C. Composition measurements are by chromatography. Over 40 binary systems have been measured.

The work done by this laboratory is primarily for the enhanced oil recovery industry; thus, systems include carbon dioxide, nitrogen, and hydrogen sulfide. In addition, work is done for the petrochemical industry. The principal investigator has a very good reputation, and has done contract work for the GPA and AIChE/DIPPR.

Principal Investigator: **Rowley, Richard L.**
Institution: Chemical Engineering Department
Brigham Young University
Address: Provo, UT 84602
Telephone: (801) 378-2590

Pure Component Properties:		
Viscosity, liquid	298 K	
Thermal conductivity, liquid and vapor	260—320 K	
Density, liquid	273—373 K	1—20 bar

Mixture Properties:
- Heat of mixing — flow calorimeter with no vapor space
- Excess volume — Mettler Paar densitometer
- LLE
- Liquid-liquid critical point
- Mutual diffusivity

Pressure measurement is made by Heise gauges. Temperature is measured by quartz crystal thermometers and thermocouples. Composition measurements are by gas chromatography, Karl Fischer, and refractometer. Cells are constructed of stainless steel, glass, and Teflon®.

Future capabilities may include critical properties, PVT data, and VLE data.

Principal Investigator: **Sandler, Stanley, I**
Institution: Chemical Engineering Department
University of Delaware
Address: Newark, DE 19716
Telephone: (302) 451-2945

Pure Component Properties:		
Vapor Pressure	300-450 K	0.1—1.5 bar

Mixture Properties:
- Excess volume — Paar high-pressure densitometer

PTxy VLE data — Stage-Muller glass still (0.1 to 1.5 bar), Ruska recirculation still (5 to 300 bar)

Pressure measurement is by Wallace Tiernan mercury manometer to 1.5 bar, Budenberg dead weight tester, and Ruska null cell. Temperature is measured by Rosemount thermometer and thermistors. Composition measurements are made by gas chromatography and high-pressure liquid chromatography. The low-pressure still is glass; the high-pressure still is stainless steel with sapphire windows.

Work can be performed by either students or hired personnel. Approximately 15 binary systems have been measured to date.

Principal Investigator: **Steele, William V.**
Institution: National Institute for Petroleum and Energy Research (NIPER)
Address: P.O. Box 2128
Bartlesville, OK 74005
Telephone: (918) 337-4323

Associate Investigators: R. D. Chirico, M. M. Strube, W. B. Collier, R. H. Harrison, I. A. Hossenlopp, A. C. Nguyen, R. Sivaraman

Pure Component Properties:

Heat of combustion		
Flash point		
Refractive index		
Heat of fusion		
Critical properties, T, P, V		
Vapor pressure	73—633 K	0.00010—350 bar
Heat capacity, solid, liquid, and vapor	4—733 K	0—350 bar
Viscosity, liquid and vapor	60—250 K	1 bar
Density, liquid and vapor	240—623 K	3—350 bar
Heat of vaporization	288—438 K	0.10—2 bar
Second virial coefficients	98—623 K	0.125—350 bar
PVT data	240—623 K	3—350 bar
Speed of sound	98—573	0.10—350 bar
Future capabilities — autoignition temperature, flammability limits		

Mixture Properties:
- PVT data
- Heats of mixing — LKB microcalorimeter
- Constant volume heat capacity, one- or two-phase
- PTxy VLE data
- Future capabilities — infinite dilution activity coefficients

Pressure measurement calibration is with dead weight gauges. An NBS-calibrated platinum resistance thermometer is the primary temperature standard used. A variety of materials of construction are available: platinum, gold, quartz, Pyrex®, stainless steel, etc. Composition measurements available are gas chromatography, liquid chromatography, IR, and gas chromatography-mass spectrometry.

This facility has always specialized in characterization and highly accurate physical property testing of hydrocarbon fluids, and has special expertise in calorimetry and PVT measurements. It may be the only remaining laboratory in the U.S. doing combustion calorimetry on pure substances. From those measurements, the heat of formation can be calculated. From measurement of phase transition heats and heat capacities, the entropy and free energy of formation also can be computed. This laboratory is thus a valuable national resource. Data measurements are made both under contract and by grant. All work is done by hired personnel who are highly skilled. Confidentiality is possible. VLE measurements are not recommended at this laboratory, but pure component thermal properties are among the best obtainable.

Principal Investigator: **Tassios, Dimitrios**
Institution: Chemical Engineering Department
New Jersey Institute of Technology
Address: Newark, NJ 07102
Telephone: (201) 736-3257

Pure Component Properties:

Vapor pressure	250—350 K	1.3 bar

Mixture Properties:
- Infinite dilution activity coefficients by gas chromatography
- PTx VLE data
- PTxy VLE data, static cell

Pressure measurement is made by U tube manometer and cathetometer. Temperature is measured by mercury thermometer. Cells are constructed of glass and appropriate only for low-pressure work. Gas solubilities also can be measured.

Principal Investigator: **Taylor, R. E.**
Institution: Thermophysical Properties Research Laboratory (TPRL)
Purdue University
Address: 2595 Yeager Road
West Lafayette, IN 47906
Telephone (317) 463-1581

Associate Investigator: R. L. Shoemaker

Pure Component Properties:
- Heat capacity, solid and liquid
- Thermal conductivity, solid and liquid

Density, solid and liquid
Thermal diffusivity
Thermal expansivity
Emissivity

The TPRL, associated with CINDAS (described in Chapter 4), measures data under either grant or contract. The normal period for completion of measurements is 2 months. With a 100% surcharge, data will be provided in 2 weeks.

Principal Investigator: **Teja, Amyn S.**
Institution: Chemical Engineering Department, Georgia Institute of Technology
Address: Atlanta, GA 30332
Telephone: (404) 894-3098

Pure Component Properties:		
Critical temperature (low residence time)		
Vapor pressure	273—450 K	0.1—1.5 bar
Heat capacity, liquid	273—500 K	10 bar
Viscosity, liquid	273—500 K	400 bar
Thermal conductivity, liquid	273—500 K	10 bar
Density, liquid and vapor	273—450 K	350 bar
PVT data	290—420 K	350 bar

Mixture Properties:
PVT data
Excess volume — vibrating tube densitometer and high-pressure pycnometer
Solubilities in supercritical gases
PTxy VLE data — recirculating
Viscosity
Thermal conductivity
Heat capacity

Pressure measurement is by Heise gauges and dead weight tester. Temperature is measured by platinum resistance thermometer. Composition measurements are by gas chromatography, density, refractive index, and high-pressure liquid chromatography. Cells are constructed of glass and stainless steel.

At this writing Dr. Teja had been appointed principal investigator for the Fluid Properties Research Inc. (FPRI). His list of equipment includes that formerly developed at Oklahoma State University for measurement of transport and thermal properties. Dr. Teja also had a contract with DIPPR to measure critical temperatures of fluids which decompose on heating.

For a small surcharge, a period of confidentiality of 1 year is possible.

Principal Investigator: **Wilson, Grant M.**
Institution: Wiltec Inc.
Address: 488 South 500 West
Provo, UT 84601
Telephone: (801) 374-6648

Associate Investigators: Paul Gillespie, Jon Owens, Ralph Wood, Louis Jasperson, Vince Wilding

Pure Component Properties:
- Critical properties
- Heat of fusion
- Vapor pressure
- Heat capacity, solid, liquid, and vapor
- Viscosity, liquid and vapor
- Density, liquid and vapor
- Heat of vaporization, calorimetric
- PVT data
- Surface tension

Mixture Properties:
- PVT data
- Heats of mixing — static (with vapor space) and flow (no vapor space)
- Excess volume — pycnometer or PVT cell
- Heat capacity
- Viscosity
- Surface tension
- Density
- Infinite dilution activity coefficients by gas chromatography
- PTx VLE data — static, still, and flow apparatus
- PTxy VLE data — static, still, and flow apparatus

Pressure measurements are made by calibrated pressure gauges and transducers, mercury manometers, and dead weight tester. Temperature is measured primarily by calibrated platinum resistance thermometers, and by calibrated thermocouples in some cases. Composition measurements are by gas chromatography, Karl Fischer, redox, and gas absorption. Cells are constructed of glass, stainless steel, Teflon®, Monel, Inconel, and Hastelloy.

Wiltec has considerable experience in measuring difficult data. Often it is the only organization willing to contract data on certain systems which are toxic or require extreme conditions. Wiltec is particularly good at analyzing difficult problems and figuring out what type of data would be most useful, and the company has a reputation for pointing out other ways to find the data if it knows them to be already measured. The organization has considerable experience with short residence time to minimize chemical reactions. Supercritical data have been measured in support of the developing supercritical extraction industry. In some cases, reaction kinetics also have been measured.

IV. SUMMARY

Experimental physical property data are always needed as the basis for a process design. For design of some processes the data base available in the literature and the correlations based on those data are sufficiently accurate. Usually, that is not the case. Whether the data are unavailable, inaccurate, or just suspect, it is necessary to make experimental measurements. There are valid reasons for choosing to contract out the data. It is usually less expensive and more reliable to do so than to measure the data in-house when the measurements are conducted at extreme conditions. Routine types of measurements may be appropriate for in-house work.

In spite of the drastic decline in experimental physical property measurement capability in the U.S., a number of people and organizations have been identified that appear to be willing to do high-quality contract work. It is hoped that in the future this list can be expanded. Sections A and B following summarize the investigators willing to measure certain properties. The principal investigators are identified by number according to the following key.

1	Atkinson, G.	10	Manley, D. B.
2	Chao, K.-C.	11	Olien, N. A.
3	Christensen, J. J.	12	Robinson, D. B.
4	Daubert, T. E.	13	Rowley, R. L.
5	Fried, V.	14	Sandler, S. I.
6	Gammon, B. E.	15	Tassios, D.
7	Hall, K. R.	16	Taylor, R. E.
8	Kobayashi, R.	17	Teja, A. S.
9	Kohn, J. P.	18	Wilson, G. M.

Pure Component Properties

Solids

Heat of fusion 5, 6, 9, 18
Heat capacity 6, 8, 10, 11, 16, 18
Heat of combustion 6, 7
Density 16
Thermal conductivity 11, 16

Liquids

Heat capacity 3, 4, 6, 7, 8, 10, 11, 16, 18
Heat of combustion 6, 7
Density 1, 5, 6, 7, 8, 9, 10, 11, 12, 13, 16, 17, 18
Thermal conductivity 5, 7, 11, 13, 16, 17
Vapor pressure 2, 4, 5, 6, 7, 8, 9, 10, 11, 14, 15, 17, 18
Critical properties 4, 6, 9, 17, 18
PVT 1, 5, 6, 7, 8, 9, 16, 17, 18
Heat of vaporization 6, 8, 18
Viscosity 5, 6, 7, 8, 9, 10, 11, 12, 13, 17, 18
Flash point 6, 9
Flammability limits 9
Refractive index 4, 5, 6, 9
Speed of sound 6, 11
Surface tension 18
Emissivity 16

Vapors

Heat capacity 3, 6, 7, 8, 10, 11, 18
Thermal conductivity 7, 11, 13
Density 6, 7, 8, 10, 11, 12, 17, 18
Viscosity 6, 7, 8, 10, 11, 12, 18

Mixture Properties

Low-pressure VLE — PTxy 2, 5, 7, 10, 11, 12, 14, 15, 17, 18
Low-pressure VLE — Ptx 2, 7, 8, 11, 15, 18
Moderate- to high-pressure VLE — PTxy 2, 3, 6, 10, 11, 12, 14, 17, 18
Moderate-pressure VLE — Ptx 2, 8, 9, 10, 11, 18
Infinite dilution activity coefficients (GLC) 2, 7, 8, 11, 15, 18
Infinite dilution activity coefficients (ebulliometry) 18
LLE 2, 13, 18
Heat of mixing 3, 5, 6, 7, 8, 10, 13, 18
Excess volume 5, 7, 13, 14, 17, 18
PVT 6, 7, 8, 9, 10, 11, 17, 18
Diffusivity 13
Solubilities in supercritical gases 17, 18

The inclusion of candidates in this list is not meant to be an endorsement or guarantee that they will perform measurements on schedule or produce accurate data. (Investigators who wish to be added to future lists should contact the author.) Clients should look carefully into the reputation of the contemplated contractor for good-quality experimental work, as well as the probability of having the proposed project completed on time. With those caveats, the greatest risk is in not having experimental backup where it is most needed.

REFERENCES

1. **Wichterle, I., Linek, J., and Hala, E.,** *Vapor-Liquid Equilibrium Data Bibliography* Suppl. II, Elsevier, New York, 1979.

Chapter 7

DEVELOPING AN IN-HOUSE MEASUREMENT CAPABILITY

TABLE OF CONTENTS

I. RATIONALE FOR IN-HOUSE DATA MEASUREMENTS

The previous chapter suggested the advantages and disadvantages of reliance on contract data measurements. The other option, use of in-house capabilities, also offers a number of advantages and disadvantages.

If there is an equipped laboratory and someone trained to run it, response time can be days instead of months. Sometimes it can take less time to actually carry out the experiments than to get approval for funds to have them done elsewhere. This is particularly true when some of the more convenient methods described below are used. If the research program is aimed not at a new chemical synthesis but at a new separation for existing chemicals, probing data may be needed to establish feasibility. Realistically, it is not easy to get money for contract research on high-risk probing projects.

Repeated measurements of the same type often can justify an in-house capability. Management can gain some control over both the financial and timing aspects of such projects. Again, response time can be a significant factor. If the equipment and trained personnel are not in place, such a project sometimes can take longer.

Security is enhanced by keeping projects in-house. This may be of concern when the data required are related to a top-secret development project. In descending order of security we would look to in-house, private contractor, university, and government laboratory research. Problems with government research would not be related to integrity of the personnel, but to access under the Freedom of Information Act.

Seldom can in-house measurements be justified on a one-time basis for some of these physical property measurements. Thus, a conscious decision has to be made that a permanent capability is desired and the company is willing to support it. This does not mean that the equipment has to be active all the time, but it must be kept in a state of reasonable readiness. It needs space, maintenance, and someone with a commitment to it.

Data that present unusual experimental requirements may be better taken under contract if there is little chance of a second use for the apparatus. Sometimes equipment becomes so sophisticated that a full-time effort is needed to maintain it. A section on calorimetry has not been included in this book because calorimetry falls into that category. Most physical property measurements look easy to someone who has not performed them. Calorimetry can be simple and straightforward, but there is a large capital investment required and service is a very difficult problem. Isothermal titration calorimeters, for example, are very sophisticated devices with complex circuitry and control systems.

The personnel problems created in trying to staff an in-house laboratory can vary depending on the work load. If the laboratory is very busy, with more than one technician at all times, it will be easy to keep it staffed. It will be possible to keep it under the part- or full-time direction of a trained professional. Technicians can be attracted because of the correct perception that work in this area is technically challenging. However, the normal circumstance of varying work requirements puts company resourcefulness to the test. Development programs, the greatest source of data needs, will come and go. This leads to problems in maintaining trained technician help. Use of a semiprofessional who can do a great deal of the data evaluation and report writing can be the best solution. Such a person usually has other skills such as computer expertise, and can move from one assignment to another so as to be available when the data are needed. Thus, the professional supervising a small laboratory does not have to spend an inordinate amount of time training new people.

II. OVERVIEW OF CHAPTER

A chapter dealing with experimental measurements could easily turn into a multivolume treatise. However, the objective of this chapter is really quite narrow. There will be an exclusive focus on vapor-liquid equilibrium (VLE) data. It is usually the most important information required for design of distillation columns. Sometimes it is also extremely important in design of multiphase reactors. Although the writer has had experience in measuring heats mixing, excess volumes, and liquid-liquid equilibrium (LLE), those topics will be ignored.

Within the framework of VLE, gas solubility can be treated. Polymer-solvent equilibrium is also a special case. However, high-pressure measurements are not treated, and solid phases are also ignored.

To introduce the problems inherent in measurement of VLE, the following discussion begins with examination of a static equilibrium experiment. The difficulties in obtaining samples and the problems with analytical measurements are both discussed. One of the better vapor/liquid recirculation cells, developed by Dr. Grant. M. Wilson, is then described.

The discussion of regular VLE measurements is followed by a discussion of PTx measurements. The vapor and liquid compositions are not measured, although the liquid composition is known by material balance. The vapor composition is computed by the Gibbs-Duhem equation. An apparatus developed by J. Gmehling is discussed in detail.

One of the most powerful laboratory techniques for measurements of VLE involves the gas chromatograph. If one component is much less volatile than the others, it can be coated on an inert support material. Then the retention times of volatile solutes passing over it can easily be measured. Thermodynamic relations enable computation of the activity coefficient at infinite dilution of the solute in the heavy solvent. This technique can be used for measurement of gas solubilities and polymer-solvent equilibria as well.

Finally, measurement of infinite dilution activity coefficients by ebulliometry is discussed. That method involves recirculation of both liquid and vapor. It can be used for pure component vapor pressures as well as ordinary VLE. The method has both the advantages and disadvantages of recirculating equilibrium cells.

III. VLE IN STATIC CELLS

VLE cells are either static or recirculating. A static cell involves no movement of the liquid or vapor except by rocking or other direct agitation of the phases. The cell with its contents must be completely purged of unwanted chemicals and gases. In such cells, temperature control and measurement are straightforward, pressure measurement is more difficult, and composition measurements are even more difficult.

A. Temperature and Pressure Measurement

Static cells must be immersed in a constant-temperature gas or liquid bath, maintained within ± 0.02°C. Care must be taken to assure that there are neither vertical nor lateral temperature gradients in the bath. The absolute temperature is measured with either a quartz crystal or platinum resistance thermometer.

Pressure measurement may be difficult, because vapor can condense in a line leading to a pressure gauge. That difficulty can be overcome by connection of pressure transducers to the vapor space of the cell. They can either give a direct reading or be used to null against a pressure on the other side of the gauge. The latter approach has the advantage that pressure standards such as dead weight testers and mercury manometers can be used. Alternatively, the line to a pressure gauge can be filled with a fluid to prevent vapor or liquid from the cell from entering the line. Both methods can suffer from problems of static head correction. At low pressures, a heated line leading directly to a heated mercury manometer reduces the problem of static head correction.

B. Mixing

Good mixing of the vapor and liquid space is needed, leading to attainment of phase equilibrium. Usually static cells either are rocked or have some form of magnetically controlled internal mixer. The use of magnetic stirring bars is usually inadequate. However, the use of magnetically controlled agitators is mechanically much more difficult. One design has been previously described.[23] The attainment of equilibrium can be determined by successive measurements of the vapor phase composition, at time separations of 1/2 hr; 2 hr are usually adequate to achieve equilibrium in a well agitated static cell.

C. Composition Measurements

One of the biggest problems with static cells is obtaining representative samples without disturbing the equilibrium. It is anticipated that methods will be developed to measure compositions through windows in the cell without disturbing the equilibrium. In the meantime, however, actual samples must be removed. Every precaution must be taken to disturb the equilibrium as little as possible.

1. Vapor Samples

Removing and accurately analyzing a vapor sample is a most difficult challenge. In the presence of a large amount of metal, adsorption can be very high. Thus, adsorption losses can change the apparent composition of a sample drastically. If the mixture constituents have similar volatility, a reasonable plan is to have a gas sampling valve mounted in a heated oven and located in extremely close proximity to the equilibrium cell. It should be connected to the cell though a bellows or other type of low-leakage valve and a vertical capillary line. At the sample valve outlet there should be another valve connected to a vacuum system. For each sample the bellows valve is closed, the vacuum valve is opened, and the sample valve is evacuated. Then the vacuum valve is closed, the bellows valve is opened to expose the sample valve to the cell vapor, and the sample is injected into the gas chromatograph or other analytical device.

Both the connecting line and the sampling valve should be maintained at a minimum temperature of 150°C, and preferably 100° hotter than the equilibrium cell itself. To avoid depleting the cell of volatile components, the sample size must be very small in relation to the volume of liquid in the cell. Computations can be made to determine the worst-case depletion of volatiles during sampling. If the sample size is about 0.5 cc, the liquid volume in the cell should be about 1 ℓ. These suggestions, of course, are subject to computations on the specific system being studied.

In spite of all these precautions, vapor sampling by this method still can be fraught with peril. Problems encountered include sample decomposition, inadequate purging of the sample system to avoid selective adsorption of heavy components, and resulting nonrepeatabilty. The researcher should determine in advance whether any of these constraints are applicable to the system proposed for study.

2. Liquid Samples

Liquid sampling difficulty depends on the cell pressure. Static cells can be used at virtually any pressure for which they are designed. Sampling can be simple with superatmospheric pressure experiments, since a capillary line can lead directly to a liquid sampling valve. However, the line must be colder than the cell in order to avoid flashing. If the equilibrium cell is below room temperature, a special sampling procedure is required to avoid fractionation in the sample line. When there is a significant difference in volatility between components, use of a liquid sampling valve is not recommended because of serious tailing.

High-pressure VLE measurements require special techniques; in fact, the cost of designing a safe cell for high-pressure measurements often can be so high as to suggest the advisability of contracting out the measurements. Nevertheless, a description of techniques is given by Renon et al.[25,26]

It is difficult to maintain a system without any leaks. The leaks are particularly troublesome in vacuum systems. In fact, with sufficient leakage into the cell there is potential for development of an explosive mixture. During vapor sampling, the presence of air can be determined. Leak testing under pressure or use of a helium leak detector can help to isolate the problems. It is important that only bellows seal or very low dead-volume valves be used, because valve stems usually have some leakage. Connections to steel cells should be welded or soldered. As a rule of thumb, anything that can leak will leak.

D. Thermodynamic Consistency Test

Whenever Ptxy measurements are made, thermodynamic consistency should be checked for the binary mixture. Techniques are described in Chapter 4. Consistency of ternary data also can be checked if the proper approach is used in fixing the liquid compositions.[23] A binary VLE point is measured. Then the third component is added and another measurement taken. Subsequent additions of the third component trace out a line of composition toward the pure component corner of the third component. The equation for testing of those data is as follows:

$$\int_{x_k=0}^{x_k} \left(\ln \frac{\gamma_k^{r+1}}{\gamma_i \; \gamma_j^r} \right) dx_i - \left(\frac{G^E}{RT} \right)_{x_k=0} + \left(\frac{G^E}{RT} \right)_{x_k} = 0 \quad (1)$$

The integration is performed at a constant ratio of components i and j.

It should be remembered that designs have been published for hundreds of VLE cells. No single cell has answered all the needs; in fact, many published designs simply do not work correctly. For that reason, a large percentage of published phase equilibrium data is of only poor to fair quality.

IV. VLE BY RECIRCULATION

Cells which recirculate the vapor, liquid, or both usually permit easier sampling. Furthermore, there is no need for degassing. Of the scores of designs proposed, many could be considered. The design developed by Dr. Grant M. Wilson, chosen for presentation here, has been proven reliable in studies done on many very difficult systems, ranging from vacuum conditions to moderately high pressures.

Wilson's still recirculates both liquid and vapor. Furthermore, the equilibrium chamber is surrounded by the boiling liquid at the same conditions of temperature and pressure.[44] In other words, it is a still within a still. The outer still, containing about 1 ℓ of liquid, serves as an adiabatic shield to the equilibrium being measured in the inner still. Figure 1 illustrates the design of this apparatus.

The operation of the still is quite straightforward. The inner and outer stills are situated

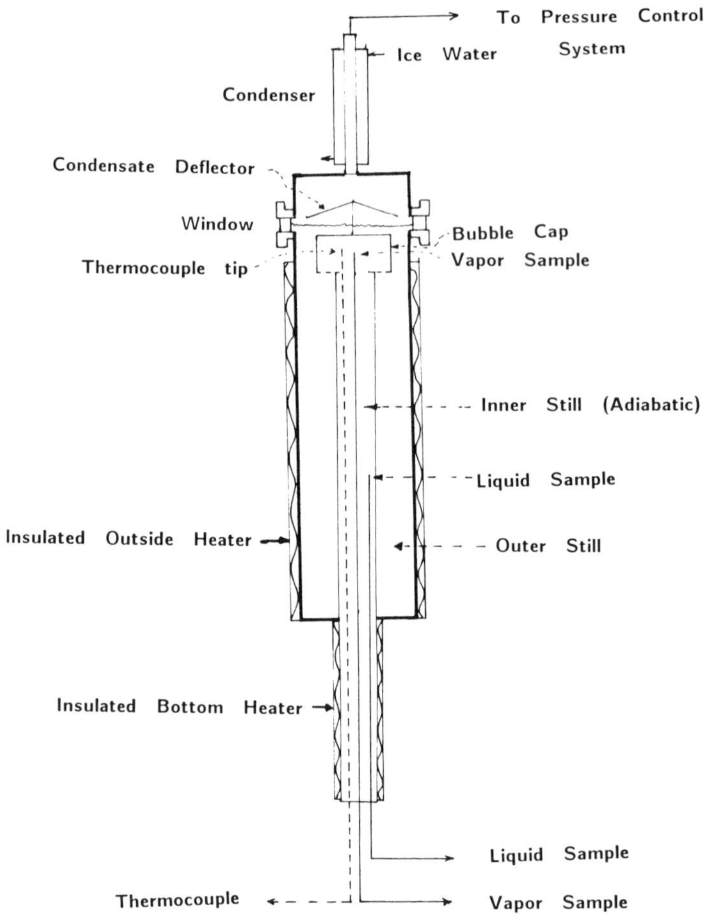

FIGURE 1. Schematic of Dr. Grant M. Wilson's high-pressure recirculating VLE apparatus.

beneath a cold condenser where equilibrium is established with the inert gas of the pressure control system. Vapor generated in the two stills is returned as cold condensate. Thus, a mixture of vapor and liquid rises from below, is condensed, and falls back into the outer cell. The two cells are connected at the bottom of the outer cell, where liquid replenishes vapor generated in the inner cell. Samples of vapor are captured at the top of the inner cell, under a bubble cap. The vapor passes out at that point for eventual condensation. The temperature is measured in the bubble cap, while liquid samples are taken just below it. Wilson's description of the overall process is as follows:

> Liquid in the inner still is also boiled by means of the insulated bottom heater. This produces vapor bubbles which rise in the vertical pipe through a distance of about 23 cm where equilibration of the vapor and liquid phases occurs in a manner analogous to the Cottrell vapor-liquid contactor in recirculation-type stills. After equilibration in the vertical pipe, the vapor bubbles become trapped in the bubble cap shown at the top of the pipe...The bubble cap serves as a means of trapping equilibrium vapor which can then be sampled and analyzed. This is done by means of the vapor sample line...It was a 0.5 mm i.d. capillary line which was connected to a special low-dead-volume sample valve...The vapor was allowed to condense in the sample line, and the sample was removed as a liquid at conditions close to ambient temperature.

The vapor and liquid samples, both in liquid form, can go to a liquid sampling valve on a gas chromatograph. Alternatively, other analytical measurements such as Karl Fischer can be used. In Wilson's work, the temperature was measured by means of a platinum resistance thermometer inserted from below into the bubble-cap vapor space. Pressure is measured easily by manometers, gauges, or dead weight testers.

Two special features of the cell shown in Figure 1 are a window for observation, and a deflector shield to prevent the cold condensate from contacting the bubble cap directly. One difficulty noted with the glass cell is initiation of boiling without excessive overheating. It is important either that agitation of some type be induced, or else that there should be a significant amount of surface area. Fritted glass surfaces are helpful in glass cells. One modification made by this author in a glass cell patterned after Wilson's idea included a liquid leg from the top of the cell to the bottom of the outer cell. It contained a check valve and appeared to work well; however, it was not tested extensively.

Wilson has pointed to a number of advantages of such a cell, which can be constructed in either metal or glass. It has exactly one theoretical plate. More intricate designs are required in some other types of stills. Steady state is achieved very quickly compared to other recirculation stills. Because of the large amount of liquid compared to the small size of vapor samples, the measured liquid compositions are very close to the charge composition.

V. VLE BY PTx MEASUREMENTS

The greatest source of error in most VLE measurements is usually the sampling and analysis of the vapor phase. However, if the gas phase nonidealities can be modeled accurately it is not necessary to measure the gas phase composition physically. This is because of the fact that the criterion for thermodynamic consistency, the Gibbs-Duhem equation, makes one of the measurements (t, p, x, y) redundant. Therefore, it was reasoned that very accurate measurement of pressure as a function of composition would be adequate. Liquid compositions are determined either volumetrically or by weight. With no necessity of sampling either phase, a static cell would be adequate. However, extremely careful degassing and high-purity components are required. Considerable success with this method has led to several viable equipment designs.

A. Experimental Apparatus

There are several different approaches to measurement of PTx data. One approach, developed by Maher and Smith,[28] uses 15 separate cells, with liquid compositions accurately determined by weighing. Degassing is accomplished by cycles of deep freezing, evacuating and melting. The cells are then attached to a manifold which permits pressure measurement. Extreme care is taken to obtain accurate pressure and temperature measurements. The level of sophistication required may be more than most companies should attempt. A combination of extreme component purity, extensive degassing, accurate control over liquid composition, and high accuracy in temperature and pressure measurements is required.

Abbott[26] and Gibbs and Van Ness[27] pioneered an approach which uses a single static cell. They degas the liquids in separate vessels and then introduce accurate measured volumes. Many points can be generated across a binary composition range. Details of a simpler approach based on the same idea have been given by Gillespie et al.[45]

Kolbe and Gmehling[46] describe in detail a similar apparatus; Figure 2 illustrates the overall setup used in the Gmehling laboratory. Liquids purified and degassed in a small distillation column are fed into the evacuated feed chambers. The piston injectors are then used to introduce accurately known quantities of liquid into the equilibrium cell. The temperature inside the bath is controlled so that the VLE data are isothermal. Water is used as the bath fluid from 0 to 90°C. From 90 to 150°C the bath fluid is glycerine. The short-term stability obtained in that bath is 0.01 K, while the long-term drift (24 hr) is 0.2 K. The pressure is measured with a dead weight tester, isolated from the cell by a differential pressure indicator

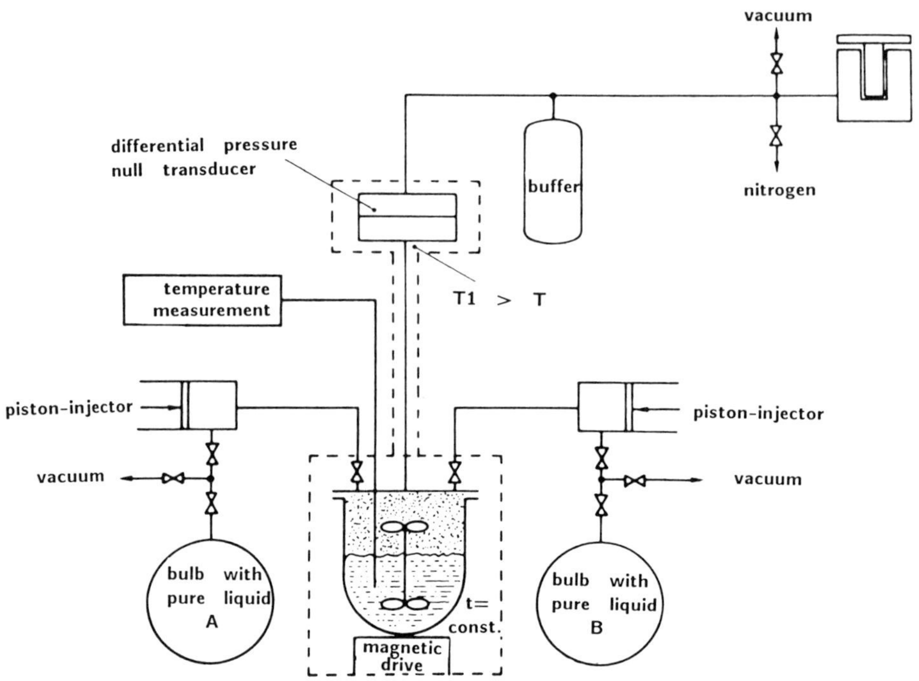

FIGURE 2. Overall equipment setup for accurate measurement of Ptx data, guaranteeing degassed liquid feeds, excellent temperature control and measurement, and accurate pressure measurement. (Reproduced from Kolbe, B. and Gmehling, J., *Fluid Phase Equilibria,* 23, 213, 1985. With permission.)

(Ruska Inc. type 2413-702). The standard deviation for that device is about 0.2 mbar; its sensitivity is 0.04 mbar. That transducer and the capillary line connected to it are kept at a temperature higher than the cell to avoid condensation. The liquid injections are made with a hand-operated piston pump whose temperature is measured at the time of each injection.

The Gmehling cell is made from thick-walled (0.5 mm) glass and can be used up to 10 atm pressure; the liquid volume is about 180 cc. A cross-section of the cell is illustrated in Figure 3. The device on the top right-hand side is the quartz crystal thermometer (Hewlett-Packard 2801A). Note the fairly sophisticated agitators, driven magnetically. The valves for the pressure measurement and introduction of liquid are built into the head to reduce dead volumes. The seals used are Teflon®. The data suggest that the apparatus is quite accurate. However, greatest errors occur in the regions near infinite dilution.

B. Data Reduction

1. Barker Method

The data reduction is almost as difficult as the experiments themselves. The Barker method assumes an expression for the Gibbs excess free energy and from that an equation for the system pressure can be computed directly.[29-31] The parameters in that equation are derived as part of the fitting procedure. The total pressure is the sum over each component of the liquid mole fraction times its activity coefficient, times its vapor pressure, times its fugacity coefficient, and divided by the system fugacity coefficient. This approach is superficially very straightforward. However, several complexities are quickly apparent. How are supercritical components treated? This is illustrated with examples by Won and Prausnitz.[32] Treatment of ternary systems is discussed in two different papers.[33,34]

Should the experimentally determined vapor pressure be used, or the value derived from a vapor pressure equation? The answer is that *only the experimentally determined vapor*

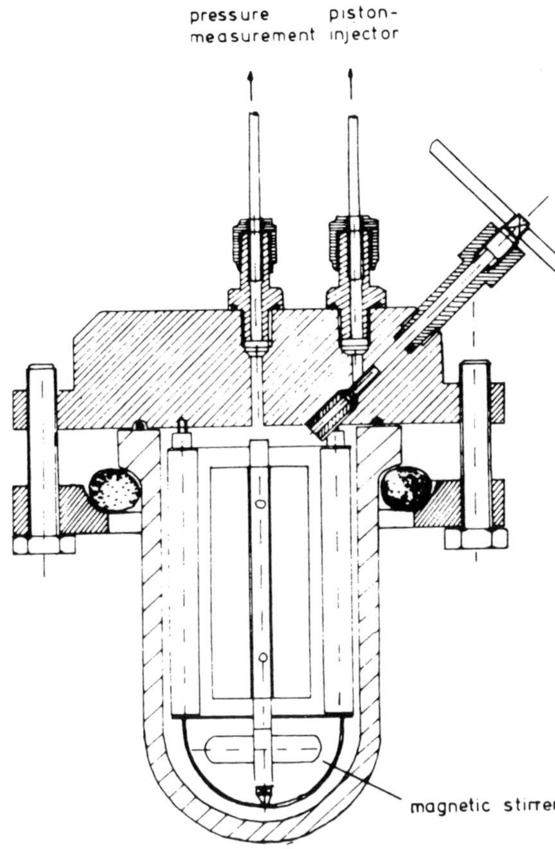

FIGURE 3. Close-up view of equilibrium cell. Note the use of a sophisticated magnetic stirrer, a thick-walled glass cell, a quartz crystal thermometer on the top right-hand side, and valves built into the cell. (Reproduced from Kolbe, B. and Gmehling, J., *Fluid Phase Equilibria*, 23, 213, 1985. With permission.)

pressures should be used. Munja et al.[35] give an example for acetonitrile + ethyl acetate. The following set of infinite dilution activity coefficients was derived for acetonitrile, depending on whether the experimentally measured vapor pressure for acetonitrile was used, or whether that value was taken from the Antoine equation.

Temperature (K)	Experimental	Antoine
313.15	1.575	1.371
353.15	1.553	1,389
393.15	1.498	1.496

Abbott and Van Ness[36] also noted the importance of choice of experimental vapor pressure in the data reduction.

Another problem in the data reduction is the choice of equation of state (EOS) for computation of the vapor phase fugacity coefficient. The effect is particularly noticeable in the computed infinite dilution activity coefficients. For example, using Barker's method with the three-constant Redlich-Kister equation, the following values were obtained.

	Component 1	Component 2
Chlorobenzene + acetone (313.15 K)		
Virial	1.617	1.598
Redlich-Kwong	1.575	1.525
Acetonitrile + ethanol (393.15 K)		
Virial	1.980	2.343
Redlich-Kwong	1.870	2.081
Acetonitrile + ethyl acetate (393.15 K)		
Virial	1.592	1.585
Redlich-Kwong	1.498	1.439

The importance of the EOS increases with increasing pressure. Uncertainties may be due to difficulties in predicting cross virial coefficients, or in definition of the interaction parameter in the Redlich-Kwong EOS. The differences indicated here would be smaller for the same systems measured at lower temperatures. It is probably wise to limit Ptx measurements to temperatures that give pressures which do not exceed 2 atm. When the large potential variance in infinite dilution activity coefficients apparent from the preceding data is considered, it is not unreasonable to ask whether interpolation based on direct measurement of the infinite dilution activity coefficients might be more accurate. It may be necessary to take a number of measurements of equilibrium pressure very close to infinite dilution in order to pin down the limiting slope. By so doing, the infinite dilution value could be determined directly by data reduction methods described elsewhere in this chapter.

2. Mixon Method

A different approach to data reduction was developed by Mixon et al.[37] It requires data points which are equally spaced with respect to composition. The equidistant spacing is achieved numerically by use of a spline fit to the existing data. No spline fit is required if the raw data points are themselves equally spaced. The predicted activity coefficient at infinite dilution is sensitive to experimental error, as the method is able to reproduce experimental pressures quite precisely. When the data in the dilute regions are accurate, the Mixon method gives a more accurate computation of the infinite dilution activity coefficient than does the Barker method.[38] The Mixon approach does not assume a given form for the activity coefficient equation; a finite difference approach is used to fit the pressure data. Overfitting can result in unreasonable activity coefficient curves, so they must be plotted to determine whether the fitting is reasonable.

3. Data Reduction Problems

When one component has a significantly higher vapor pressure than the other, the computed activity coefficient at infinite dilution can be in significant error for the less volatile component. Usually, it is significantly underestimated. This is illustrated for the acetone + isopropylbenzene system in an unpublished paper by B. D. Smith and co-workers. The "true" values are based on the slope at infinite dilution.

	Acetone	Isopropylbenzene
"True" value	2.463	4.849
Mixon method	2.466	2.805
Barker method		
Van Laar	2.261	2.428
Wilson	2.287	2.454
NRTL	2.432	2.585
UNIQUAC	2.263	2.437
Redlich-Kister, five constants	2.513	2.591

These values suggest the difficulties involved in obtaining accurate values at infinite dilution from reduction of Ptx data. In the midrange the activity coefficients are probably close to the true values. However, it is at infinite dilution where most of the trays are usually required in distillation columns. Therefore, it is often necessary to measure Ptxy data to obtain the true values. The author is aware of a case where a very respected thermodynamicist had a student recalculate the vapor compositions for the ammonia + water system. The experimentally determined vapor compositions were ignored. Subsequent careful measurements by Wiltec Inc. for the Design Institute for Physical Property Data of the American Institute of Chemical Engineers showed that the computations were in very serious error.

Although it would be possible to take good Ptx data in an industrial laboratory, it is not recommended. Pressures must be accurate to within less than 0.1% accuracy, and mole fractions must be accurate to within 0.0005 or less. This requires state-of-the-art equipment, and the results still may not be accurate enough. If this procedure is used, it probably should be supplemented with measurements at infinite dilution.

It would be wise to use both the Mixon and the Barker methods when reducing the data. It would not make sense to use an activity coefficient equation in the Barker method other than the one intended for use in the process simulation.

VI. INFINITE DILUTION ACTIVITY COEFFICIENTS BY GLC

A. Introduction

The importance of activity coefficient at infinite dilution was discussed in Chapter 3. It is often sufficient to pin down the ends of the activity coefficient curve and then interpolate with activity coefficient equations. Use of a gas chromatograph for measurement of infinite dilution activity coefficients is a very efficient means of taking equilibrium data. It is applicable to measurement of the partial pressure of volatile solutes in nonvolatile solvents. In an attempt to collect all published activity coefficients at infinite dilution, it was discovered that 88% of them had been measured by the chromatographic retention time method.[1] The ebulliometric method is described in the next section. The stripping method, not described here, is explained in Reference 2.

The basic principle of the chromatographic retention time method is easily understood. A liquid solvent is coated on an inert column support with high surface area. When a solute is injected into the column, it partitions itself between the gas and liquid phases. An infinite partition into the gas phase, as with air over most supports, gives the limiting value. Something which partitions primarily into the liquid phase will take much longer to elute than air. The difference in retention time between the air peak and the solute peak (net retention time) can be quantitatively related to thermodynamic properties. The net retention time is a function of the amount of solvent on the column, the vapor pressure of the solute, and its activity coefficient at infinite dilution in the solvent. The entire procedure is quite simple and can give a considerable amount of data in a short amount of time. Within a single day it is possible to construct a column and measure accurate data on a range of solutes at various temperatures. Such data have been checked against data measured by other methods and found to be comparable. The fundamental principles have been described in References 3 to 10.

B. Column Preparation

The packing support should not have any appreciable interaction with the solute. A good column support choice is Chromosorb® W, acid washed but not DMCS treated. Where water or acids are to be used, Chromosorb® T is a better choice. When in doubt concerning a support, put bare support into a column and check to be sure that solute retention times are the same as for air. Typical mesh size would be 60/80 or 80/100.

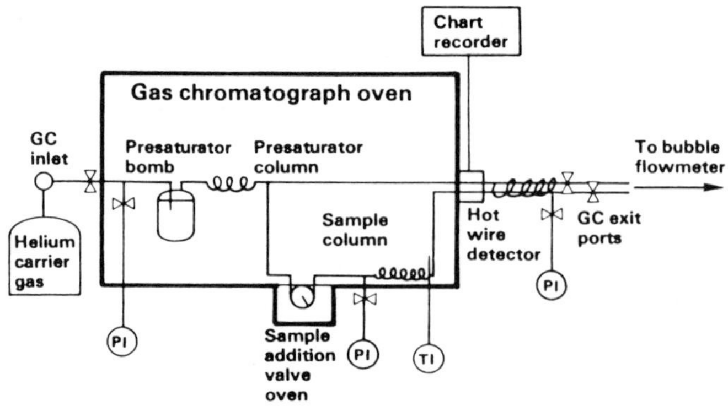

FIGURE 4. Gas chromatograph setup for measurement of infinite dilution activity coefficients when the solvent on the column is partially volatile.

The column size will be between 1/8 and 1/4 in. O.D., with normal wall thickness. The disadvantage of small-diameter columns is their high-pressure drop. The disadvantage of larger-diameter columns is a tendency to wider peaks and more difficulty in measurement of the time to the peak maximum. If there is a tendency for solvent loss, a presaturation bomb and column are normally used, as illustrated in Figure 4. In this circumstance it is desirable to minimize pressure drop by use of the large column diameter and adjustment of carrier gas flow rate. Otherwise, the solvent may be in equilibrium at the column inlet but not at the outlet; solvent loss from the column then would be inevitable.

Why would column solvent loss be a problem? The reason is that the amount of solvent on the column must be known exactly. The uncertainty in quantity of solvent will be reflected directly in uncertainty of the measured activity coefficient. Without presenting the full derivation for the equations of equilibrium, it is intuitively obvious that the more liquid that is present in the column, the more tendency there will be for a solute to stay in the column and net retention time will increase proportionately.

If the solvent has a vapor pressure less than 1 mmHg at room temperature, the coating of a column is simple. Weigh a quantity of packing more than sufficient to fill a column; this would typically be greater than 20 g. Then add the solvent to the solid and measure the new weight. Typical coating loadings would be 10 to 50%. The important quantity derived from these measurements is the weight percent of solvent on the support. If the quantity of liquid on the support is too low, there is a greater chance for support interaction effects, however small, to affect the results. Also, there will be more pressure drop than necessary for exposure to the needed amount of liquid. If too much liquid is used, the mixture will not be free-flowing, a condition for proper loading of the column.

Before packing, the column should be washed with toluene and acetone to remove any oils. The column needs to be dried and then assembled but not curled prior to addition of the support. One end should be plugged with glass wool to prevent solid loss during packing. Metal plug fillings are added to each end. Then the empty column with fittings is weighed. Thereafter, the end without glass wool is opened and connected to a funnel with flexible tubing. With adequate vibration the column is then filled with the support. The fitting is replaced and the column is reweighed. The weight difference can be cross-checked against the amount of support measured. There would of course be a very small loss on the surface of the funnel and the tubing. After getting the weight, glass wool should be used to protect the packing from spilling out of the top end. Then the caps are reapplied and the column is curled around a piece of pipe. Finally, the column is installed. A matching column containing approximately the same amount of dry packing is placed on the other side of the detector system.

A word is in order concerning volatile solvents. All the same procedures are followed, with the addition of precautions to minimize volatilization. After weighing the mixture of support and solvent, the coated support is chilled with dry ice. It is kept at dry ice temperature up to the very moment of insertion into the column. The author found that this procedure gave reproducible results with acetonitrile.

The procedure also must be modified when the solvent being coated on the support has high viscosity. In that case, it will not be possible simply to mix it into the support. The heavy solvent first must be dissolved in a volatile solvent, quite often toluene. The dilute mixture then is mixed in with the support, after which the support must be dried. Thorough drying can pose a problem, and sometimes the heavy solvent has impurities which bleed from the column during operation. The answer to the latter problem was accomplished by packing a column with uncoated support and weighing. Then vacuum was used to pull a toluene-diluted solution of heavy solvent through the column. The column was dried and then conditioned overnight above operating temperature. Finally, the column was reweighed.

C. Operation

1. Flow Rate

The flow rate can be varied to given reasonable retention times as the temperature is changed. An ideal net retention time is about 3 to 15 min. The other consideration in setting the air rate is pressure drop. Preferably, the inlet pressure will not exceed 1.3 times the outlet pressure. Under normal circumstances the outlet pressure will be atmospheric. Back pressure is used only when the solvent losses from the column at operating temperature would be a problem. Flow rates are measured, in either event, at atmospheric pressure and temperature using a bubble flowmeter. Room temperature and pressure are always required in order to compute the moles of gas flow per unit time.

2. Pressure

The inlet pressure at the entrance to the column can be measured with a handheld pressure gauge having a needle for insertion into the injection port. A more accurate and preferable approach is to tee into the line at the entrance to the column. A capillary line then connects that spot to a pressure transducer. If the column is operated under back pressure, pressure also should be measured before the outlet valve. Preferably, pressure is measured before rather than after the detector. With volatile solvents it is always possible to partially clog the detector. The pressure at the exit of the column, not the detector, is the required information. Fortunately, under normal circumstances the difference is negligible.

3. Temperature

Most gas chromatograph ovens have very good air circulation, resulting in a uniform temperature. However, the thermocouples in such ovens do not even approach the needed accuracy. The minimum accuracy required is ±0.1°C. Preferably, that would be reduced to ±0.02°C. There are two ways to obtain such accuracy. The simplest is to use a quartz crystal thermometer. The older, but still reliable approach is to use an NBS-calibrated platinum resistance thermometer. If the uniformity of temperature in a chromatograph oven is found to be inadequate, the entire column can be placed in a constant-temperature liquid bath. This procedure in fact was observed in one industrial laboratory.

If absolute values of activity coefficients are desired, the effect of temperature errors can be computed easily. The percentage error is based simply on the percentage change in vapor pressure with the temperature increased by the estimated uncertainty. An error of 1° can result in substantial error. If absolute values are not needed, as for instance in screening of extractive distillation solvents, the temperature error will not be too significant. In that case, the simple ratio of net retention times gives the relative volatility. Vapor pressures do enter in if the data are to be used for computation of liquid extraction potential. For VLE computations such approximations are inadequate and it is important to know the absolute value of the activity coefficient at infinite dilution.

4. Retention Times

The most accurate way to measure net retention time is to ignore the time at which the injection is made, then take the difference in time from the maximum in the air peak to the maximum in the solute peak. One of the problems is in obtaining a good but not excessive air peak. It is best if the peaks do not exceed about 30 on the scale, with attenuation of the detector at its minimum value. The point is that the quantity of solute must indeed be small to justify the contention that it is at ''infinite'' dilution.

Thermal conductivity detectors appear to give better results than do hydrogen flame detectors. The problem with the latter is the lack of response to an air peak. Methane could be used to establish the minimum retention time, but it will be absorbed to a measurable extent in most solvents. Therefore, thermal conductivity detectors are recommended whenever possible.

There are three methods of injection which have been used successfully in the Amoco laboratories.

1. With use of gas sampling valves kept in a valve oven at about 200°C the resulting peaks are quite sharp. The valves are particularly useful when the column is operated under pressure. In that circumstance the use of syringes and septums becomes very difficult.
2. Liquid microliter syringes can be used; however, timing must be done from the first injection. Air must be injected separately to find a time that would be subtracted from total retention time.
3. Small gas syringes are very effective. With solutes of moderate to high volatility the syringe is used to sample the vapor above the liquid, simultaneously picking up some air. Since absolute quantities of sample are of little importance, this approach works very well.

D. Data Reduction

A generalized thermodynamic expression relating all the measured data to activity coefficients is given in Equation 2. It will be seen that the gas phase nonidealities are correlated in terms of second virial coefficients. However, other EOS also could be used. The equation uses an average pressure, and is thus applicable even when the column is operated under back pressure.

$$\gamma_2^\infty = \frac{(RT_{col} + BP_{col})\, m_1 \bar{P} T_{room}\, \hat{\phi}_{2,P\epsilon}}{Q \Delta t P_{atm} T_{col}\, \phi_{2,P_2*} P_2{}^* M_1} \tag{2}$$

where M_1 = molecular weight of solvent, R = 62,365 cm^3 mmHg/gmol K, $P_2{}^*$ = P^*_{solute} (mmHg), m_1 = g of solvent on column, Δt = total retention time minus air peak, $\bar{P}$ = average column pressure, (mmHg) and T = temperature of column or room (K).

$$\hat{\phi}_{2,P} = \exp\left[\frac{BP}{RT} + \frac{P}{RT}(2y_1B_{12} + 2y_3B_{23})\right]$$

$$\phi_{2,P_2*} = \exp\left(\frac{P_2{}^*\, B_{22}}{RT}\right)$$

$$\epsilon = \exp\left(\left(\frac{P_2^* - P_{col}}{RT}\right) V_2^L\right) \qquad (2) \text{ cont'd.}$$

$B = y_1{}^2B_{11} + y_3{}^2B_{33} + 2y_1y_3B_{13}$. Component 1 = solvent, component 2 = solute, component 3 = helium.

When the outlet pressure is atmospheric and there is substantial pressure drop through the column, the average pressure may need further refinement. That is obtained by Equation 3.

$$\bar{P} = P_0 \frac{2}{3}\left[\frac{(P_i/P_0)^3 - 1}{(P_i/P_0)^2 - 1}\right] \qquad (3)$$

Henry's constants for supercritical gases also are obtained by Equation 2; however, the vapor pressure is set to unity. The Henry's constant is obtained instead of the activity coefficient. It is in pressure units, appropriate to the units used in the data reduction.

E. Data Testing

For *n*-hexane in squalane, the literature gives two data points; 0.651 and 0.666. Gas-liquid chromatography data in the literature gave 0.684, while the author's data gave 0.666. This is excellent agreement between the types of data. Similar comparisons are given by Eckert and co-workers[6] on four different types of systems. These included carbon tetrachloride in 1,2 dichloroethane, acetone and cyclohexane in benzene, and toluene in ethanol. The data obtained by the chromatographic method were especially consistent with infinite dilution values measured by the ebulliometric method described later.

F. Undefined Oils

Heavy oils, oil additives, and other undefined chemical mixtures are commonplace in the oil and chemical industries. Very often, they are produced or refined in the presence of light solvents such as hexane or methanol. Usually, those solvents must be stripped out at the end of the process. Definition of the volatility is essential to the process design, and the assumption of ideality is usually false. The chromatographic method is ideally suited to evaluation of such systems.

The molecular weight of heavy oils varies over a range; however, that is not a problem in the data reduction. An estimated average molecular weight is used for the material coated on the column. The computations of K values based on the derived activity coefficients will be correct provided that the molecular weight used in subsequent computations for the high-molecular-weight product is kept the same. The activity coefficients will not have true thermodynamic rigor, but the results will be correct.

The techniques described here are applicable up to about 10 bar. At higher pressure, more sophisticated techniques are required. These are described by Kragas and co-workers,[11] who have an enviable history of producing high-quality data.

G. Polymer-Solvent Equilibrium

1. Introduction

Polymer devolatilization also requires accurate equilibrium data which can be effectively measured using the gas chromatograph technique. Because of essentially zero vapor pressure, it is possible to carry out these measurements at relatively high temperatures; the author once took data up to 250° C. However, polymers allow much lower rates of mass transfer than do ordinary solvents, so special techniques are required. Otherwise, the assumption of equilibrium in the column will not be valid.

2. Column Preparation

The target concentration of polymer on a solid support should not exceed 10%. To compensate for the smaller loading, more packing is needed and column length can be increased. There is no problem with bleedoff of the polymer so the higher pressure drop resulting from use of a long column is not a problem. The following description is typical of the approach that should be used in column preparation:

The column packing was prepared by dissolving 2.5 grams of polystyrene in 500 ml of reagent grade toluene in a 1-liter 3-neck flask. The toluene was refluxed and stirred under a nitrogen blanket. When the polymer was entirely dissolved (about 45 minutes), the heat was removed. After the reflux had stopped, 50 g of column support was added to the solution. Stirring was continued until the slurry had reached room temperature. It was then poured into a glass evaporating dish and gently purged with nitrogen to evaporate toluene. The slurry was stirred every half hour. When the support seemed dry and free flowing, it was placed in a vacuum oven at 70 degrees C for eight hours. The coated support was stirred every half hour during that time. A sample of the packing was analyzed for total percent carbon (4.11%). The percent carbon was divided by the ratio of carbon to total polymer, 0.923. This number was multiplied by the total weight of packing loaded into the column to obtain column loading in grams of polymer.

3. Operations

It is important to make sure that mass transfer is not affecting the results. Therefore, the first step is to measure the activity coefficients at the lowest operating temperature. This is then repeated with different flow rates. When the flow rate is too high, equilibrium is not obtained and the apparent volatilities are too high. A critical flow rate will be found below which flow rate has no effect on the activity coefficient. Failure to observe this precaution with polymers can result in erroneous data.

4. Data Reduction

The standard equations for reduction of data presented previously in this chapter are used, with a slight modification, for polymers. Solute molecular weight is entered into Equation 2 instead of the polymer molecular weight. This gives a weight fraction-based activity coefficient (Γ_1). This weight fraction activity coefficient is more meaningful than mole fraction-based activity coefficients when dealing with polymers. Techniques for obtaining the partial molar heats of mixing and the Flory-Huggins interaction parameter from these measurements over a range of temperatures are given in Reference 12.

5. Activity Coefficients at Finite Concentration

It is often adequate to interpolate from infinite dilution to finite concentration with polymers. However, it is also possible to measure the activity coefficients directly, using a gas chromatograph. This technique was pioneered by Guillet[13] and refined at the Amoco Research Center by Brockmeier et al.[14] They examined compositions up to 10 wt% solvent. This is a more complex experiment than measurements at infinite dilution, involving measurement of elution rates on a plateau of continuously added solute.

VII. INFINITE DILUTION ACTIVITY COEFFICIENTS BY EBULLIOMETRY

A good general-purpose recirculating still was described previously for ordinary VLE measurements. A more specialized still can be used to obtain both activity coefficients at infinite dilution and regular VLE data. It is an ebulliometer, which has recirculation of both liquid and vapor. The pumping of both phases simultaneously is accomplished by overflowing a boiling mixture upward through a small tube. The vapor and liquid travel together and impinge on a thermowell in an equilibrium chamber. The technique was discovered by Cottrell[16] in about 1910. It was expanded by Swietoslawski and Romer[17] and with elaborations by many investigators such as Eckert and co-workers. (see Thomas et al.[41]).

The advantage of ebulliometers is that they provide a continuous supply of vapor that can be sampled, if necessary. Degassing is accomplished continuously, with inerts being purged out of the condenser. Ebulliometers can handle liquids with wide ranges of volatility, but do not work well with low-volatility solvents. Thus, the gas chromatography technique previously described is complementary to ebulliometry; it handles well the systems with low-volatility solvents.

A. Modified Malanowski Still

Two different stills are described by Malanowski[18] and Rogalski and Malanowski.[19] Both are ebulliometers, one designed for determination of activity coefficients at infinite dilution and the other designed for phase equilibria. A modified design which permits both types of measurements is shown in Figure 5.

The ebulliometer shown has considerable capabilities, both in the atmospheric pressure region and under vacuum. The modifications were made at Amoco Chemicals Corp. where it was used for a number of VLE measurements. It also was used by Prof. Malanowski during a sabbatical leave in the U.S. The ebulliometer can be fabricated by Reliance Glass Works, 17 Gateway Road, Bensenville, Ill. 60106 [(312) 766-1816].

Liquid is introduced on the top left-hand side or through a septum on the liquid sampling port. The liquid level is allowed to rise in the liquid reservoir until it is about half full. Liquid thus fills the bottom of the ebulliometer as well, including the boiler. A heating rod is inserted into the boiler, and it boils away a mixture of liquid and vapor. They are separated in the equilibrium chamber, where the liquid is returned to the liquid reservoir. The vapor goes to a condenser and is then recycled. The procedure now will be described in more detail.

The boiling chamber at the bottom is a critical part of the apparatus. The surface of the boiler should be coated with very fine fritted glass to obtain good boiling. It is necessary to vaporize liquid steadily, without ''bumping''. By so doing, a continuous stream of liquid and vapor are forced upward through the tube; this is the so-called ''Cottrell pump''. It may be helpful to heat the tube to get the proper proportion of vapor to liquid. The system works well at atmospheric pressure. At vapor pressures under 0.1 atm there can be problems with unsteady boiling. The care with which the fritted glass is applied can affect the lower pressure limit of the apparatus. It may be helpful to apply the fritted glass to the concave surfaces of the boiler, as well as to the convex surface shown in the drawing. Alternatives to use of fritted glass which have proven effective include high-speed stirring at over 1000 rpm. Another possibly fruitful approach would use ultrasonic mixing.

The mixture of vapor and liquid impinges on a thermowell. A spiral ribbon on the exterior of the thermowell causes the liquid to spend some time in contact with the tube. In this chamber the liquid and vapor are in equilibrium. An evacuated chamber which surrounds the equilibrium chamber helps to reduce heat losses. The entire equilibrium chamber would be insulated during operations. The interior diameter of the thermowell is determined such that a quartz crystal thermometer probe will fit inside with a small clearance. Silicone oil helps to improve heat transfer. A tip at the bottom of the thermowell directs liquid away from the vapor receiving line. The bottom of the vapor receiving line is down toward the bottom of the chamber, and any droplets of liquid would have to change direction by 180° to enter that line.

In the transit from the equilibrium chamber to the condenser the vapor is heated to prevent condensation. The condenser itself is designed with cooling on the outside and the inside to prevent vapors from being lost to the pressure control system. There is a drop counter at the bottom of the condenser; it gives a semiquantitative measure of the rate of vapor generation. Condensate flows down past the vapor sample port. The liquid holdup there should be as small as possible; the condensate is sampled by syringe through a Teflon® septum.

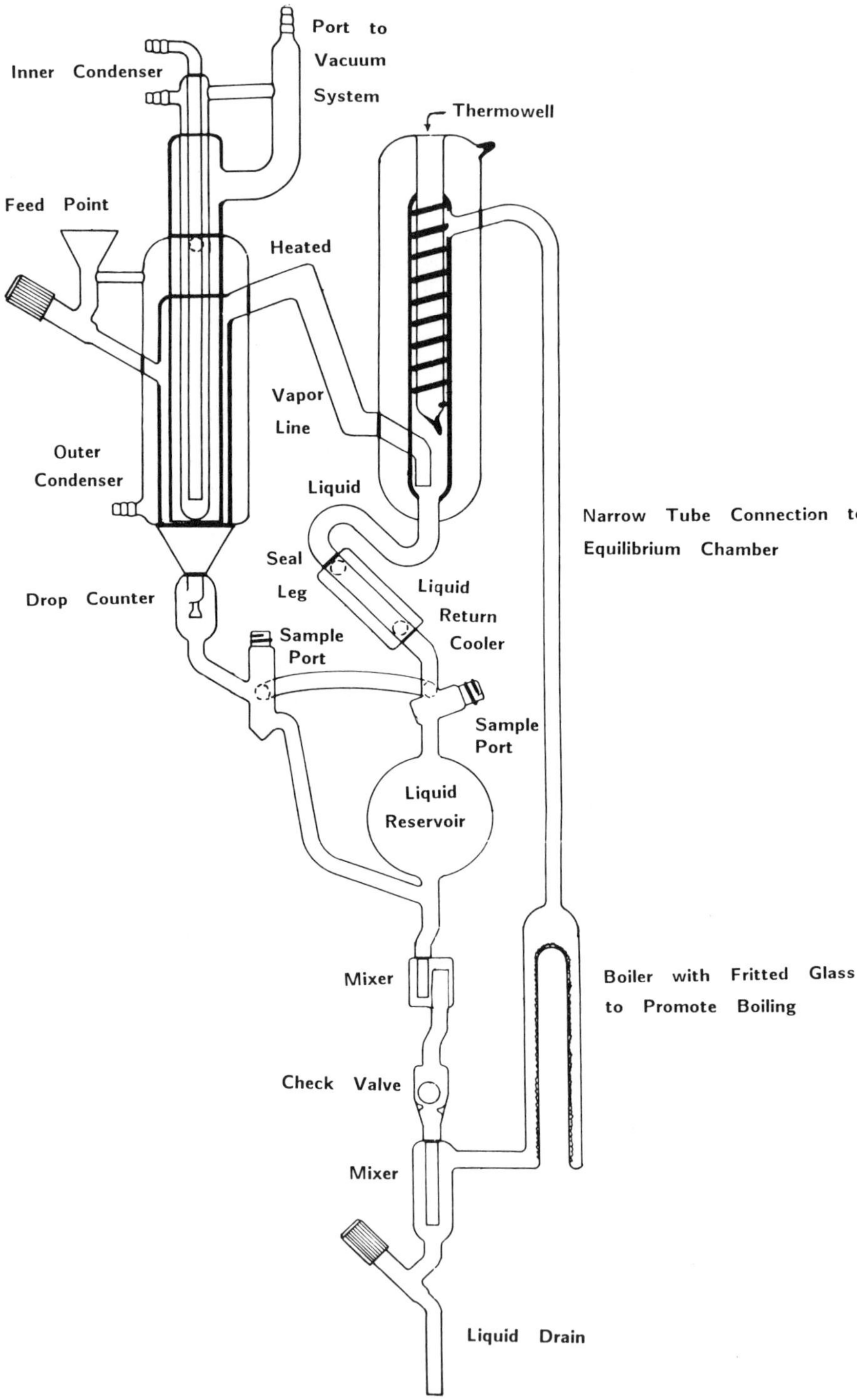

FIGURE 5. Modified Malanowski ebulliometer, used for infinite dilution activity coefficients as well as ordinary VLE data. (Reproduced with permission of Reliance Glass Works, Bensenville, Ill., [(312) 766-1816].)

Stopcock grease on the top of the septum helps to prevent air leakage. The condensate thereafter joins the liquid being returned to the boiler. The mixed stream goes through a mixer, past a check valve, and through another mixer back to the boiler.

The liquid leaving the equilibrium chamber goes through a bend which provides a liquid seal. This results in the need for a small connection between the vapor condensate side and the liquid return side in order to balance the pressure. That line can be very small. The need for the liquid seal has not been proven, and the ebulliometer may well perform properly without it. Below the liquid seal there is a cooler, which is needed to prevent flashing when the condensate meets the returning liquid at the bottom of the liquid reservoir. Liquid compositions can be measured using liquid withdrawn before the liquid reservoir.

The boiler temperature is irrelevant, except as it affects the boil-up rate. It is important that the temperature measured in the equilibrium chamber be accurate; therefore, a quartz crystal thermometer is a good choice. Alternatively, a calibrated platinum resistance thermometer could be used. Syringes used for sampling should be cold, to prevent any possible flashing. The pressure is measured in the pressure control system. A variety of accurate electronic pressure gauges are available, and they can be checked against a mercury manometer. The author has found that it is easier to control the pressure by hooking the system to a large surge tank kept at room temperature in a liquid bath than to control pressure actively with a manostat. Through use of a digital pressure gauge, small pressure fluctuations can be recorded easily. They can be used with the vapor pressure equation for the solvent to make small corrections to the measured temperature. Thus, the technique of having a reference ebulliometer may have been rendered obsolete by the progress of electronics. Those who do not trust electronic pressure readings will be more happy with a matched ebulliometer filled only with the solvent. In that case, quartz crystal thermometer probes should be particularly valuable; the thermometers can register the temperature difference directly.

B. Activity Coefficients at Infinite Dilution

The fundamentals of measuring activity coefficients at infinite dilution by ebulliometer were described by Null.[20] The advantage of measuring activity coefficients at infinite dilution is that it avoids sampling problems. The important measurement is the change in temperature with added solute at constant pressure: (1) A known quantity of approximately 100 mℓ is added into the ebulliometer. (2) The vapor pressure of that pure material is determined and checked against literature values for accuracy. (3) A small amount of solute is added through the liquid sampling septum. The syringe is carefully weighed before and after the addition. The amount of the second component added is sufficient to determine a measurable temperature change. The temperature should be stable within 1/2 hr. (4) Addition of a number of increments, up to perhaps 5 mol %, gives a temperature change per mole fraction which can be fitted numerically or graphically. Using the convention that component 1 is the solute, and the major component in the ebulliometer is component 2, the following equation is applicable for reduction to infinite dilution activity coefficients at infinite dilution:

$$\gamma_1^\infty = \exp\left[\frac{(B_{11} - V_{1L})(P_2^* - P_1^*)}{RT}\right]\left(\frac{1}{P_1^*}\right) \times$$

$$\left\{P_2^* - \left(\frac{\partial P_2^*}{\partial T}\right)_{x_1}\left(\frac{\partial T}{\partial x_1}\right)_P^\infty\left[1 + P_2^*\frac{(B_{22} - V_2^L)}{RT}\right]\right\} \tag{4}$$

where B_{ii} = second virial coefficient (cm^3/mol) and ∞ = infinite dilution condition. In this equation the slope of the vapor pressure for the solvent is determined from its vapor pressure equation. The derivation of the equation is given in the supplementary material for Reference 21.

Liquid compositions are obtained by computation rather than measurement. However, it is intuitively obvious that a highly volatile solute will tend to partition preferentially into the vapor phase. This would render the true liquid concentration of solute somewhat lower than its expected value. Therefore, a correction must be made to account for that effect; the correction is not needed for ordinary VLE data because the liquid composition is measured directly. For a total feed to the ebulliometer with a mole fraction of solute q_i, the corrected liquid phase composition is given by

$$x_i = q_i(1 + f)/(1 + y_i f/x_i) \tag{5}$$

The coefficient f determines the ratio of the number of moles in the stream of equilibrium vapor and in the equilibrium liquid. According to Malanowski,[18] it is not a function of the system studied, but of the vapor rate. There is a reasonably wide range of vapor rate through which the equilibrium temperature is not affected. Thus, the value of f determined by measurements on a well-defined and previously measured system will be generally useful; the vapor rate must, however, be kept approximately the same as in the test run.

C. Other Equipment Designs

The apparatus just described is adequate but probably not optimal. Another apparatus was described by Eckert et al.[22] based on the same principles. It had the disadvantage of not being simultaneously useful for regular VLE measurements. The Eckert ebulliometer was tested further at E. I. du Pont de Nemours & Co., where it was discovered that an improved boiler was essential. The design uses a high-speed (2000 rpm) agitator setup which prevents significant superheating in the boiler. The size of the flask containing the liquid is apparently significant and needs more elaboration. Presumably, the same modification would improve operation of the Malanowski still. The designer, L. E. Scott, reports much more satisfactory temperature control.

There are hundreds of designs for ordinary dynamic VLE devices. One is available commercially from Cal-Glass [(714) 546-7250]. It is based on modifications to a design developed at C. F. Braun.[42,43] The author has had no personal experience with the cell and so cannot suggest how reliable it might be. A commercial ebulliometer recommended by Dr. Amyn S. Teja is manufactured in West Germany.

Fischer
5309 Meckenhaim, Bonon Industrial Park
Kottenforst, West Germany
0 22 25/4044

It is not clear whether either of these latter two ebulliometers has an adequate design to give reliable data on a wide variety of systems.

VIII. SUMMARY

Routine physical property measurements are often best conducted in-house. The cost is not necessarily lower than contracting out the measurements. However, it gives more control over timing. Very specialized experiments, with costly setup charges, are best given to contractors. A description has been given of various types of apparatus which could reasonably be assembled by an industrial company to measure VLE data at low to moderate pressure.

Both static cells and vapor recirculation devices were described. Neither is without pitfalls. The biggest problem with static cells is obtaining accurate samples, particularly of the vapor.

Mixing and attainment of equilibrium can be a problem. With recirculation stills there are problems with unsteady boiling and in having the steady state actually represent the true VLE. A recirculation still designed and operated by Dr. Grant M. Wilson was described and recommended. Data taken by such methods should be thoroughly tested for thermodynamic consistency.

Another technique which does not require composition measurements is determination of the vapor pressure for defined mixture compositions. The vapor composition is determined by use of the Barker or the Mixon method. There are computational difficulties, and even with well-behaved systems the different methods give different results. The errors increase with the system pressure due to uncertainties in modeling the gas phase nonidealities. Two different types of static measurement were noted.

Accurate infinite dilution activity coefficients are particularly helpful. The gas chromatography approach, measuring retention times of volatile solutes over nonvolatile solvents, is especially useful. Once a column is coated with a solvent, the incremental effort to screen new solutes is very small. The method also can be used to measure Henry's constants. Even undefined solvent fractions and polymer-solvent equilibrium can be studied.

Ebulliometers which recirculate both liquid and vapor can be used to obtain activity coefficients for either volatile or nonvolatile solutes in volatile solvents. They are well suited to measurement of infinite dilution activity coefficients, since the change in temperature with a change in composition at constant pressure is the only measured variable. The same apparatus can be used to measure ordinary VLE over the entire composition range, for both binary and multicomponent mixtures. Composition measurements are required for measurements at finite concentrations.

REFERENCES

1. **Bastos, J. C., Soares, M. E., and Medina, A. G.,** Selection of solvents for extractive distillation. A data bank for activity coefficients at infinite dilution, *Ind. Eng. Chem. Process Des. Dev.*, 24, 420, 1985.
2. **Santacesaria, E., Berlendis, D., and Carra, S.,** Measurement of activity coefficients at infinite dilution by stripping and retention time methods, *Fluid Phase Equilibria*, 3, 167, 1979.
3. **Alessi, P., Kikic, I., and Tlustos, G.,** The determination of activity coefficients at infinite dilution by gas-liquid chromatography, *Chim. Ind.*, 52, 1197, 1970.
4. **Kwantes, A. and Rijnders, G. W.,** The determination of activity coefficients at infinite dilution by gas-liquid chromatography, in *Gas Chromatography 1958*, Desty, D. H., Ed., Academic Press, Orlando, Fla., 1958, 125.
5. **Porter, F. E., Deal, C. H., and Stross, F. H.,** The determination of partition coefficients from gas-liquid partition chromatography, *J. Am. Chem. Soc.*, 78, 2999, 1956.
6. **Eckert, C. A., Newman, B. A., Nicolaides, G. L., and Long, T. C.,** Measurement and application of limiting activity coefficients, *AIChE J.*, 27, 33, 1981.
7. **Tassios, D. P.,** Infinite dilution relative volatilities through gas-liquid chromatography, *Ind. Eng. Chem. Process Des. Dev.*, 11, 43, 1972.
8. **Bogeatzes, A. S. and Tassios, D. P.,** Prediction of gas-liquid equilibrium through gas-liquid chromatography, *Ind. Eng. Chem. Process Des. Dev.*, 12, 274, 1973.
9. **Everet, D. H. and Stoddart, C. T. H.,** The thermodynamics of hydrocarbon solutions from G.L.C. measurements, *Trans. Faraday Soc.*, 57, 746, 1961.
10. **Martire, D. and Pollara, L.,** *J. Chem. Eng. Data*, 10, 40, 1965.
11. **Kragas, T. K., Pollin, J., Martin, R. J., and Kobayashi, R.,** A high-pressure-high-temperature chromatographic apparatus for the measurement of vapor-liquid equilibria, *Fluid Phase Equilibria*, 16, 205, 1984.
12. **DiPaola, G. and Guillet, J. E.,** Estimation of polymer solubility parameters by gas chromatography, *Macromolecules*, 11, 228, 1978.
13. **Guillet, J. E.,** *J. Macromol. Sci. Chem.*, 4, 1669, 1970.

14. **Brockmeier, N. F., McCoy, R. W., and Meyer, J. A.,** Gas chromatographic determination of thermodynamic properties of polymer solutions. I. Amorphous polymer systems, *Macromolecules,* 5, 465, 1972.
15. **Masuoka, H., Murashige, N., and Yorizane, M.,** Measurement of solubility of organic solvents in polyisobutylene using the piezoelectric-quartz sorption method, *Fluid Phase Equilibria,* 18, 155, 1984.
16. **Cottrell, F. G.,** On the determination of boiling points of solutions, *J. Am. Chem. Soc.,* 41, 721, 1910.
17. **Swietoslawski, W. and Romer, W.,** Sur un nouvel appareil ebullioscopique, *Bull. Acad. Sci. Pol. (A),* 1924, 59, 1924.
18. **Malanowski, S.,** Experimental methods for vapour-liquid equilibria. I. Circulation methods, *Fluid Phase Equilibria,* 8, 197, 1982.
19. **Rogalski, M. and Malanowski, S.,** Ebulliometers modified for the accurate determination of vapour-liquid equilibrium, *Fluid Phase Equilibria,* 5, 97, 1980.
20. **Null, H. R.,** *Phase Equilibrium in Process Design,* John Wiley & Sons, New York, 1970.
21. **Palmer, D. A.,** A practical synthesis of experimental and correlation techniques for modeling highly nonideal systems with many components, *Ind. Eng. Chem. Process Des. Dev.,* 23, 259, 1984.
22. **Eckert, C. A., Newman, B. A., Nicolaides, G. L., and Long, T. C.,** Measurement and application of limiting activity coefficients, *AIChE J.,* 27, 33, 1981.
23. **Palmer, D. A. and Smith, B. D.,** Thermodynamic excess property measurements for acetonitrile-benzene-*n*-heptane system at 45°C, *J. Chem. Eng. Data,* 17, 71, 1972.
24. **Renon, H., Richon, D., Laugier, S., Legret, D., Meskel-Lesavre, M., Fontalba, F., Guillevic, J.-L., and Rousseaux, P.,** The need of high-pressure vapor-liquid equilibrium data in industrial applications, Proc. 62nd Ann. Convention, Gas Processors Association, 228.
25. **Guillievic, J.-L., Richon, D., and Renon, H.,** Vapor-liquid equilibrium measurement up to 558 K and 7 MPa: a new apparatus, *Ind. Eng. Chem. Fundam.,* 22, 495, 1983.
26. **Abbott, M. M.,** Measurement of vapor-liquid equilibrium, in *Phase Equilibria and Fluid Properties in the Chemical Industry,* ACS Symp. Ser. No. 60, Storvick, T. S. and Sandler, S. I., Ed., American Chemical Society, Washington, D.C., 1977, 87.
27. **Gibbs, R. E. and Van Ness, H. C.,** Vapor-liquid equilibria from total-pressure measurements. A new apparatus, *Ind. Eng. Chem. Fundam.,* 11, 410, 1972.
28. **Maher, P. J. and Smith, B. D.,** A new total pressure vapor-liquid equilibrium apparatus. The ethanol + aniline system at 313.15, 350.81, and 386.67 K, *J. Chem. Eng. Data,* 24, 16, 1979.
29. **Barker, J. A.,** Determination of activity coefficients from total pressure measurements, *Aust. J. Chem.,* 6, 207, 1953.
30. **Abbott, M. M. and Van Ness, H. C.,** An extension of Barker's method for reduction of VLE data, *Fluid Phase Equilibria,* 1, 3, 1977.
31. **Abbott, M. M. and Van Ness, H. C.,** Vapor-liquid equilibrium. III. Data reduction with precise expressions for G^E, *AIChE J.,* 21, 62, 1975.
32. **Won, K. W. and Prausnitz, J. M.,** High-pressure vapor-liquid equilibria. Calculation of partial pressures from total pressure data. Thermodynamic consistency, *Ind. Eng. Chem. Fundam.,* 12, 459, 1973.
33. **Abbott, M. M., Floess, J. K., Walsh, G. E., Jr., and Van Ness, H. C.,** Vapor-liquid equilibrium. IV. Reduction of P-x data for ternary systems, *AIChE J.,* 21, 72, 1975.
34. **Martinez-Ortiz, J. A. and Manley, D. B.,** Direct solution of the isothermal Gibbs-Duhem equation for multicomponent systems, *Ind. Eng. Chem. Process Des. Dev.,* 17, 346, 1978.
35. **Plank, C. A., Olson, J. D., Null, H. R., Muthu, O., and Smith, B. D.,** Reduction of total-pressure vapor-liquid equilibrium data. Common pitfalls encountered, *Fluid Phase Equilibria,* 6, 39, 1981.
36. **Abbott, M. M. and Van Ness, H. C.,** An extension of Barker's method for reduction of VLE data, *Fluid Phase Equilibria,* 1, 3, 1977.
37. **Mixon, F. O., Gumowski, B., and Carpenter, B. H.,** Computation of vapor-liquid equilibrium data from solution vapor pressure measurements, *Ind. Eng. Chem. Fundam.,* 4, 455, 1965.
38. **Munja, S., Muthu, O., Khurma, J. R., and Smith, B. D.,** Reduction of total-pressure vapor-liquid equilibrium data. Comparison of data reduction methods, reliability of infinite dilution activity coefficient values, and effect of the equation of state assumed, *Fluid Phase Equilibria,* 12, 29, 1983.
39. **Tremper, K. K. and Prausnitz, J. M.,** *J. Chem. Eng. Data,* 21, 265, 1976.
40. **Horvath, M. J., Sebastian, H. M., and Chao, K. C.,** Gas chromatograph method for the determination of gas solubility in liquids, *Ind. Eng. Chem. Fundam.,* 20, 394, 1981.
41. **Thomas, E. R., Newman, B. A., Nicolaides, G. L., and Eckert, C. A.,** Limiting activity coefficients from differential ebulliometry, *J. Chem. Eng. Data,* 27, 233, 1982.
42. **Hipkin, H. and Myers, H. S.,** Vapor-recirculating equilibrium still, *Ind. Eng. Chem.,* 46, 2524, 1954.
43. **Triday, J. O. and Veas, C.,** Vapor-liquid equilibria for the system cyclohexane-*tert*-butyl alcohol, *J. Chem. Eng. Data,* 30, 171, 1985.
44. **Freeman, J. R. and Wilson, G. M.,** High temperature vapor-liquid equilibrium measurements on acetic acid/water mixtures, in Experimental results from the Design Institute for Physical Property Data. I. Phase equilibria, Benson, M. S. and Zudkevitch, D., Ed., *AIChE Symp. Ser.,* 81(244), 14, 1985.

45. **Gillespie, P. C., Cunningham, J. R., and Wilson, G. M.,** Total pressure and infinite dilution vapor liquid equilibrium measurements for the ethylene oxide/water system, in Experimental results from the Design Institute for Physical Property Data. I. Phase equilibria, Benson, M. S. and Zudkevitch, D., Ed., *AIChE Symp. Ser.*, 81(244), 26, 1985.
46. **Kolbe, B. and Gmehling, J.,** Thermodynamic properties of ethanol + water. I. Vapour-liquid equilibria measurements from 90 to 150°C by the static method, *Fluid Phase Equilibria,* 23, 213, 1985.

Section IV: Applications

Chapter 8

TACKLING DIFFICULT PROBLEMS

TABLE OF CONTENTS

I. INTRODUCTION

Previous chapters have given short examples of the use of techniques, equations, and apparatus. This chapter organizes some of those techniques in the solution of two fairly difficult problems which might be considered intimidating. These two examples illustrate the fact that technically difficult problems can be solved when approached from a fundamental and organized point of view.

The first example, the shorter of the two, relates to a combined vapor-liquid and reaction equilibrium. Its solution does not require use of a formal free energy minimization program but the result is the same. The second problem describes the development of a phase equilibrium model for a highly nonideal system with many defined components. The actual system studied was butadiene extraction.

Both examples make use of predictive methods, described in Chapter 3. The first example shows how to expand the applicability of those predictive methods to solve a particular practical problem. Both examples take advantage of the literature, described in Chapter 4. The correlation methods of Chapter 5 are brought into play as well. Contract data measurements, discussed in Chapter 6, are utilized. In-house data measurements described in Chapter 7 are also used in a particularly challenging application.

II. SIMULTANEOUS CHEMICAL AND PHASE EQUILIBRIUM

A. Introduction

Material and energy balances were needed for a liquid phase phthalic anhydride plant. The unit operations included a simultaneous reaction and phase equilibrium, recovery operations, three phase flashes, and purification towers. Inasmuch as the project was in the process development phase, considerable interaction between the design effort and the pilot plant work was anticipated, therefore a process simulation appeared to be justified. An in-house process simulator was to be used.

The following obstacles were encountered in development of the simulation:

1. There were at least 14 components, including gases, water, ortho-xylene, phthalic acid (1,2 dicarboxy benzene), phthalic anhydride, and reaction intermediates.
2. There was a chemical equilibrium between phthalic acid, phthalic anhydride, and water. Appropriate Benson method parameters described in Chapter 2 were not available for the anhydride group.
3. There were almost no mixture phase equilibrium data.
4. There were great variations in temperature, pressure, and composition.
5. Actual reaction rates were sensitive to the water, acid, and anhydride concentrations.

B. Problem Definition and Strategy

Fortunately, it was clear that the gases present in the process would have low solubility in the liquid. The presence of water, particularly, was a clue that there would be solubilities low enough to use an activity coefficient approach rather than an equation of state (EOS). Furthermore, operations would not be near the critical point for any constituent in the mixture. Thus, use of the unsymmetric convention for gases, as described in Chapter 5, was appropriate. Because of the low gas solubility, the activity coefficient for gas in the reference solvent (water) could be set equal to unity.

Henry's constants for gas solubility in water were obtained from the literature. The Yen-McKetta correlation[1] was used to predict Henry's constants for gases in the organic solvents. Partial molar volumes for the pressure correction factor were estimated from literature values.

Table 1
ASOG INTERACTION PARAMETERS DERIVED FOR ACID AND ANHYDRIDE SYSTEMS

	CH_2	Benzene	Acid	Anhydride	OH
CH_2	1.0	0.734	1.1333	0.7703	0.305
Benzene	1.24	1.0	1.1372	0.9841	0.534
Acid	0.034	0.0442	1.0	0.6205	2.352
Anhydride	0.0249	0.1876	0.5369	1.0	2.04
OH	0.0147	0.045	0.1960	0.300	1.0

Note: All of these groups contribute 1 "size" group, except for the anhydride which contributes 2 size groups. Water contains 1 size group and 1.4 interaction (OH) groups. The temperature dependence of these interaction parameters has not been determined.

Over the range of pressures encountered (up to 30 bar), the pressure correction affects the equilibrium constant by only 2%.

The liquid phase activity coefficients were correlated with the Chien-Null multicomponent expansion of the Van Laar equation.[2] It appeared to be the best alternative at a time before publication of the UNIQUAC equation. Other data required were the vapor pressure constants and critical properties needed for the Redlich-Kwong EOS; the latter was used to calculate gas phase fugacity coefficients.

C. Activity Coefficients by Group Contribution Method

The most difficult problem in this development was the estimation of activity coefficient parameters. The group contribution approach was used to estimate binary parameters for mixtures of minor importance. At the time, the only procedure which was sufficiently developed was the ASOG method, described in Chapter 3. However, there were insufficient interaction parameters available; acid and anhydride group parameters had to be determined for this project. In choosing experimental data for development of such interaction parameters it is desirable to use components in which the new functional groups play a large role. In other words, acetic acid is a better choice for determination of acid interactions than dodecanoic acid, where the influence of the acid group is diminished by a large number of methyl groups. Fortunately, appropriate data were available.

Cyclohexane + acetic acid	CH_2/acid
Acetic acid + water	Water/acid
Toluene + acetic acid	Benz/acid
Benzene + acetic acid	Benz/acid
Cyclohexane + acetic anhydride	CH_2/anhydride
Acetic acid + acetic anhydride	Acid/anhydride

The actual values[14] derived from that study are shown in Table 1.

Using the group contributions from these 6 systems, the vapor-liquid equilibrium (VLE) for 26 new systems was predicted. This illustrates the way in which a small number of experimental data can be expanded in usefulness using the group contribution method.

The next step was to test the predictive capabilities of the analytical solutions of groups (ASOG) method with the new group interaction parameters. In Figure 1 the activity coefficient predictions are shown for the water + propionic acid and water + formic acid systems. Neither of these systems was used in the data base and yet the predictions are quite accurate. It should be emphasized that these are predictions, not correlations. Even the negative deviations from ideality exhibited by the water + formic acid system were handled

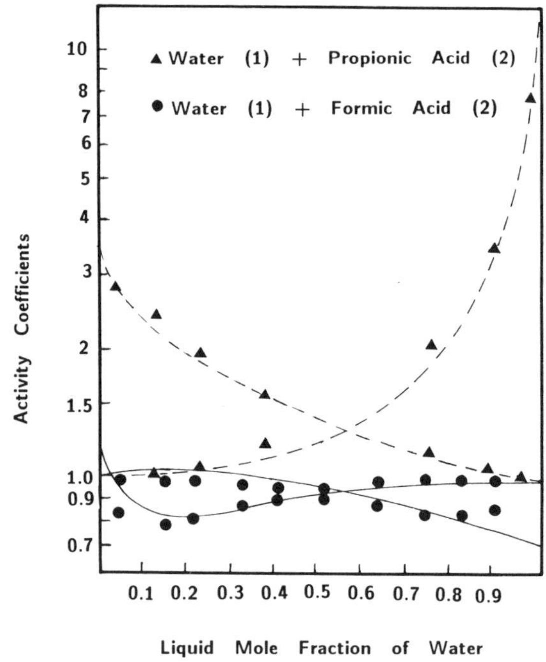

FIGURE 1. ASOG prediction of activity coefficients for the water (1) + propionic acid (2) and water (1) + formic acid (2) binary VLE. The parameters were based on water + acetic acid. Note the excellent prediction of activity coefficients both above and below unity for propionic and formic acids. (Adapted from Palmer, D.A., *Chem. Eng.*, June 9, 80, 1975.)

quite well. Total agreement in that system was not possible because the data were not completely consistent thermodynamically. As a matter of interest, it should be noted that in derivation of the activity coefficients for these alkanoic acids, gas phase dimerization was taken into account. The methods used to do this were described in Chapter 5 in the section on reactive systems.

In Figure 2 benzoic acid + phthalic anhydride, a VLE system of more direct interest to the process development, is illustrated. The only data available were x-y data from the literature. They were predicted extremely well using the ASOG method. Had full data been available, it would have been possible to compare the activity coefficients, a more sensitive measure of predictive capabilities. Figure 3 shows a similar prediction for the ortho-toluic acid + phthalic anhydride system. It will be noted that a fair extrapolation was made with respect to temperature. Yet, even the azeotrope was predicted quite well.

D. Experimental Measurements

At this point in the development, key elements needed to simulate the overall system were missing. Both the chemical equilibrium and the physical interaction between an anhydride group and a water group were undetermined. It takes little imagination to understand the absence of literature data on the reaction and VLE in such a system. To remedy this problem, experimental data on the heat of hydration of phthalic anhydride were measured under contract to Amoco Chemicals Corp by Dr. Grant M. Wilson. He found it to be −5200 cal/g mol. He also measured the activity coefficient of water in phthalic anhydride at infinite dilution, and found it to be 2.45. Data measured at the Analytical Laboratories of the Amoco Research Center gave information on the acid-anhydride equilibrium vs. water concentration.

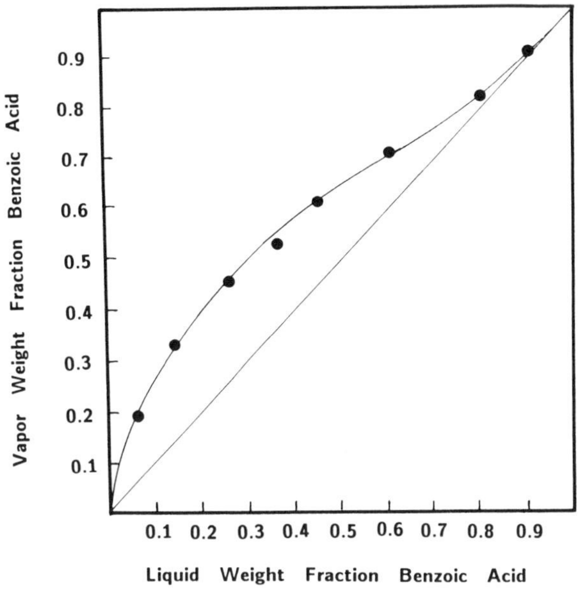

FIGURE 2. ASOG prediction of VLE for benzoic acid (1) + phthalic anhydride (2) at 100 mmHg pressure, compared to data by BASF in a private publication. (Adapted from Palmer, D.A., *Chem. Eng.*, June 9, 80, 1975.)

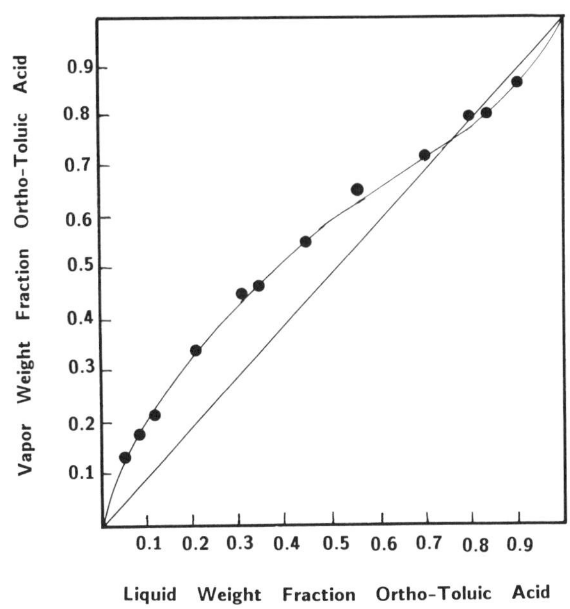

FIGURE 3. ASOG prediction of VLE for ortho-toluic acid (1) + phthalic anhydride (2) at 100 mmHg pressure, compared to data by BASF in a private publication. (Adapted from Palmer, D.A., *Chem. Eng.*, June 9, 80, 1975.)

Those data were taken in a cell under pressure, using Laser-Raman spectroscopy for the analytical measurements. None of these data have been published, but the results of the analysis are included below.

All the data were utilized in a regression analysis with the determination that the equilibrium constant at 437 K is 0.426. The activity coefficient for phthalic anhydride in water was found to be 135. The equilibrium constant varied with temperature in accordance with Equation 1.

$$\ell n \frac{K_1}{K_2} = \Delta \frac{H^r}{R} \left(\frac{1}{T_1} - \frac{1}{T_2} \right) \quad (1)$$

The compositions were computed from the equilibrium constant and the activity coefficients. The activity coefficients were computed with the Van Laar equation, based on group contributions and the specific data previously mentioned.

$$\frac{x_{PAN}\ x_W}{x_{ACID}} = \frac{K\ \gamma_{ACID}}{\gamma_{PAN}\ \gamma_W} \quad (2)$$

Further, these data yielded the anhydride/water interaction parameters for the ASOG method which are included in Table 1. With those final parameters, a complete correlation was possible for the heat of reaction, activity coefficient, and chemical equilibrium data. The predictive capability of the correlation was checked using equilibrium data at other temperatures which were not part of the original data base. This instilled confidence needed to proceed with the reaction simulation, where the chemical and phase equilibrium affect each other.

E. Computation Sequence

The following diagram shows the computation sequence and the iterative method used to solve this problem.

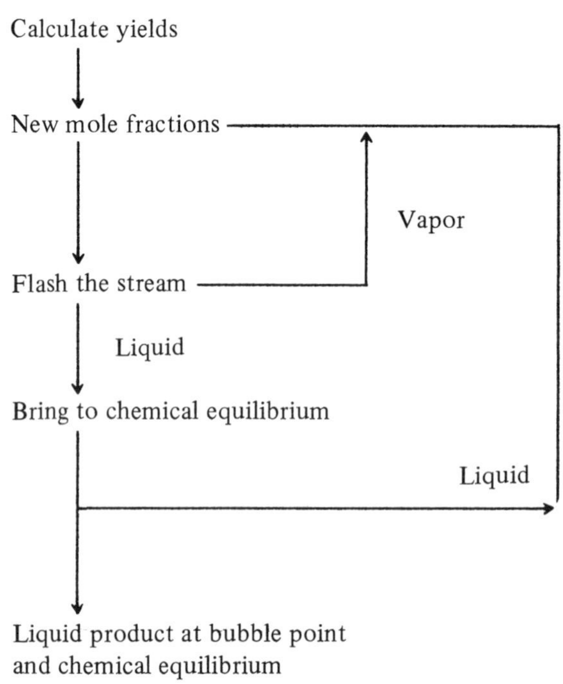

Table 2
COMPARISON OF PREDICTED WITH EXPERIMENTAL REACTOR K VALUES IN MIXTURE OF PHTHALIC ANHYDRIDE AND REACTION INTERMEDIATES

	Predicted	Experimental
Water	1.66	1.67, 1.80
ortho-Xylene	0.32	0.32, 0.36

After calculation of the yields, giving new mole fractions, the stream was flashed, giving a vapor and a liquid. The latter was brought to chemical equilibrium. Then the liquid and vapor were mixed, computationally, and flashed once more. This procedure was repeated a number of times until there was convergence. The result was a liquid product at the bubble point in chemical equilibrium, and a vapor stream in equilibrium with the liquid. The computations could have been done by hand but were actually included as a subroutine in a chemical process simulator program.

There was of course concern as to whether the circuitous route taken to arrive at this predictive scheme would model reality; therefore, experimental data were taken in the pilot plant. Table 2 presents the comparison of experimental and predicted K values for water and ortho-xylene in the mixture of phthalic anhydride and reaction intermediates; the agreement shown is excellent.

Once the simulation appeared to model the reactor operation correctly, development of the process simulation for the rest of the process was straightforward. Some experimental VLE data and vapor pressures were measured to represent the key distillation columns. However, very little additional fine-tuning was necessary in the physical properties package.

III. A PRACTICAL SYNTHESIS OF EXPERIMENTAL AND CORRELATION TECHNIQUES FOR MODELING HIGHLY NONIDEAL SYSTEMS WITH MANY COMPONENTS

A. Introduction

There are a great many systems in the chemical industry which are composed exclusively of nonelectrolytes and which operate at moderate pressures. However, those simplifications quite often are negated by the presence of highly nonideal mixtures and very many components. In fact, highly nonideal mixtures are sometimes created deliberately, in order to separate close-boiling components. Extractive distillation is the quintessential example of that. In this section the development of a complete phase equilibrium model for the butadiene extraction system will be described. It uses an acetonitrile + water mixture as the solvent. The techniques described are useful beyond the range of extractive distillation. Liquid extraction, azeotropic distillation, and the handling of other nonideal components are approached similarly.

B. Problem Definition and Strategy

The feedstock environment in the chemical industry changes rapidly. Many ethylene plants are designed to be able to handle feedstocks varying from ethane and propane down to naphthas. Some steam crackers can even handle gas oil. When the feed to the unit changes, the feed to the butadiene extraction section can swing drastically. In order to respond in optimal fashion to feed changes to the extraction unit, a process simulation process package was desired.

Table 3
TYPICAL BUTADIENE UNIT FEED COMPOSITION

Major components (mol %)		Minor components (mol %)	
cis-2-Butene	4.3	2-Butyne	0.01
trans-2-Butene	6.8	Propadiene	0.02
n-Butane	4.5	1-Butyne	0.12
1-Butene	16.8	1,2 Butadiene	0.12
Isobutene	24.4	Propyne	0.28
1,3 Butadiene	41.2	Isobutane	0.57
		Vinyl acetylene	0.62
		Isopentane, 1-Pentene, 3-Methyl-1-butene	0.22

Details of the feed composition can be obtained from Table 3. The acetonitrile + water solvent is added to this feed in the extractive distillation column.

Chapter 3 describes how a polar solvent interacts with hydrocarbon molecules in such a way that the volatility of more saturated molecules is enhanced compared to that of less saturated molecules. Thus, a separation can be effected by extractive distillation, removing butadiene from the bottom of the tower with the solvent. (Liquid extraction is not chosen for this separation because the separation factors are not particularly high and because the feed components all boil in the same narrow range. Azeotropic distillation was not chosen because of the high volatility of the hydrocarbons and the additional fuel requirement that would be necessary by comparison with extractive distillation.) The paraffins and mono-olefins are removed from the top of the extractive distillation tower. There are other towers for recovery of the solvent, purification of the butadiene, and purging of the acetylenes. The process[3] has been described by Shell Development Co., the licensor.

It is infeasible to reduce the problem to a set of key components and a couple of key pairs, for the reason that it is necessary to know the volatilities of all the hydrocarbons in the mixed solvent used in the extractive distillation tower. Futhermore, since activity coefficients are large, the components cannot be ordered with respect to volatility based simply on vapor pressures and judgment. In fact, the volatility order changes from solvent to solvent. In each of the towers there is a different set of conditions which must be taken into account by a phase equilibrium model. The process variations thus require a fairly thorough approach to definition of the phase equilibrium.

Because all molecules in the system are below their critical points throughout the process, an activity coefficient approach was chosen to model the nonideal molecular interactions. An EOS is used in this approach, but only for modeling of the modest gas phase nonidealities. The virial EOS was selected, with predictions of virial coefficients made by the Tsonopoulos correlation.[4] The critical evaluation of C_4 hydrocarbons completed by the Thermodynamics Research Laboratory at Washington University was used to obtain the vapor pressures, liquid densities, and virial coefficients. The properties of acetonitrile and water were found in the literature, using the tools described in Chapter 4.

The key parameters to be found and correlated were the activity coefficients. The UNIQUAC equation, described in Chapter 5, was used to correlate the activity coefficients with respect to composition and temperature. Two parameters are required for each binary pair. Types of data that can be used to derive the parameters have been discussed in Chapters 3, 4, 5, and 7. From those options, thermodynamically consistent VLE data over the composition range were used. Additionally, activity coefficients at infinite dilution and solubility data for partially miscible mixtures were employed to obtain the equation parameters.

The choice of a correlating equation is always a compromise, and the UNIQUAC equation certainly is not perfect. Its principal deficiency is inconsistency between predictions made from vapor-liquid and liquid-liquid data,[5] introducing a minor source of error in the computations that follow. There were no infinite dilution activity coefficients less than unity, so the potential problem of multiple sets of UNIQUAC parameters was not encountered.

Even with simplified measurements, the experimental determination of 153 pairs of binary interaction parameters is unrealistic for the single project. Therefore, some type of prediction method had to be chosen for representation of the less important binary pairs. The UNIFAC method was selected because more interaction parameters were available in the literature at the time of this project than for the ASOG method.

It is evident from the list of typical feed concentrations shown in Table 3 that the most important hydrocarbon/hydrocarbon interactions that must be considered are between the major components. Interactions of minor components with each other should have just a small effect, and interactions of minor components with major components will have an intermediate impact. Of most importance are the interactions of the hydrocarbons with the solvent system of water and acetonitrile. The solubilities of most of the hydrocarbons in water had been published. The interactions with acetonitrile had been published only indirectly, with insufficient information to derive the activity coefficients.

Accordingly, the following approach was taken.

1. Use the literature data as much as possible. This was particularly helpful in finding interactions of the major hydrocarbons with each other. Solubility data were used for hydrocarbon + water interactions.
2. Use the UNIFAC predictive scheme to fill in missing hydrocarbon + hydrocarbon interactions. Acetylenic interactions could not be predicted because the group parameters were unavailable.
3. Make rough estimates of interaction parameters for acetylenes with themselves.
4. Measure the activity coefficients at infinite dilution for mixtures of acetonitrile with hydrocarbons. These are the interactions of most importance in the system. The reasons for emphasizing infinite dilution activity coefficients were discussed in Chapter 3. To review, note that

   A. The activity coefficient is usually at its largest value at infinite dilution.
   B. In separations, the greatest number of trays is usually required to achieve high purity at the extremes of concentration.
   C. There is less uncertainty in activity coefficients obtained by interpolation with thermodynamically consistent equations, such as UNIQUAC, than there is in extrapolation of ordinary VLE-type data to infinite dilution.
   D. In most cases it is simpler to make measurements at infinite dilution because vapor composition measurements can be avoided.

C. Evaluation of Literature Data

Data from the literature on mixtures of interest were found by screening the bibliographies of Wichterle et al., the *American Petroleum Institute Technical Databook-Petroleum Refining,* Horsley's "Azeotropic Data", and the "Bulletin of Thermodynamics and Thermochemistry" going back to 1972. Solubility reviews by Battino and Clever were also used. References are given in Chapter 4. The useful DECHEMA series had not been published at the time of this study.

The raw PTxy data points extracted were reduced to activity coefficients and fitted to the UNIQUAC equation. The data-fitting procedure included statistical information that was helpful in screening and eliminating a number of bad points. UNIQUAC parameters can be made temperature-dependent, and temperature-dependent parameters were obtained for the acetonitrile + butane and acetonitrile + isobutane systems. A careful analysis of systems where the temperature dependence was not handled well with two parameters usually showed that the data, rather than the correlation, were at fault.

Where possible and appropriate, the Gibbs-Duhem area ratio test for thermodynamic consistency was applied to the data. Another test is the fit of the data by the UNIQUAC equation. As discussed in Chapter 4, an accurate fit by that thermodynamically consistent equation is a sufficient condition to establish consistency of the data.

One of the systems of most consequence was water + acetonitrile. Data considered came from six references. Although most of these data meet requirements of thermodynamic consistency, they are still in substantial disagreement with each other. This is particularly true in the region of 10 to 20 mol % water where some processes operate. Depending on which author one believes, the volatility of water compared to acetonitrile can vary by a factor of two. The data of References 7 and 8 appear to be the best.

D. Experimental Strategy

Two different techniques were used to measure activity coefficients at infinite dilution without making composition measurements. The first method was ebulliometry, applicable when the major component is volatile. The other method involved measurement of retention times in a gas chromatograph, used for systems where the solute is more volatile than the solvent. Both methods were described in Chapter 7. The two techniques are complementary and were used in this study to measure the activity coefficients at infinite dilution for acetonitrile mixed with each of the hydrocarbons.

E. Measurements with Ebulliometer

The activity coefficients for acetonitrile in hydrocarbons were measured at infinite dilution in an ebulliometer. Chapter 7 presented a description of how the ebulliometer measures a small change in temperature with addition of a small amount of solute. To get an accurate measurement of the small amount of solute added, a liquid sampling valve was installed in the liquid return line. Its volume was found by filling with mercury and emptying a number of times. The ebulliometer had to be constructed in stainless steel because of operation under pressure. In those experiments a reference ebulliometer was used and temperature differences were measured with thermistors. Temperature differences were referenced to the ebulliometer in which no solute was added. A precision of 0.03 mmHg on pressure measurements was determined using a pressure transducer.

Several solute additions were made, and curves were plotted through values of temperature change vs. composition change. The use of multiple points, typically ranging from a 0.1% to 2% concentration of solute, gave curves in which the slope and uncertainty in the slope could be determined. The data reduction is handled by Equation 4 in Chapter 7.

F. Measurements with Gas Chromatograph

Activity coefficients for the hydrocarbons in acetonitrile were measured at infinite dilution in a gas chromatograph. Relative to the usual simplicity of gas chromatographic measurements described in Chapter 7, this system posed a special problem. The solvent was itself quite volatile. The usual solution when studying such a system is to presaturate the carrier gas with solvent so that loss out of the column itself can be minimized. However, operations were planned at temperatures as high as 75°C, where at atmospheric pressure the solvent would form a very appreciable fraction of the vapor coming out the end off the column. This would given an impractically high loss rate, so the whole system was operated under pressure, usually about 6 to 7 bar. At these pressures the gas phase nonidealities are significant and must be taken into account. The appropriate data reduction procedure is given in Equation 2 in Chapter 7.

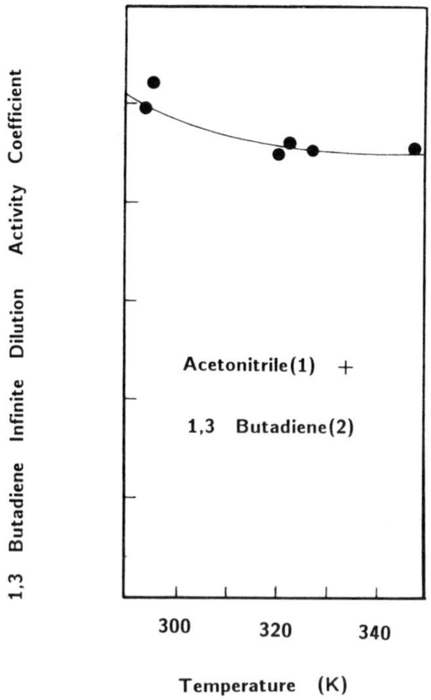

FIGURE 4. Variation of activity coefficient of 1,3 butadiene in acetonitrile vs. temperature, illustrating precision of the measurements. Data were measured at Amoco Chemicals by gas chromatography.

It is important to know the exact amount of solvent within the column. Because of the solvent volatility, special measures were required to determine it. The columns themselves were 3/16 in. O.D. × 6 ft long and were 316 stainless steel. The packing was Chromosorb® W, acid washed, 80/100 mesh. To a weighed amount of solid Chromosorb® was added sufficient acetonitrile to give a packing containing 40 wt % solvent. The mix was still free-flowing. The mixture was cooled with dry ice to avoid vaporization and inserted into the columns through a funnel; vibration was used to avoid bridging. Acid-washed glass wool was used to retain the support in the column. The weight of support in the column was determined by weighing the column with fittings before and after solid addition. A material balance showed very little solvent loss.

Ideally, there would be no pressure drop through the column. Unfortunately, there always is, even with optimization of the column size and gas flow rates. During the runs the pressure drop was usually 4.4% of the column pressure. As a consequence of this pressure drop, there was a continuous loss of solvent, even though a bubbler and presaturator column were used. To compensate for this loss, isobutane was used as a marker, being injected before and after the series of other hydrocarbon compounds. Following the last injection, the column was weighed to determine the amount of solvent remaining. With the ratio of isobutane retention times, the initial amount of solvent would be computed; then the solvent on the column at a given point in time during the run could be calculated. Although this procedure was quite straightforward, it was likely that nonreproducible losses of solvent during shutdowns were a significant source of error. Therefore, the experiments were repeated with different columns to assure accuracy.

Using different columns and different temperatures, the activity coefficient was measured for butadiene. The extent of scatter is suggested by Figure 4. Comparisons between the

Table 4
COMPARISON OF RELATIVE VOLATILITIES OF C_4 HYDROCARBONS WITH AND WITHOUT AN 80-MOL % ACETONITRILE 20-MOL % WATER SOLVENT

	No Solvent	Solvent
1,3 Butadiene	1.00	1.00
cis-2-Butadiene	0.72	1.35
Isobutylene	0.90	1.83
1-Butene	0.90	1.96
n-Butane	0.86	2.84
Isobutane	0.93	3.63

Table 5
EFFECT OF HYDROCARBON UNSATURATION ON ACTIVITY COEFFICIENTS IN PURE ACETONITRILE-VOLATILITY RELATIVE TO 1,3 BUTADIENE IN ACETONITRILE

Double bonds	Molecule	Relative volatility
0	*n*-Butane	3.7
1	*trans*-2-Butene	2.0
2	1,3 Butadiene	1.0
3	2-Butyne	0.76
4	Vinyl acetylene	0.52

relative volatilities measured here and those reported by Pavlov et al.,[6] who used a similar technique, indicated an average difference of 5%. That is an extraordinary agreement and suggests the value of infinite dilution activity coefficient measurements. If ordinary VLE measurements had been extrapolated to infinite dilution, differences between the two sets of measurements would probably have averaged 30 to 50%.

G. Correlation of the Data

The UNIQUAC equation was used to correlate all the literature and experimental data. As previously mentioned, the UNIFAC correlation was used to fill in where there were missing data of only moderate importance. In almost all instances, the built-in temperature dependence of the equation was capable of handling well the temperature dependence of the activity coefficients.

The effect of the solvent mixture on the unit feed is well illustrated by Table 4. The volatilities relative to butadiene are all fairly close to unity in the absence of a solvent, whereas the feed at infinite dilution in the solvent has volatilities ranging up to 3.63 for isobutane. The component most difficult to separate is *cis*-2-butadiene.

The interaction between acetonitrile and the hydrocarbons is a strong function of extent of unsaturation. This is seen very clearly in Table 5, which presents the C_4 molecules in order of increasing unsaturation. Relative to butadiene, the volatility of *n*-butane is 3.7, which contrasts with a volatility of 0.52 for vinyl acetylene.

H. Testing the Correlation

Data from a number of sources not included in the data base were compared against K values or relative volatilities predicted by the model. The prediction of equilibrium pressures was very successful, comparing with a classic study of mixtures of C_4 hydrocarbons.[9] The K value predictions were good except for *cis*-2-butene. Since there was little bias error, a substantial part of the discrepancy was attributed to experimental error. Table VI in Reference 10 gives details of this comparison.

Black[11] has published PTx data in graphical form for the butadiene, 1-butene, and *n*-butane mixtures with acetonitrile. The curves are not linear because of the nonideality of the systems. Figure 5 compares predictions from this model with the data on 1,3 butadiene. There is a very small discrepancy compared to the data, in that the predictions are slightly high at $x = 0.4$, although probably within experimental accuracy. Figure 6 presents the 1-butene + acetonitrile comparison. Once again the prediction is within probable experimental accuracy. There are square points on Figures 5 and 6, representing Romanian data at 50°C.

A substantial number of PTx data were measured in Romania[12] on C_4 hydrocarbons with

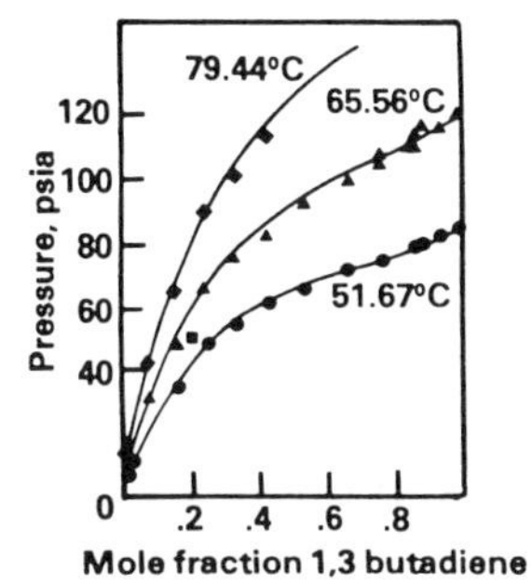

FIGURE 5. Comparison of the UNIQUAC prediction with total pressure data measured by Black for the 1,3 butadiene + acetonitrile binary system. UNIQUAC parameters were derived from Amoco Chemicals measurements of the infinite dilution activity coefficients.

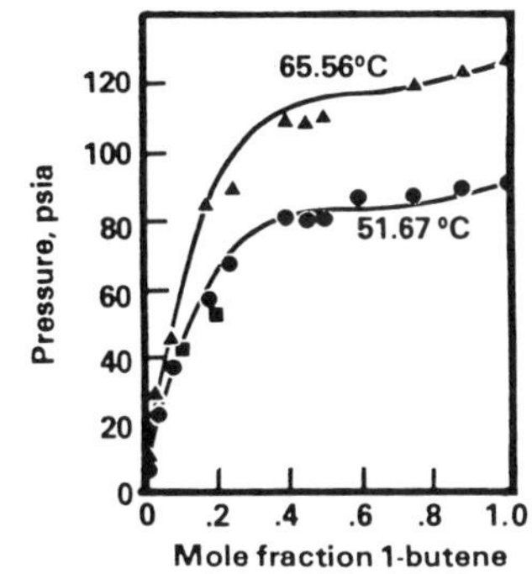

FIGURE 6. Comparison of the UNIQUAC prediction with total pressure data measured by Black for the 1-butene + acetonitrile binary system. UNIQUAC parameters were derived from Amoco Chemicals measurements of the infinite dilution activity coefficients.

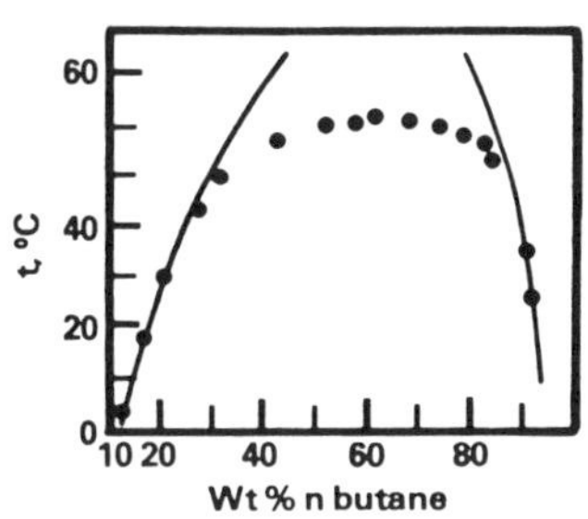

FIGURE 7. Prediction of liquid-liquid phase envelope for *n*-butane + acetonitrile binary system. UNIQUAC parameters were obtained from Amoco Chemicals measurements of the infinite dilution activity coefficients.

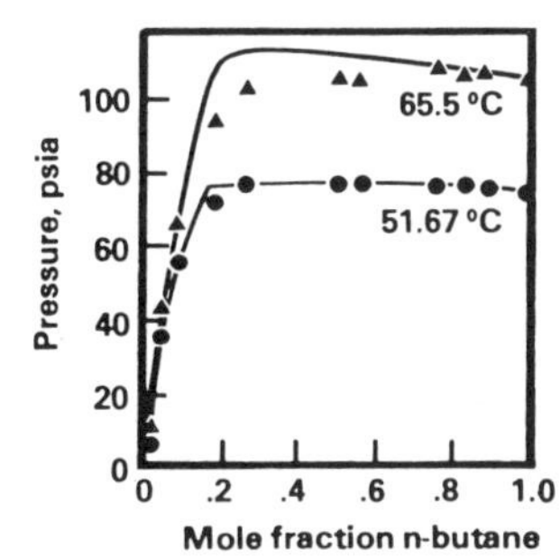

FIGURE 8. Comparison of the UNIQUAC prediction with total pressure data measured by Black for the *n*-butane + acetonitrile binary system. UNIQUAC parameters were derived from Amoco Chemicals measurements of the infinite dilution activity coefficients.

the acetonitrile + water solvent. These data would be most useful if they were accurate; however, a careful examination of the graphically presented data indicates that there were probably serious errors. Taken at 50°C, these data give excessively high pressures for butadiene relative to the Black data and the correlation. It appears that the authors must have had an impure sample of butadiene. Confirming that suspicion is the observation that with reference to butadiene, all relative volatilities for other components are 10 to 14% below those listed by Pavlov[6] at identical conditions. The discrepancy relative to the Amoco Chemicals data is slightly greater. Therefore the Romanian data were rejected.

Figure 7 presents the prediction of liquid-liquid equilibrium (LLE) for the *n*-butane + acetonitrile system. The accuracy with which the UNIQUAC equation predicts these equilibrium data of Pavlov[6] based only on vapor-liquid data is remarkable. The overprediction of the critical solution temperature is not surprising, given difficulties experienced by all such molecular models in handling the prediction of phase envelopes in the region near the critical point. Figure 8 shows the prediction of VLE for the same system. The failure to fit

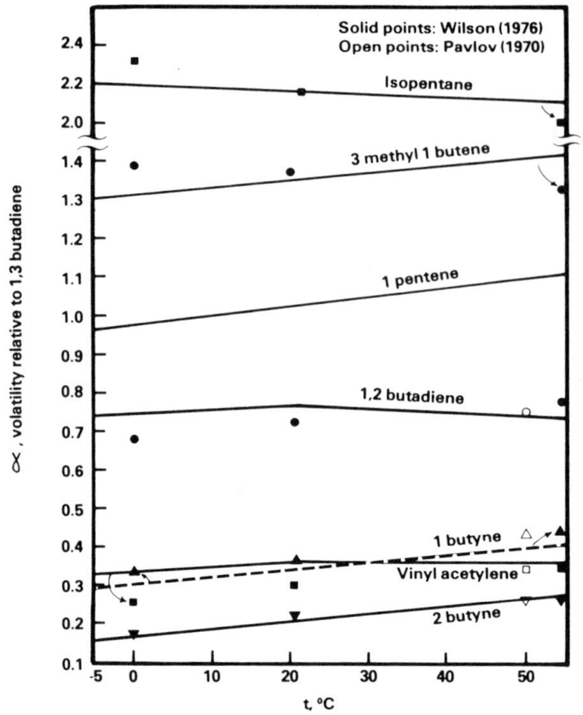

FIGURE 9. Comparison of UNIQUAC predictions of relative volatility with data by Dr. Grant M. Wilson and Pavlov. Solvent is 80 mol % acetonitrile and 20 mol % water. Note excellent agreement between Wilson and Pavlov data. UNIQUAC parameters were derived from Amoco Chemicals measurements of the infinite dilution activity coefficients, plus solubilities of the hydrocarbons in water.

the data at 65°C while representing the data at 52°C could be expected from consideration of Figure 7.

Wilson and Oscarson[13] measured the volatilities of various C_4 hydrocarbons at infinite dilution in a mixture of 80 mol % acetonitrile and 20 mol % water, previous to Amoco's in-house study. They used a gas chromatographic technique, with a temperature range from 0 to 54°C. A comparison of our predictions and the Wilson measurements of relative volatility is given in Figure 9. Note that the data base for the correlation was made up entirely of binary data. The Wilson data are for ternary systems. The open points show data by Pavlov which tend to verify the Wilson data on four compounds. The predictions for this multicomponent system are viewed as quite good, especially considering the variety of data sources built into the correlation.

A large amount of multicomponent data which appeared to be of reasonable accuracy were measured by Pavlov.[6] Activity coefficients measured by gas chromatography were not presented directly in the literature. Rather, the data were presented as relative volatilities which could be used to check the correlation but not to build a model. Figures 10 and 11 compare the data and our prediction for anhydrous acetonitrile at varying hydrocarbon concentrations. Figures 12 and 13 do likewise, with a solvent concentration of 10 wt % water and 90% acetonitrile. The 5% concentration comparison is illustrated in Reference 10. The agreement between the correlation and the Pavlov data at infinite dilution is remarkably good. The prediction of the concentration dependence is also reasonable. There is, however, a problem with isobutane, *n*-butane, and propadiene when water is added,

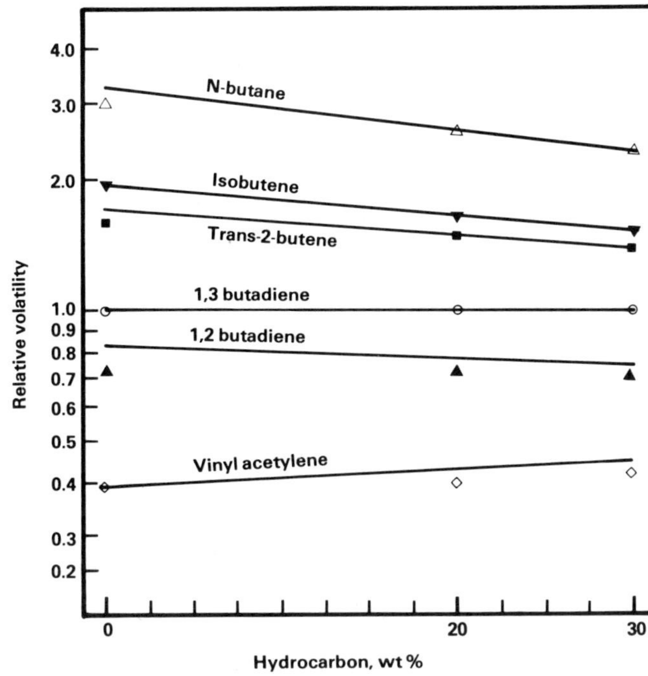

FIGURE 10. Amoco Chemicals prediction of the relative volatility of various hydrocarbons in anhydrous acetonitrile as a function of hydrocarbon concentration. Compare against the experimental data points by Pavlov.

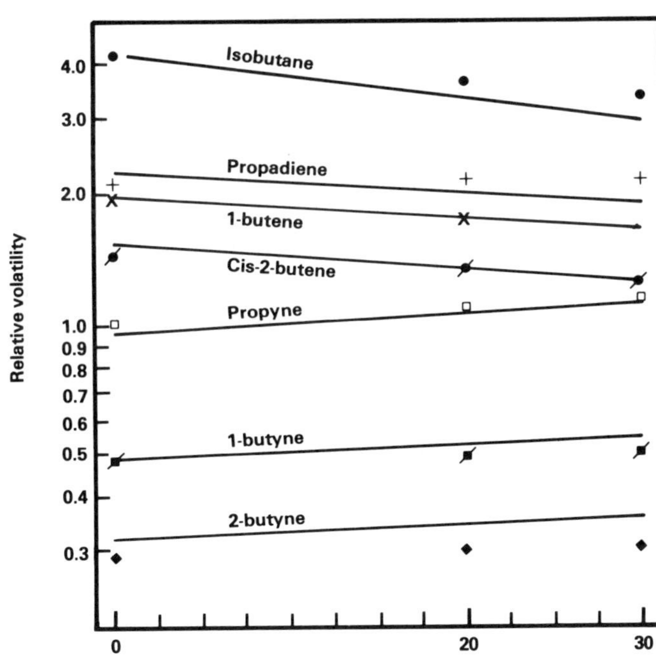

FIGURE 11. Amoco Chemicals prediction of the relative volatility of other hydrocarbons in anhydrous acetonitrile as a function of hydrocarbon concentration. Compare against experimental data points by Pavlov.

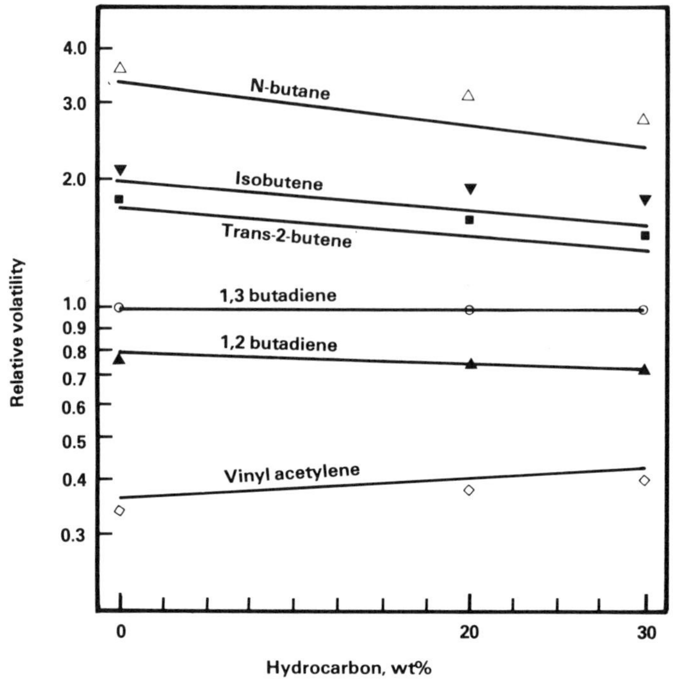

FIGURE 12. Amoco Chemicals prediction of the relative volatility of certain hydrocarbons in a commercial concentration of solvent-acetonitrile with 10 wt % water. Compare against the experimental data points by Pavlov.

which may be due to incorrect solubility data. Alternatively, it may be due to previously noted problems in predicting VLE from liquid-liquid data with UNIQUAC. Overall, however, it appears that quite reasonable predictions resulted from UNIQUAC and the data base used.

IV. SUMMARY

The principles discussed in the preceding chapters of this book have been applied to the two industrial examples presented here. It was necessary to measure experimental data to obtain the chemical equilibrium for the anhydride/acid/water system. In that particular case, a knowledge of the effects of pressure and temperature on the reaction equilibrium was most helpful in adjusting conditions to optimize reaction rate. Nevertheless, it was not necessary to measure the chemical and phase equilibrium data before testing the reaction. Those properties were actually measured during the process development phase.

The phase equilibrium data and the model developed by Amoco Chemicals Corp. for the Shell butadiene process were both assembled subsequent to commercial operation of that process by various licensees. However, the Shell Development Co. made physical property measurements and developed its own model prior to development of that process. This is an example of the type of system in which the process itself is a separation and requires accurate knowledge of the phase equilibrium before process development. For many processes, the physical property and phase equilibrium measurements can be deferred until the development phase.

The phase equilibrium characteristics of an 18-component butadiene extraction system were modeled by use of the UNIQUAC equation for activity coefficients. Representation of the phase equilibrium was challenging because the solvent is a mixture of acetonitrile

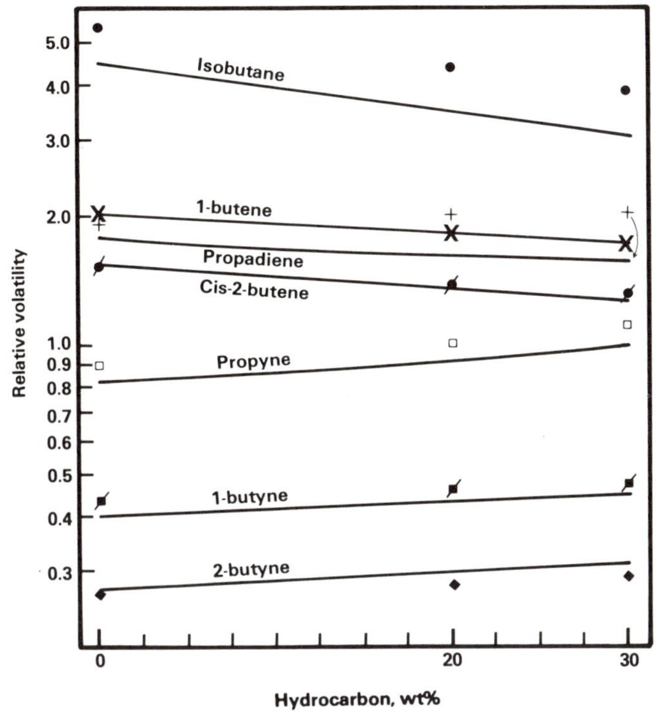

FIGURE 13. Amoco Chemicals prediction of the relative volatility of certain hydrocarbons in a commercial concentration of solvent-acetonitrile with 10 wt % water. Compare against the experimental data points by Pavlov. Note the excellent prediction of the relative volatility of the key component that must be separated, *cis*-2-butene.

and water. Solubility data were used to find interactions of hydrocarbons with water. The activity coefficients of hydrocarbons with acetonitrile were measured at infinite dilution with both ebulliometry and gas chromatography. Other interactions were obtained from literature data or predicted with the UNIFAC correlation. For the acetylenes, appropriate UNIFAC parameters were not then available, so educated guesses were made. The small concentration of the acetylenes suggested that errors in those estimates would not be of great consequence to the process simulation. Altogether, 153 pairs of UNIQUAC parameters were developed. Table 6 shows the source of parameters for each of the binary pairs. This approach permitted successful prediction of available binary and multicomponent data.

The strategy employed in developing the extractive distillation phase equilibrium model is applicable to similar systems as well. It is not the specific model that is of importance. If the solvent had been nonaqueous and nonvolatile, the problem would have been much more simple. The correlation would have been more straightforward, and the experimental work would certainly have been easier.

The methods described here are not a panacea for all separations problems. Some problems will require use of the EOS approach. Different types of data may be required. However, the concepts are similar and the approach is basically the same. Literature information was used as much as possible, filling in with predictive methods and having in-house and contract data taken on the key areas of concern. Although this may seem self-evident, many organizations have shunned the fundamental approach for a "quick and dirty" test of a system in an actual distillation column. Although there is value in such tests, the empirical approach is more costly in the long run, takes more time, is less effective, and gives no legacy of lasting information.

Table 6
SOURCES OF BINARY INTERACTION PARAMETERS FOR ACETONITRILE RECOVERY UNIT SIMULATION

	1	2	3	4	5	6	7	8	9	10	11	12	13	14	15	16	17	18
1																		
2	L																	
3	A	L																
4	A	L	L															
5	A	L	U	U														
6	A	L	L	U	U													
7	A	L	U	U	U	U												
8	A	L	U	U	U	U	U											
9	A	L	U	U	U	U	U	U										
10	A	L	U	U	U	U	U	U	U									
11	A	L	U	U	U	U	U	U	U	U								
12	A	L	L	U	U	L	U	U	U	U	U							
13	A	L	U	U	U	U	U	U	U	U	U	U						
14	A	L	G	G	G	G	G	G	G	G	G	G	G					
15	A	L	G	G	G	G	G	L	L	G	G	L	G	G				
16	A	L	G	G	G	G	G	G	G	G	G	G	G	G	G			
17	A	L	G	G	G	G	G	G	G	G	G	G	G	G	G	G		
18	A	L	G	G	G	G	G	G	G	G	G	G	G	G	G	G	G	

Note: Component designation: 1, acetonitrile; 2, water; 3, butane; 4, isobutane; 5, isopentane; 6, 1-butene; 7, isobutene; 8, *trans*-2-butene; 9, *cis*-2-butene; 10, 3-methyl-1-butene; 11, 1-pentene; 12, 1,3 butadiene; 13, 1,2 butadiene; 14, propadiene; 15, vinyl acetylene; 16, 1-butyne; 17, propyne; 18, 2-butyne. Legend: L, literature data; A, Amoco Chemicals Corp. data; U, UNIFAC prediction; G, educated guess.

REFERENCES

1. **Yen, L. C. and McKetta, J. J.,** A thermodynamic correlation of non-polar gas solubilities in polar, non-associated liquids, *AIChE J.*, 8, 501, 1962.
2. **Null, H. R.,** *Phase Equilibrium in Process Design,* Robert E. Kreiger, Huntington, N.Y., 1980, 54.
3. **Evans, H. D. and Sarno, D. H.,** Acetonitrile as an extractive distillation solvent, *7th World Petroleum Congress Proceedings,* Vol. 5, Elsevier, Barking Essex, England, 1967, 259.
4. **Tsonopoulos, C.,** An empirical correlation of second virial coefficients, *AIChE J.*, 20, 263, 1974.
5. **Lobien, G.,** Activity Coefficients at Infinite Dilution for Partially Miscible Liquid Systems, Ph.D. dissertation, University of California, Berkeley, 1980.
6. **Pavlov, S. Y., Gorshkov, V. A., Kaikina, T. G., Bushin, A. N., and Skorikova, V. V.,** Separation and purification of butadiene with the aid of extractive fractional distillation using ACN, *Khim. Prom. (Kiev),* 46, 810, 1970.
7. **Maslan, F. D. and Stoddard, E. A.,** Acetonitrile-water liquid-vapor equilibrium, *J. Phys. Chem.*, 60, 1146, 1956.
8. **Nagahama, K. and Kirata, M.,** *Bull. Jpn. Pet. Inst.*, 18, 79, 1976.
9. **Carmichael, L. T., Kwang, K. C., Berry, V. M., and Sage, H. H.,** Phase behavior in a six-component hydrocarbon system, *J. Chem. Eng. Data,* 7, 331, 1962.
10. **Palmer, D. A.,** A practical synthesis of experimental and correlation techniques for modeling highly nonideal systems with many components, *Ind. Eng. Chem. Process Des. Dev.*, 23, 259, 1984.
11. **Black, C.,** Multicomponent vapor-liquid equilibria from binary data, *Ind. Eng. Chem.*, 51, 211, 1959.
12. **Antonescu, I., Lipovics, R., and Milhailescu, M.,** *Rev. Chim. (Bucharest),* 6, 740, 1975.
13. **Wilson, G. M. and Oscarson, J. L.,** Private report to Amoco Chemicals, Chemical Engineering Department, Brigham Young University, Provo, Utah, August 25, 1976.
14. **Palmer, D. A.,** Predicting equilibrium relationships for maverick mixtures, *Chem. Eng.*, June 9, 80, 1975.

Chapter 9

THERMODYNAMICS OF PROCESS OPTIMIZATION MADE EASY

TABLE OF CONTENTS

I. INTRODUCTION

Capital and operating cost are usually thought of as trade-offs in process optimization. However, that is not always correct. Assume that too much heat is put into a process because of an inefficient design. That excess heat then must be removed. There is wasted capital in transferring the heat in and out, in addition to the capital cost for excess utilities. The chemical process industries are prodigious consumers of energy. The energy crises of the 1970s stimulated many successful efforts to identify and reduce energy losses. In just 5 years the industry reduced energy consumption by 30% relative to the value of goods produced.

With the efforts previously made, further improvements are less easily achieved. While many proposals can be made for improvement of energy utilization in a chemical complex, what is their relative merit? What impact would they have on the overall system? Have all the major potential improvements been identified, or has something been overlooked? Is an apparent savings on one unit counterbalanced by an unrecognized cost somewhere else? Are the methods used to optimize a chemical plant well founded, or do the methods of economic analysis employed lead to nonoptimum solutions? In the worst case, do they lead to wrong investment decisions?

II. EXERGY EFFICIENCY

There are pulse combustion furnaces sold into the home heating market which are advertised as having close to 100% efficiency. Is that claim justified? The answer is that it is true, relative to the accounting methods that traditionally have been used. However, in another sense, it is extremely inefficient. Heat at the flame temperature of methane is converted to warm air just a few degrees above room temperature. The loss in ability to do work is very large. This simple example leads to consideration of the next generation in analysis of energy efficiency, exergy efficiency.

The concept of exergy analysis was introduced about a century ago. Regardless of what it is called — essergy analysis, available work, second law analysis, or exergy analysis — most of the literature in the field has been developed since 1975. The basic ideas were presented by Monsanto in an American Institute of Chemical Engineers (AIChE) short course, and Exxon personnel expanded on those ideas more explicitly in another AIChE short course.[5] At the 1980 AIChE annual meeting, there was a general discussion of the subject from both an academic and industrial perspective. Also in 1980, a workbook was published on exergy analysis.[7]

Exergy analysis measures the ability to do work, and gives more information than the energy evaluation methods commonly employed. This is not to say that the concepts previously have been entirely unutilized. For example, most companies use 10,000 Btu/kW as a standard conversion factor. Actually, the true conversion factor is 3413 Btu/kW. This distinction implies that a unit of work, represented by electricity, is about three times as valuable as a unit of heat. In fact, it takes roughly 3 Btu of heat to generate 1 Btu of work. This is a consequence of the second law of thermodynamics. Engineers also recognize that low-pressure steam is not worth as much as high-pressure steam, but that intuition seldom has been quantified.

Exergy (available work) analysis provides the means to present these arguments quantitatively. Analyses can be made without doing an efficiency evaluation of specific pieces of equipment such as turbines and without introducing other thermodynamic cycles. Exergy analysis facilitates examination of losses within processes, whereas an energy analysis based on the first law of thermodynamics looks only at the system boundaries. Proposals for cogeneration of steam and electricity ultimately are related to exergy analysis.

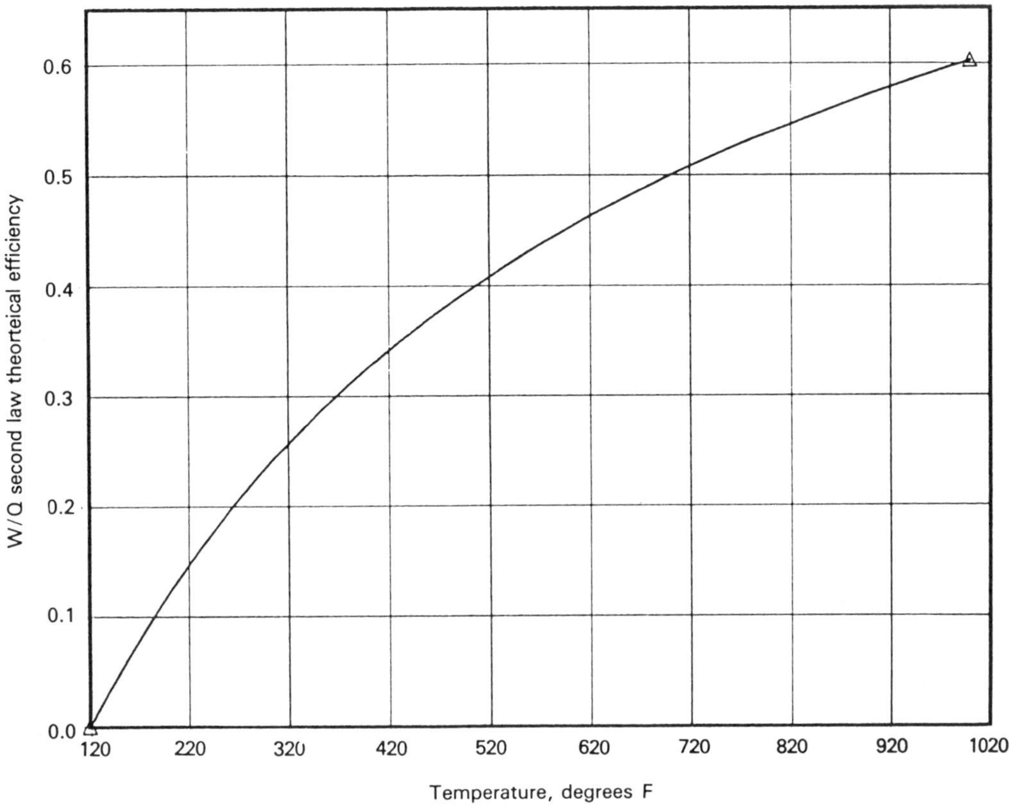

FIGURE 1. Theoretical maximum recovery of heat as work: effect of temperature on second law of thermodynamics efficiency. Base temperature is 120°F, lowest temperature reasonably achieved with cooling water.

The first law of thermodynamics indicates that energy is neither created nor destroyed. While it deals with quantity, it ignores the quality of energy. The second law of thermodynamics considers quality and shows that the potential for work of a heat source increases with temperature. The second law efficiency, in conversion of heat to useful work, is limited by Carnot's law:

$$\eta_{II} = 1 - \frac{T_o}{T} \tag{1}$$

where T_o is the heat sink (ambient) temperature. In these studies it is concluded that a reasonable ambient temperature base is 580° R, which can be attained with cooling water and a reasonably sized heat exchanger. Figure 1 shows graphically the effect of temperature on the theoretical second law efficiency. Figure 2 is similar, but was developed for subambient temperatures where there is concern about the coefficient of performance for refrigeration. At absolute zero, the work required to remove heat goes to infinity. In addition to showing that the value of heat increases with temperature, the second law implies that large quantities of heat at ambient temperature are worthless. The most valuable form of energy is work; heat can only be partially converted to work, being limited by Carnot's law and other irreversiblities.

Exergy is defined by

$$Ex = \Delta H - T_o \Delta S \tag{2}$$

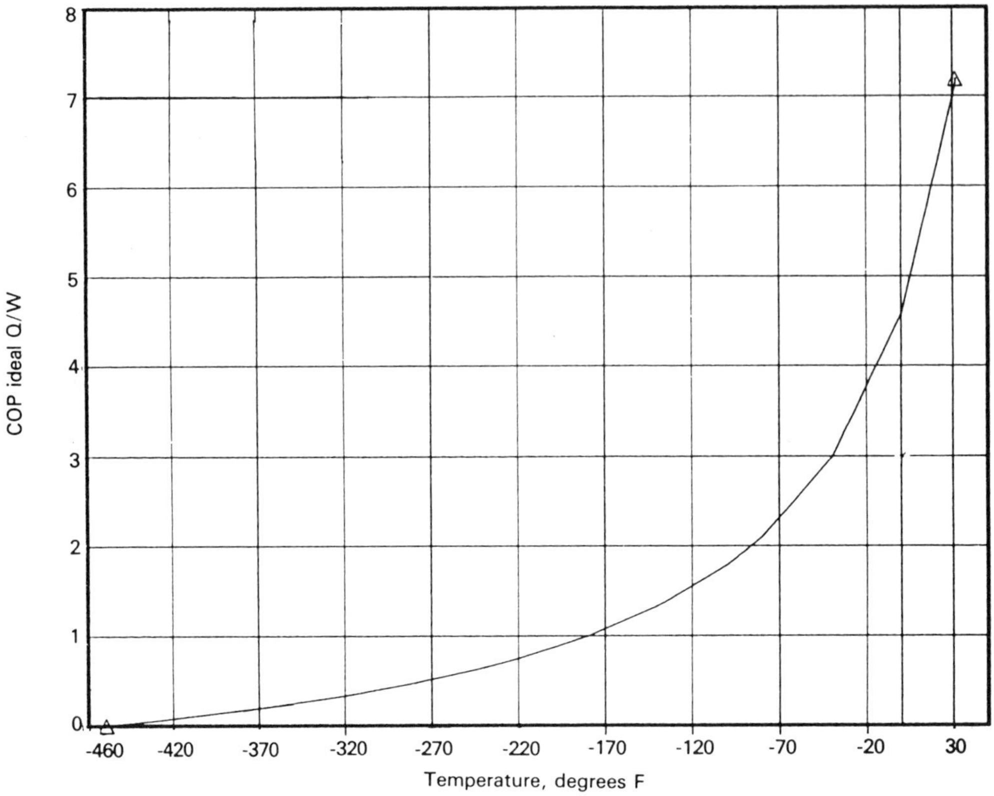

FIGURE 2. Ideal refrigeration performance, coefficient of performance (COP) vs. temperature. The base temperature is 100°F. Note that the coefficient of performance goes to zero at absolute zero where the cost of refrigeration essentially becomes infinite.

Exergy is a quantitative measure of the ability to cause change. By contrast with energy, exergy can be destroyed. This occurs with irreversibilities which increase the entropy of the system. Exergy, rather than entropy, is used in analysis, since entropy gives ambiguous information when there are simultaneous changes in temperature and pressure.

When exergy is used in efficiency evaluation, losses can be identified that are not readily apparent. For example, it is obvious that reducing steam pressure through a valve is a loss of ability to do work. In fact, there is a 38% loss when pressure is reduced from 1325 to 80 psia. However, it is not so obvious that there is a loss of exergy when 1325 psia condensate is depressurized. It might even seem advantageous to drop the pressure of condensate; for every 100 lb of such condensate reduced to atmospheric pressure, 42 lb of steam are generated. Actually, 57% of the ability to do work is lost. That is a real cost, which ultimately translates into more fuel into the boiler or electricity consumption.

III. EXERGY ANALYSIS OF PROCESS UNITS

A process exergy analysis can be used for identification of the biggest losses in a process, either by section or by pieces of equipment. It also can stimulate creative changes to the process which would enhance the overall process efficiency. Unfortunately, some presentations of exergy analysis suggest excessive complication and difficulty. For example, in one Ph.D. thesis the author stated, "In summary, it is concluded that a Second Law analysis in the context of the chemical process design is (1) difficult to produce and (2) difficult to interpret."[6] That statement is correct, assuming that a complicated approach is used. However, it appears possible to simplify the problem considerably. The following discussion gives straightforward rules and procedures that should not impose inordinately on the engineer's time.

The approach recommended is straightforward. First, determine the losses due to streams leaving at nonambient conditions. Also compute heat exchanger exergy losses, pressure reduction exergy losses, and mixing or quenching operation contributions to exergy loss. This method will be explained shortly. The next step is analysis of the results. Has a high percentage of the exergy losses been identified? What are the major exergy losses in the process? Are the losses actually created somewhere other than where they appear? Can a number of losses be consolidated and solved with a single piece of hardware? What can be done to reduce the remaining losses? Having identified reasonable alternatives for the reduction of major losses, comprehensive economic analysis can begin.

A. Overall Process Exergy Consumption

Some advocates of exergy analysis would begin with an overall analysis of the process and a computation of the process efficiency. The author has not found this to be particularly helpful. Rather, it is recommended that major process losses first be identified. However, since the overall analysis is described so frequently, an example of that approach has been included here for completeness.

The example is based on the terephthalic acid process developed by Amoco Chemicals Corp. and licensed world-wide. In this process, para-xylene mixed with acetic acid solvent is oxidized in the liquid phase with air. The operating pressure will be between 250 and 450 psig. The catalyst system is cobalt and manganese with a bromine promoter. The product precipitates during reaction but is kept in suspension by agitation. The reactor product is cooled by evaporation in crystallizers as the pressure is reduced in succeeding stages. The crude product is recovered in centrifuges. It is then dissolved in water, hydrogenated over a catalyst to remove aldehyde impurities, and recrystallized, recovered, and dried. Meanwhile, mother liquor in the oxidation section is stripped, and the acetic acid + water mixture is distilled. A schematic of the process, with exergy losses in each section as computed by ICI,[2] is given in Figure 3.

The ICI approach recognizes that all the heat generated by the reactor cannot be converted to work. This follows from the fact that exothermic reactions cannot be run at infinite temperature, nor is it usually practical to run endothermic reactions at room temperature. Therefore, for exothermic reactions Linnhoff[1] subtracts the "inevitable" loss, which is defined by

$$(T_o/T_r)\ \Delta G^o_{T_r} \tag{3}$$

Townsend of ICI[2] found that the major losses are in the reaction and solvent dehydration sections. The author's own analysis led to the same conclusion, but in addition discovered serious problems in standard layouts of the utility distribution system. The problem in the reactor area is that heat cannot easily be withdrawn at reactor temperature due to solids in the reactor.

B. Exchanger Exergy Loss Calculations

Exergy is lost in exchangers to the extent that the exchange is irreversible. The irreversibility is a function of temperature differences and heat transferred. The exergy change is calculated for both the hot and cold sides of the exchanger by Equations 4 and 5.

$$\Delta Ex_{HS} = \Delta Q\left[1 - \frac{T_o}{T_{HS}}\right] \tag{4}$$

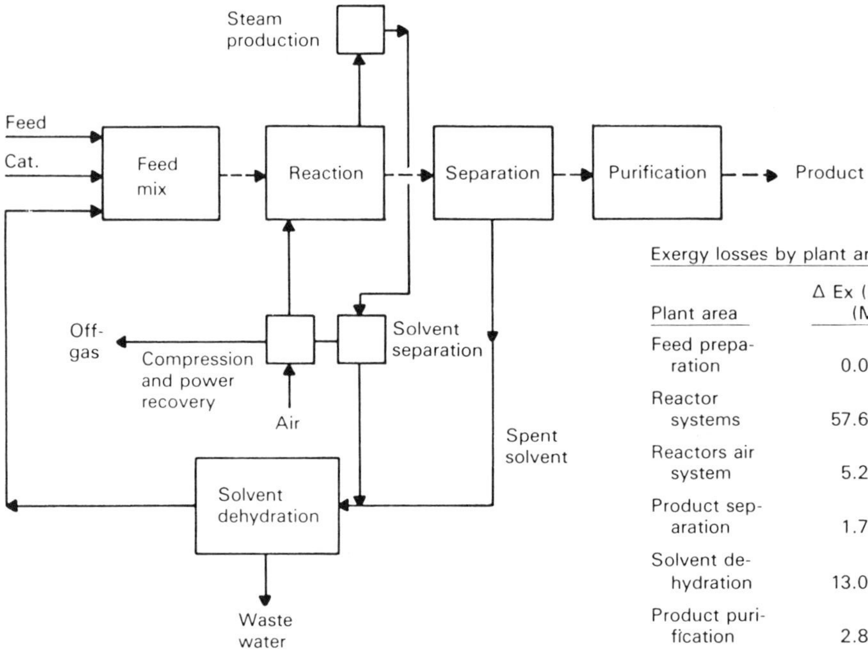

Exergy losses by plant area

Plant area	Δ Ex (Irrev.) (MW)
Feed preparation	0.04
Reactor systems	57.64 (9.1 avoidable)
Reactors air system	5.24
Product separation	1.76
Solvent dehydration	13.05
Product purification	2.89

FIGURE 3. Diagram of the ICI terephthalic acid process and the ICI evaluation of exergy losses.

$$\Delta Ex_{CS} = \Delta Q\left[1 - \frac{T_o}{T_{CS}}\right] \tag{5}$$

The lost exergy is the summation of these two exergy equations, with an exergy gain (negative exergy loss) on the cold side. An example is now presented, dealing with a stream being heated with high-pressure steam.

The computation, for a heat duty of 40 MM Btu/hr, is

$$\Delta EX_{HS} = 40\left[1 - \frac{580}{1041}\right] = 17.71 \text{ MM Btu W}$$

$$\Delta EX_{CS} = -40\left[1 - \frac{580}{858}\right] = -12.96 \text{ MM Btu W}$$

$$\Delta EX_{Lost} = 17.71 - 12.96 = 4.75 \text{ MM Btu W}$$

The efficiency of this exchanger is 73%, from

$$\eta = 12.96/17.71 = 0.73$$

Using this approach, it is an easy matter to compute exergy losses for each of the exchangers in a plant. Both their efficiency and magnitude of the loss will be evident. One of the greatest losses is often the condensation of overhead products. Rejection of exergy to cooling water or the environment in general, as in an air-cooled condenser, is a total loss. If the condensing temperature is greater than that of water, and pressure is above atmospheric, consideration should be given to use of a turbine plus condenser. The capital cost is high for such an option, and for steam it is particularly high because of large vapor volumes. The justification is the value of electricity generated or saved. It is often preferable to find an opportunity to use the overhead from one column to reboil another, but turbines are an option that often should be considered.

C. Vent Streams Losses

Vent streams are quite often a source of large exergy loss but are sometimes overlooked in the search for energy savings. In addition to the raw material value, if any, in vent streams, there is often pressure reduction and mixing of the stream at higher than ambient temperature with the atmosphere. The computation of exergy loss is carried out as follows:

1. Find the heat of condensation (negative) at the weighted average condensing temperature, T_{av}.
2. Find the sensible heat (negative) for cooling the stream to ambient temperature, ΔH_s.
3. Find the entropy change from

$$\Delta S = \underset{\text{(negative)}}{\Delta H\left[\frac{\ln(T/T_o)}{T - T_o}\right]} + \underset{\text{(positive)}}{R \ln(P/P_o)} + \underset{\text{(negative)}}{\frac{\Delta H^v}{T_{av}}} \tag{6}$$

4. Finally, based on Equation 2, the exergy change is calculated from

$$\Delta Ex \text{ (negative)} = \Delta H^v + \Delta H_s - T_o \Delta S \tag{7}$$

D. Analysis of Separation Processes

There is a finite amount of exergy required to separate mixtures due to the increase in entropy when the mixtures were first created. This ideal exergy requirement is calculated quite easily. First, a material and energy balance is developed around the unit operation. Next. the activity coefficients are found or predicted. For isomeric mixtures such as xylenes they can be set equal to unity. The exergy for each stream is calculated from the composition and activity coefficients, using

$$Ex = RT_o \Sigma x_i \ell n(1/x_i\gamma_i) \tag{8}$$

The minimum overall exergy requirement is

$$\Delta Ex_{min} = -[\Sigma Ex_{prod} - \Sigma Ex_{feed}] \tag{9}$$

The actual exergy input is calculated from the steam and electricity input exergies minus the condensate exergy leaving the process. (Utility exergy computations will be discussed subsequently.) Then the second law efficiency of the separation is calculated from

$$\eta_{II} = \frac{\Delta Ex_{min}}{\Delta Ex_{actual}} \tag{10}$$

This approach was used to analyze the efficiency of the acetic acid + water separation. It is a common separation in the chemical industry, since acetic acid is a large-volume chemical and in most solvent uses a dehydration step is needed for recycle. With straight distillation at a conventional reflux ratio, the efficiency is 6.7%. Use of an entrainer could improve that efficiency to 8.5%. That is not much of an improvement, considering the added complexity of an azeotropic distillation process.

The reason for such low efficiencies is that the heat of vaporization is usually much higher than the exergy of mixing. Distillation continues to be the workhorse separation process in the chemical industry because other separation methods have their own set of problems. Crystallization is potentially more attractive, but when refrigeration is required a new set of exergy losses is introduced.

When analyzing distillation processes, it would be well to consider the following useful, but not universally applicable, rules of thumb:

1. Introduce the feed on the tray having a composition closest to that of the feed, to minimize mixing exergy loss.
2. Consider whether the reflux ratio has been reduced sufficiently by addition of trays.
3. If the column has a large temperature rise, consider inter-reboilers using lower-pressure steam. If refrigeration is needed for the condenser, consider intercondensers that could use cooling water.
4. Maximize heat reuse through multiple-effect distillation.
5. Consider heat pumps to avoid refrigeration condensing, especially when there is a very small temperature difference across the column.
6. Consider condensing turbines for overhead streams condensed with cooling water, particularly when the column operates under pressure and the stream is totally condensable.

E. Mixing/Quenching

Exergy losses due to mixing are not always obvious. One such source is a hot stream going to a scrubber. Some scrubbers reject heat to cooling water, so the entire exergy value of the stream is lost. Crystallizers are another source of exergy loss. Not only is there a loss at the crystallizer due to a temperature difference between the coolant and the process fluid, there is also a loss due to warmer liquid being quenched in the crystallizer.

IV. ALLOCATION OF ENERGY COST

Energy cost is at the heart of most process optimization in research, engineering, or operating plants. The cost of energy is expressed as cents per kilowatt, dollars per thousand pounds of steam, or dollars per million Btu. The author intends to show that current steam accounting systems used by most companies are in fundamental error. Most companies value steam strictly on a Btu basis, so that variations in pressure level are of little economic significance. It is proposed that steam delivered at pressures lower than generation pressure should be priced significantly below the value commonly assigned. It is also proposed that high-temperature condensate should be given a value in corporate accounting and optimization systems. Typically it carries zero or little value. These are not new proposals, but they still need to be implemented.

Incorporation of these simple changes will give a quantitative basis for cost allocation which harmonizes with intuition. More importantly, these changes will affect future investment decisions. Application of the recommendations in this section can provide a foundation upon which normal investment decisions will be implicitly based on exergy considerations. Also, it will be easier to identify the places in energy distribution systems and chemical processes where irreversible energy losses are incurred.

A. Exergy Pricing Basis for Steam

The earlier discussion on exergy leads to the simple conclusion that steam should be priced based on exergy rather than energy. To appreciate the reason for this, consider three simple examples.

1. Heat is needed at 150°F for warming a waste treatment pond. With current pricing methods it would be reasonable to generate the steam at low pressure and charge the waste treatment facility full price. However, the steam could be generated at high pressure, with superheat, with work extracted in a back pressure turbine before exhausting the steam at atmospheric pressure for warming of the pond. The cost of the steam to the waste treatment facility would now be reduced by the value of work produced.
2. Heat is needed at 300°F for a distillation column. In this case some power can be extracted before exhausting the steam, but it is less. Obviously, the steam is now worth more, because less work could be extracted along the way.
3. Heat is required at 550°F to run a high-pressure dissolver. In this case, 1300 psi steam is used directly, and no work is extracted. Obviously, steam in this form is worth more than in the two previous cases.

Does it make sense to charge as much for energy close to ambient temperature as for energy at high temperature where there is no potential for recovery of work? That is essentially what is done all over the world today.

It is worth citing an example of the irrationality of enthalpy as opposed to exergy pricing. In a large chemical complex, steam was required to run a large styrene distillation column. It was needed at only 75 psig. The steam was available at over 400 psig, so it was used first in a different unit to run a back pressure turbine. The second unit justified its investment on the basis of "free" energy. Meanwhile, the styrene unit was charged the full energy basis for the steam it received. With the demand for energy savings ideas, it was decided to look at vapor recompression for the styrene column. Installation of a compressor now appeared justified. But was it really?

This is just one more example of how isolated projects get consideration when the net effect may not be that which was desired. This was pointed out in an editorial in *Hydrocarbon Processing*.[6] It gave an example of a major energy conservation project at a Gulf Coast refinery, with projected savings of 5% of fuel fired. Subsequent to completion, tests failed to show those savings. The editor pointed out that the problem was in use of the wrong basis for assignment of utility values: "After more than a decade since the drastic increase in OPEC oil prices, a realistic, technically sound pricing technique for low level energy sources is badly overdue. As long as industry continues to assign the wrong benefits to energy conservation, everybody will lose . . . And the credibility of the engineering profession will suffer because it gets the blame when the payout isn't there."

One suspects that the sum total of all energy savings claimed would exceed the total energy used! It can be argued that entire site-wide steam balances would solve the problem, but that seldom can be justified for individual projects. Fortunately, exergy analysis provides a tool which can be helpful in assuring that projected savings are real.

Figure 4 shows the variation of exergy for steam and condensate as a function of temperature. The values are generated from steam tables which list both enthalpy and entropy. It is interesting to note that the values for both decline to zero at the chosen reference temperature of 120°F. This agrees with our intuition that energy is worthless in a plant once it drops to the reference temperature. The exergy of high-pressure steam is about one third of its enthalpy, a result of the inefficiencies created by the second law of thermodynamics in conversion of heat to work. If steam is generated at a lower pressure or temperature in the boiler than would be technically and economically feasible, the cost per unit of exergy is increased significantly. Figure 4 also indicates that on an exergy basis, high-pressure condensate is worth more than low-pressure steam.

The cost of exergy is equal to the cost of absorbed heat times the energy input to the steam, divided by the exergy of the steam. For example, with energy available at $4.50/

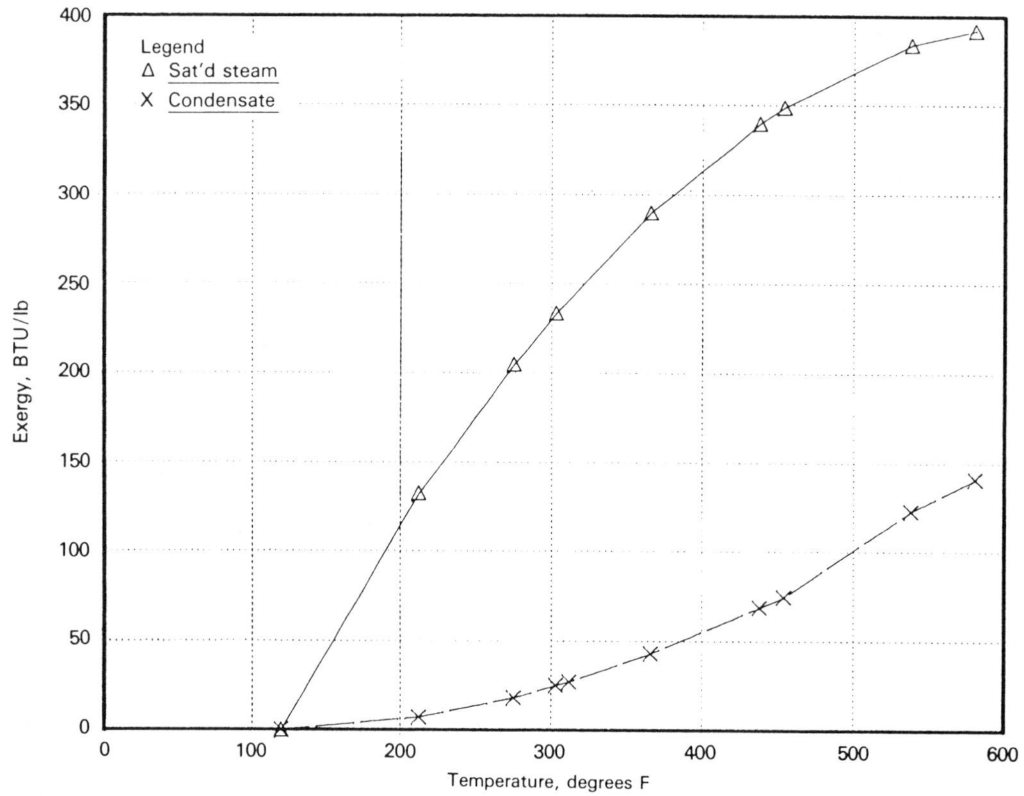

FIGURE 4. Exergy values of steam and condensate, with base temperature of 120°F.

MM Btu net and 85% furnace efficiency, the absorbed energy cost is \$5.29. The exergy cost for 1325 psia steam at 660° is computed from the following equation (enthalpy of the steam is 1267 Btu/lb and enthalpy of boiler feedwater at 120 degrees is 88 Btu/lb):

$$(1267 - 88) \text{ BtU} * \$5.29/434 \text{ BtU W} = \$14.37/\text{MM Btu W}$$

Neglecting the cost of pumping or of the cold boiler feedwater, the high-pressure steam cost is \$6.24/Mlb. Steam at 70 psia has an exergy of 234 Btuw, so its value is just ratioed down directly, and is \$3.36. Saturated condensate at 1300 psia is worth \$2.03/Mlb. Steam and condensate at 120°F are worth practically nothing.

This method is thus very simple. Table 1 gives a list of the values compiled from steam tables for one plant with multiple-pressure steam headers. It can be used as a model to develop exergy pricing for plants with different situations based on local energy costs and different boiler pressure. The cost of exergy, at a fixed fuel price, will be greater for those plants generating steam at low pressure than for plants generating it at high pressure.

B. Comparison of Exergy and Enthalpy Steam Pricing

Table 2 compares energy and exergy pricing policies for a process using energy at different temperature levels. In this example, the net heat input to the process is 1000 Btu at the boiler. There, steam is generated at 440 psia. It flows to Process #1 which produces steam at 165 psia and returns condensate at 440 psia. An example of such a process would be a high-temperature distillation where steam is generated in the overhead condenser. As shown, the energy from Process #1 is used in Process #2, and at a still lower temperature in Process

Table 1
EXERGY VALUES FOR AN ADVANCED STEAM SYSTEM

Nominal steam pressure (psig)	Actual steam pressure (psia)	Vapor exergy (Btu W/lb)	Condensate exergy (Btu W/lb)	Condensation exergy (Btu W/lb)	Condensation energy (Btu/lb)
1300	1340, 650°F	429	—	—	
1300	1340	392	141	251	585
425	440	349	75	274	770
262	375	340	69	271	788
175	165	290	43	247	857
75	80	243	27	216	901
60	70	234	25	209	908
30	45	205	18	187	929
0	15	133	7	126	970
Vacuum	1.7	0	0	0	1025

#3. In Process #3 the heat is rejected to cooling water. Table 2 shows the number of pounds of condensate and steam flowing between each of the processes. Also indicated is the exergy flow in each stream. The net energy consumption of Processes #1 and #2 is zero, so enthalpy pricing currently in use would result in no charge to those processes. The entire cost would be allocated to Process #3. By contrast, the exergy approach results in the cost being divided between the three processes. The total cost will be the same, but rational pricing will help avoid subsequent irrational decisions on capital expenditures related to energy conservation in any of the three plants.

The consequences of a shift in pricing policies to an exergy basis will be as follows.

1. Breakdown of steam and condensate to atmospheric pressure will be discouraged.
2. Cost reduction programs conserving high-pressure steam and condensate will be encouraged at the expense of programs that save low-pressure steam.
3. A quantitative method for screening energy conservation proposals will be available.
4. Steam production at high pressure, with possible cogeneration, will receive more attention.
5. Hidden process efficency losses will be identified.
6. Allocation of true costs will give a better measure of how thoroughly the exergy losses have been identified.
7. More attention will be given to losses in steam distribution systems.

As attempts are made to reduce exergy losses in new or existing processes, real reductions in boiler fuel consumption should result. The attractiveness of exergy-based steam pricing is that it is easily accomplished. With both enthalpy and entropy listed in the steam tables, computation of the exergy is simple. It is an invaluable first step to imposition of rationality on energy conservation programs. It also provides a fairer basis for allocation of energy costs between process units.

V. EXERGY ANALYSIS OF A STEAM DISTRIBUTION SYSTEM

It can be frustrating to see the best process optimization efforts sidetracked during a phase of detailed design often left to mechanical engineers. Steam distribution system losses can be far more important than has been previously thought. This principle is illustrated through assessment of an actual plant. The same pressure levels for steam as detailed in Table 1 are applicable. Based on data collected by plant personnel, it was calculated that the exergy of

Table 2
COMPARISON OF EXERGY AND ENERGY PRICING POLICIES FOR A UTILITY SYSTEM

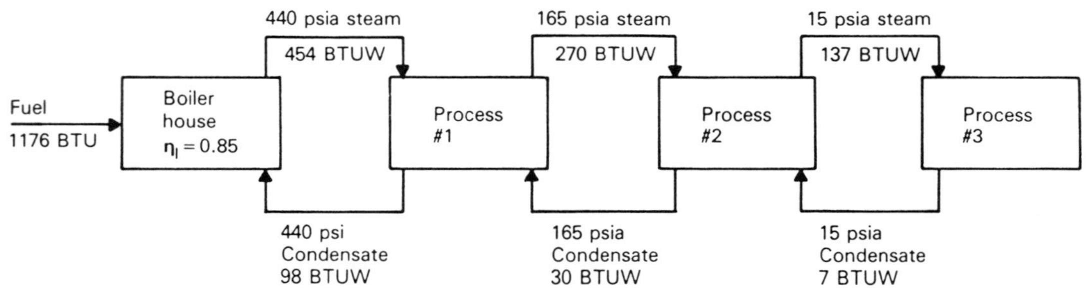

	Boiler house	Process #1	Process #2	Process #3
Lb steam in		1.30	1.11	1.03
Lb steam out	1.30	1.11	1.03	0
Lb condensate in	1.30	1.11	1.03	0
Lb condensate out		1.30	1.11	1.03
Exergy	356 Btu	116	110	130 = 356
Energy	1000 Btu	0	0	1000
Cost, exergy basis	$\frac{1.486¢}{1000 \text{ Btu W}}$	0.173¢	0.163¢	0.193¢ = 0.529
Cost, energy basis @ 4.50/MM Btu, 85% efficiency	$\frac{0.529¢}{1000 \text{ Btu}}$	0	0	0.529¢

1300 psia steam from the boiler amounted to 94 MM Btu W. There was also 99 MM Btu W in the form of 75 and 30 psig steam from the reactor condensers. Exergy in the boiler feedwater return amounted to 5 MM Btu W. All the uses of exergy were identified, such as reboilers, steam tracing, etc., and a first law balance was made. The uses of exergy were subtracted from the input of 193 MM Btu W to find the distribution system loss by difference. It amounted to 36% of the exergy exported from the boiler! This dramatically increases the effective cost of exergy consumed. An effort was made, therefore, to identify the specific losses in the distribution system.

The largest loss came in letdown of high-pressure condensate (15.2 MM Btu W). This could be rectified by pumping the high-pressure condensate back to the boiler. In retrofit situations this can create problems, so it is best implemented in the original plant design. Flashing of 75 psig condensate cost as much in that plant as pressure reduction of steam. The condensate flashing could have been eliminated by installing pumps to return such condensates to their appropriate steam drums. Venting of low-pressure steam, related to letdown of high-pressure condensate, was also very costly. Overall, it was possible to balance the observable exergy losses in the distribution system with the loss previously computed.

An overall system optimization, using certain of the ideas just presented, showed that fuel consumption could be dropped by $2.5 MM $/yr. This could be accomplished with investment in some pumps and insulated condensate return lines. It was assumed that 25 MM Btu W out of a total of 34 MM Btu W in distribution system losses could be saved. To optimize the design of a steam system without the benefit of an overall plant optimizer (in some cases those optimizers fail to consider the most important options), the following suggestions could be considered.

1. Compute actual exergy cost for steam based on boiler operation.
2. Develop a steam and condensate balance for the plant.
3. Compute the exergy consumption for each known process use.
4. Compute the exergy input from the boiler house.
5. Find distribution system losses by difference.
6. Try to close the balance by identification of major losses due to venting, condensate breakdown, and steam breakdown.
7. Consider the effect of pumping condensate back to its own level, elimination of venting, and use of turbines where necessary.
8. Test the effect of the new modifications on the overall unit balance and boiler fuel consumption.

VI. THERMODYNAMIC ANALYSIS OF RANKINE CYCLES

Rankine cycles are used increasingly as a tool to reduce exergy losses. Their application is most appropriate where heat is rejected under conditions at which it is not needed. Reference 4 lists a number of practical applications of Rankine cycles. Beyond uses in chemical plants, there are a number of unconventional sources of energy now being tapped. Both dry and wet geothermal steam are being used in the western U.S. Because of the corrosive compounds in that steam, secondary cycles using organic fluids are used to drive the turboexpanders. Until now, relatively inefficient hydrocarbons have been chosen as the Rankine cycle fluid. Geothermal brines available at about 330°F (166°C) are used at Casa Diablo Hot Springs to provide the heat for an organic Rankine cycle. At Beatrice, Neb., hot compressor exhaust was used to drive a Rankine cycle and provide a 10% increase in compressor capacity without adding more fuel. In Hawaii, an ammonia fluid Rankine cycle was used in the successful test of a process to recover solar energy from the ocean (Ocean-Thermal Energy Conversion).

Heat from reactors is usually available at a temperature fixed by optimum reaction conditions rather than by energy recovery considerations. The heat will be converted to useful steam wherever possible. However, what is the optimal solution when that steam is not needed or is required at higher temperatures? A liquid phase phthalic anhydride plant operating without acetic acid solvent has the kind of energy imbalance just alluded to. Distillation of the product requires steam or hot oil at temperatures higher than that of the reactors. Yet, the reaction is highly exothermic. The first approach is to look to other units in the same plant as a potential consumers of low-pressure steam. Failing that, a Rankine cycle is sometimes a viable alternative. Because of familiarity with steam systems, the first thought might be to generate steam anyway and send it through a condensing turbine. However, steam has a poor cycle efficiency because of mechanical problems. It often may be found that development of an organic fluid Rankine cycle could be more attractive. The power generated would be converted to electricity or used directly to run the process air compressor.

A. Factors in Choice of a Rankine Cycle Fluid

It has long been known that for fixed-temperature heat sinks the Rankine cycle is one of the best second-law-limited methods of converting heat at moderate temperature into work. It is also known that different fluids will give different cycle efficiencies. This is due to thermodynamic properties and a variety of other factors. How can one choose, from the large number of options, a chemical that will do the best overall job? This is not a trivial problem, as some researchers have discovered. Since most investigators have assumed that a complete equation of state (EOS) is necessary for each fluid, the scope of chemicals considered has been restrictively small. It is shown here that use of simple thermodynamic principles can eliminate the need for such an EOS and permit rapid testing of many fluids.

The shape of the liquid-vapor envelope is also important, since condensation of over 10% on expansion can be troublesome. Superheating on expansion can reduce efficiency or result in a more complicated cycle. Mechanical considerations are also important. Excessive pressure-drop ratios and excessive wheel tip speeds, related to enthalpy drop, must be avoided. Further, the turbine size should not be excessive. This eliminates some otherwise interesting fluids which have low vapor pressures at the turbine outlet conditions.

Cost, availability, corrosivity, toxicity, and knowledge of the pure component physical properties are added important criteria for selection. Finally, the chemical chosen must have good stability at operating conditions.

B. Thermodynamic Analysis

It has long been thought that screening of fluids as Rankine cycle candidates required either a Mollier diagram or a computer simulation of the P,H,T surface, either of which is ultimately based on an EOS. Thus, most such analyses have been limited to nonpolar fluids because of the problems encountered in modeling polar fluids with such equations. Exceptions, of course, are water and ammonia, which have been well known for many years because of their industrial importance. Because of lack of understanding of the manner in which various physical properties interact to result in cycle efficiencies, contradictory guidelines are found in the literature.

There is no disagreement that the theoretical maximum efficiency of a Rankine cycle is given by the Carnot efficiency

$$\text{eff} = \frac{T_2 - T_1}{T_2} \tag{11}$$

where the temperatures are absolute and T_2 represents the high-temperature heat sink. It is understood that the Carnot efficiency must be reduced by the turbine inefficiency. What is

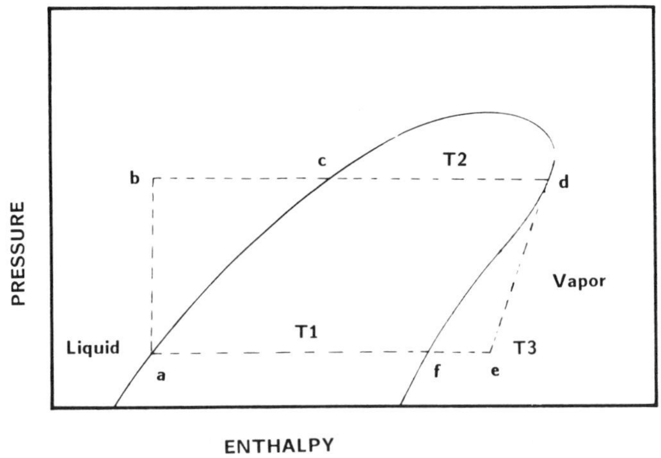

FIGURE 5. Generalized pressure-enthalpy diagram for a Rankine Cycle.

less well known is the reason why different fluids appear to reduce the ideal cycle efficiency by significant but differing amounts.

Actually, different fluids give different cycle efficiencies because of the Carnot law. Recognition of that fact and of its significance permits accurate predictions of cycle efficiencies using only liquid heat capacity and heat of varporization data.

The following P-H diagram, Figure 5, suggests some of the characteristics of a typical Rankine cycle. In Step a-b the liquid is pumped to high pressure. In Step b-c the sensible heat is added, bringing the fluid to its vapor pressure. This step represents the major inefficiency. For each increment of heat added in that step, the Carnot law must be observed. To illustrate, assume the simple case where a fluid has a liquid heat capacity which is constant with temperature. The average unit of heat is put into that fluid at a temperature midway between the condensing and boiling temperatures. The cycle efficiency for that portion of the heat input can be only 50% of the Carnot efficiency calculated for the high and low temperatures of the process. Step c-d represents vaporization, the most second-law-efficient process step. Expansion for work recovery, Step d-e, can terminate inside or outside of the two-phase envelope. In the latter case there is some added inefficiency because the fluid is superheated and not all the heat is removed at the temperature of the low-temperature sink. The temperature at point e is designated T_3. Step e-f represents removal of superheat. If it is accomplished by regeneration, heat interchange with the cold condensate, there is a recovery efficiency in the cycle. Step f-a represents condensation at the lower temperature. This temperature is always higher than the coolant temperature, to permit design of a reasonably sized condenser.

Based on the preceding description, it is now possible to write a simple equation that represents the isentropic efficiency for any given fluid between any two temperature levels. The liquid and vapor heat capacities and the heat of vaporization are required as a function of temperature. In Equation 12 the liquid heat capacity is usually much more important than the vapor heat capacity.

Table 3
NUMBER OF TURBINE STAGES REQUIRED WITH DIFFERENT RANKINE CYCLE FLUIDS IN TURBOEXPANDER WITH 6/1 EXPANSION RATIOS

	Stages (*n*)	Isentropic efficiency (%)
Fluorocarbon 11	3	72.2
Chlorobenzene	3	80.4
Carbon disulfide	3	84.7
Acetone	4	77.5
Methylene chloride	4	79.1
Acetonitrile	5	83.9
Benzene	5	79.0
Ethylene dichloride	5	81.6
Methanol	7	81.3
Ethanol	10	79.4
i-propanol	12	76.5
Water	12	90.5
n-propanol	15	78.4

$$\text{eff} = 1 - \frac{\dfrac{T_1(\Delta H^v)_{T_1} + \int_{T_3}^{T_1} C_p^v\, TdT}{(\Delta H^v)_{T_1} + \int_{T_3}^{T_1} C_p^v\, dT}}{\dfrac{T_2(\Delta H^v)_{T_2} + \int_{T_1}^{T_2} C_p^L\, TdT}{(\Delta H^v)_{T_2} + \int_{T_1}^{T_2} C_p^L\, dT}} \tag{12}$$

If the lower-temperature heat sink, T_1, were at absolute zero, the isentropic efficiency would be unity. Also, if the heat capacities were zero, the cycle would be maximally second-law-efficient. In many cases there is no or little superheat on expansion, so the numerator reduces to T_1. From Equation 12 it is evident that one characteristic of a good Rankine cycle fluid will be a high ratio of heat of vaporization at the boiling temperature to the average liquid heat capacity.

Results of the simple computations with Equation 12 were compared against data from enthalpy/entropy tables for carbon disulfide, benzene, and fluorocarbon 11. Agreement was within 1% for carbon disulfide and was exact for benzene and fluorocarbon 11.

A list of efficiencies for 54 chemicals is given in Reference 3. Over a temperature range from 50 to 175°C, the efficiency relative to Carnot varies from a low of 64% for isopentane to a high of 90.5% for water. In fact, steam usually is used, although it is not necessarily optimal because of other problems related to its unique set of physical properties.

C. Mechanical Considerations

Pressure drop in radial reaction turbines usually must be limited to a 4/1 ratio. Beyond that, the efficiency declines and the loss is 5% at a pressure ratio of 11/1. However, adding stages adds to the capital cost and reduces efficiency as well. Table 3 shows the number of stages that would be required in a turboexpander operating with 6/1 expansion ratios. The

Table 4
RELATIVE TURBINE WHEEL SIZES

Methylene chloride	0.87
Fluorocarbon 11	1.00
Acetone	1.1
Carbon disulfide	1.1
Acetonitrile	1.4
Benzene	1.6
Ethylene dichloride	1.6
Chlorobenzene	3.9

computations are based on an upper temperature of 175°C and a lower temperature of 50°C. The chemicals chosen are from among those screened which had the highest isentropic efficiency relative to Carnot.

The data in this table suggest that the alcohols and water require too many stages. The only viable alcohol would be methanol. These conclusions are reinforced by consideration of the relative tip speed. To stay within the tip speeds of normal rotors (950 ft/sec) and still have efficient expansion, the enthalpy drop should not exceed 24 cal/g. For water, it is 180 cal/g. For methanol it is 79. For the other chemicals the expansion ratio is controlling.

The *relative wheel size* affects the capital cost. It is governed by the enthalpy change during expansion, molecular weight, and vapor pressure at the condenser temperature. The relative turbine wheel sizes for two fluids derived from turbine scaling laws are

$$\frac{D_1}{D_2} = \left(\frac{\Delta H_2}{\Delta H_1}\right)^{3/4} \sqrt{\frac{MW_2\ P_2^*}{MW_1\ P_1^*}} \tag{13}$$

where ΔH_2 = isentropic enthalpy change on expansion.

The relative turbine sizes are compared in Table 4 for those fluids which survived the previous criteria. Fluorocarbon 11, often considered for Rankine cycles, is used as the reference and has a value of 1.0. It is clear that very low-pressure substances are impractical, even if they are ideal in other ways.

Carbon disulfide can be deleted from the list because of its very high toxicity. That leaves methylene chloride, fluorocarbon 11, acetone, and acetonitrile as alternatives to steam cycles. For all of them, the capital cost of the expander should be lower. Using this same approach, it is likely that other candidate compounds can be identified.

D. Concluding Remarks

This analysis of the Rankine cycle gives results which dovetail with the exergy analysis previously presented. It is clear that the problem with efficiency in such cycles is not entirely due to expander inefficiency. A larger portion of the inefficiency is due to exergy loss as the liquid is reheated to the bp temperature. The solution to the problem is clear, if not always practical. It is to find a stream that needs to be cooled sensibly. By balancing the heating and cooling requirements through sensible heat transfer with another stream, the Rankine cycle efficiency can be dramatically improved.

VII. SUMMARY

Chapter 8 gave examples of application of the principles described in this book to problems involving phase equilibrium. Certainly, phase equilibrium problems are usually a very important application of thermodynamic principles. Chapter 9, on the other hand, has shown that thermodynamic principles also can be applied to process optimization. Often, this is done in connection with pure component properties. Even the single-component utility system, steam distribution, is amenable to analysis and improvement.

It is recommended that steam pricing be shifted from an enthalpy to an exergy basis. This would formalize the intuition of most engineers to the effect that high-pressure is worth more than low-pressure steam. By formalizing this evaluation of steam values and by giving appropriate value to condensate, it is much easier to identify losses in process plants. Furthermore, it is easier to determine whether proposed process improvements really are economical; some energy conservation projects are ill-conceived and fail to save money because their justification rests on the unrealistic enthalpy pricing policy for steam that is used in most companies.

Exergy can be calculated simply from enthalpy and entropy data. However, entropy data are readily available for only a few compounds, so simplified equations have been presented to handle most process situations and take entropy implicitly into account. From the exergy analysis it was possible to develop a number of generalized rules that can be used when making quick process appraisals.

Use of heat at moderate temperatures sometimes can be a problem, particularly when it is unavoidably generated by an exothermic reaction. Conversion of such energy into power is sometimes a practical solution. However, steam turbines must be large and need many stages, due to the physical properties of steam. For that reason, there is interest in organic fluids for Rankine cycles. Fluids differ in isentropic efficiency because of the part of the cycle where the liquid working fluid is heated to the vapor pressure. Equations were presented which permit computation of efficiencies based simply on heat of vaporization and heat capacities. Methylene chloride, fluorocarbon 11, acetone, and acetonitrile were identified as potential cycle fluids.

REFERENCES

1. **Linnhoff, B.,** Thermodynamic Analysis in the Design of Process Networks, Ph.D. thesis, University of Leeds, 1979.
2. **Townsend, D. W.,** Second Law analysis in practise, *Chem. Eng.*, 628, 1980.
3. **Palmer, D. A. and Sirovich, B. E.,** Selection of a Rankine cycle fluid for recovery of work from heat at a moderate temperature, *IECEC Proc.*, 1500, 1978.
4. **Holm, J.,** Energy recovery with turboexpander processes, *Chem. Eng. Prog.*, 63, July 1985.
5. **Rathore, R. N. S. and Kenney, W. F.,** *Thermodynamic Analysis for Improved Energy Conservation*, AIChE Today Series, American Institute of Chemical Engineers, New York, 1980.
6. **Evans, F.,** The elusive energy-saving dollar, *Hydrocarbon Process.*, July, 81, 1982.
7. **Sussman, M. V.,** *Availability (Exergy) Analysis — A Self Instruction Manual*, Mulliken House, Lexington, Mass., 1980.

Index

INDEX

A

B

C

F

G

H

I

J

K

L

M

N

O

P

Q

R

S

T

U

V

W

X

Y

Z